Marine Navigation 1: Piloting

Titles in the
Fundamentals of Naval Science Series

Seamanship

Marine Navigation 1: Piloting

Marine Navigation 2: Celestial and Electronic

Introduction to Shipboard Weapons

Introduction to Naval Engineering

Marine Navigation 1:
Piloting

Second Edition

Richard R. Hobbs
Commander, U.S. Naval Reserve

Naval Institute Press
Annapolis, Maryland

Fundamentals of Naval Science Series

Library of Congress Cataloging Publication Data

Hobbs, Richard R.
  Marine navigation.
  (Fundamentals of naval science series)
  Includes index.
  Contents: —1. Piloting
  1. Navigation.   2. Pilots and pilotage.
3. Nautical astronomy.   4. Electronics in naviga-
tion. I. Title. II. Series.
VK555.H67   1981   623.89   81–9538
ISBN: 0–87021–358–X (v. 1)

                                          AACR2

To My Parents

Whose guidance set my life
on a steady course

Contents

Foreword

Throughout the history of warfare at sea, navigation has been an important basic determinant of victory. Occasionally, new members of the fraternity of the sea will look upon navigation as a chore to be tolerated only as long as it takes to find someone else to assume the responsibility. In my experience, such individuals never make good naval officers. Commander Hobbs has succeeded in bringing together the information and practical skills required for that individual who would take the first step down the road toward becoming a competent marine navigator.

At the outset of this book, the author stresses the necessity for safe navigation, but there is another basic tenet of sea warfare that this book serves. The best weapons system man has ever devised cannot function effectively unless it knows where it is in relation to the real world, where it is in relation to the enemy, and where the enemy is in relation to the real world. Not all defeats can be attributed to this lack of information, but no victories have been won by those who did not know where they were.

W. P. MACK
VICE ADMIRAL, U.S. NAVY

Although these words were written by Admiral Mack for the first edition of this book almost ten years ago, they are certainly no less valid today than they were then. The modern student of navigation could well take the last sentence to heart, not only as it applies to the practice of navigation, but also to life in general.

THE AUTHOR
ANNAPOLIS, 1981

Preface to the Second Edition

Marine Navigation 1: Piloting was originally written as a text to introduce midshipmen attending the U.S. Naval Academy and the various NROTC colleges and universities to the shipboard navigation department organization and the principles of piloting as practiced on board ships of the U.S. Navy. Therefore, the reader will find a definite Navy orientation in much of the material presented in this volume. Nevertheless, because prime emphasis has been placed on setting down the basic knowledge required for the safe navigation of any vessel in piloting waters, most of the information herein will be equally as applicable to navigators of private and commercial craft as to navigators of Navy surface warships.

Piloting refers to the safe navigation of a surface vessel in and near coastal and intracoastal waters. It is done primarily with reference to land and seamarks, as opposed to taking celestial observations or using electronic navigational aids and systems. Most of the piloting techniques discussed in this text are not new; in fact, many were developed by early seafarers well before the time of Christ, but modern instruments and navigational aids have made attainable a degree of accuracy undreamed of by the ancients. If the inexperienced student navigator can master the time-tested practical techniques of piloting presented in this book, he may be confident that he possesses the basic knowledge required to navigate any surface vessel safely in the piloting environment. Moreover, after some practical experience, the student should have sufficient background to proceed on to other more advanced techniques and navigational texts if he so desires.

The subject of navigation at sea beyond piloting waters by use of celestial and electronic navigation is covered in detail in the second volume of this series, entitled *Marine Navigation 2: Celestial and Electronic*. Its presentation assumes the student will be familiar with the basic techniques and nomenclature introduced in this first volume.

While much of the material presented in *Marine Navigation 1* is based upon similar topics covered in standard references such as *The American Practical Navigator* (Bowditch) and *Dutton's Navigation*

and Piloting, it is the author's hope that through a logical step-by-step approach, much of the "mystery" associated with the more technical approaches of these and other navigational texts can be eliminated. In order that familiarity might be gained with the many publications that support the practice of navigation during piloting, excerpts from many of the navigation publications produced and distributed by the Defense Mapping Agency appear throughout. Wherever possible, specific year dates have been deleted from these excerpts and from example problems using them, in order to lessen the sensitivity of this text to the passage of time from the date of publication. When dealing with actual or theoretical problems and situations other than those presented as examples in this text, the student is cautioned to remember always to obtain and use the correct edition covering the specific dates in question.

The author is greatly indebted to Commander F.E. Bassett, USN, who conceived the original edition of this text, and under whose guidance and assistance it was prepared, and to Lieutenant Commanders J.D.L. Backus, RN; J.L. Roberts, USN; F.A. Olds, USN; and Mr. E.B. Brown of the Defense Mapping Agency, who made many suggestions incorporated in it.

Special thanks for their help with this second edition are gratefully extended to Colonel E.S. Maloney, USMC (Retired), and Colonel W.P. Davis, USA (Retired), for their very thorough review and many suggestions for improvement of the initial draft, and to Lieutenant Commander David Russell, RN, and Lieutenant Commander Royal Connell, USN, of the U.S. Naval Academy Navigation Department, Commander G.F. Hotchkiss, USCG, of the U.S. Coast Guard Academy, Lieutenants T.S. Harden and D. Prince, USN, of the University of Utah NROTC Unit, and Mr. Frank Flyntz of the Maritime Institute of Technology, for their thoughtful suggestions for improvement of several sections, and to Sue Thompson, USNI staff photographer, for several photos of nav aids used herein. Finally, special gratitude is due Beverly Baum, Cynthia Taylor, and Carol Swartz, USNI design manager, assistant designer, and editor, repectively, who have succeeded in creating a superb book in spite of the best efforts of the author. Thanks to you all.

Marine Navigation 1: Piloting

The Art of Navigation

The word "navigate" is derived from the Latin words *navis,* meaning ship, and *agere,* meaning to move or direct. Navigation is generally defined as the process of directing the movements of a vessel from one place to another. For the contemporary navigator, however, this definition is incomplete, as it lacks two essential modifying terms—the words *safely* and *efficiently.* In today's world of rampant inflation and increasingly serious energy shortages, the cost of replacing a vessel lost through negligent navigation can often be completely prohibitive, quite apart from the consideration of any attendant injuries or deaths among the vessel's crew or loss of cargo. Revenue losses caused by inefficient navigation with consequent increases in fuel bills and other operating costs can be almost if not equally as severe over time. Thus, modern navigation may be more properly defined as the process of directing the movements of a vessel safely and efficiently from one place to another.

It is often said that modern navigation is both an art and a science, with ancient navigators initiating the practice of navigation as an art, and modern man developing it into a science. Anyone who has seen the chart of a professional navigator after a day's work in a difficult operating area would certainly be impressed with the artistry displayed in the navigational plot. When one considers the wide range of electronic aids and other sophisticated devices routinely used by the modern navigator, the scientific aspects become equally as apparent. In the last analysis, it is rather difficult to differentiate between navigation as an art and navigation as a science. In fact, the art of navigation probably represents one of the first instances of the practical use of science by mankind. The annual migration of early man from one hunting ground to another with the changing seasons is an established archaeological fact. Modern man calls this type of movement from one place to another over land areas by reference to landmarks *land navigation.* When man extended his wanderings to coastal waters and rivers by means of primitive boats and river craft, the division of navigation by land or seamarks now known as *piloting* was born.

As man continued in his quest for knowledge of unknown territories, he began to venture to sea beyond the range of piloting aids, necessitating the development of a new form of navigation: *dead reckoning*. In its initial form, this process was simply concerned with keeping a record of the estimated distances and directions traveled so as to enable the mariner to return to familiar surroundings. As the length and duration of ocean voyages became ever more extended, certain instruments were developed to assist in the determination of course, speed, depth, distance traveled, and ultimately, the position of the vessel at sea. Among these primitive instruments were the lodestone and other early forms of the magnetic compass, the chip log, the hand leadline, the hourglass, and eventually the backstaff, quadrant, early chronometers, the astrolabe, and the early sextant (Figure 1–1).

As improvements continued to be made in navigation instruments, improved techniques of using them were also developed, which ultimately culminated in the mid-1800s in determining position at sea by observing celestial bodies, or *celestial navigation*. With the de-

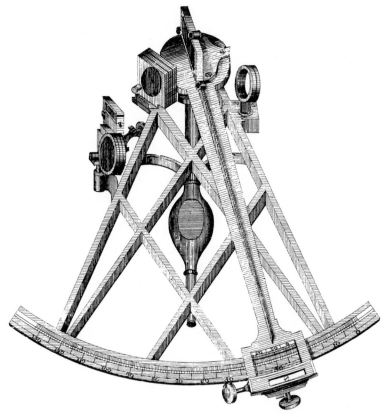

Figure 1–1. An early sextant.

velopment of radio and the electromagnetic wave in the early 1900s came *radionavigation* and the subsequent development of more sophisticated means of position-finding now referred to as *electronic navigation systems*. With the advent of the airplane came the practice of *air navigation;* the navigation of a submarine fostered new techniques of *submarine navigation;* and the launching of man into space necessitated a new branch of navigation called *space* or *astral navigation.*

Marine surface navigation, as it is currently practiced, is subdivided into two basic areas—piloting, and celestial and electronic navigation—with the latter fast becoming a third major subdivision. The remainder of this book is concerned with the routine practice of piloting in waters contiguous to U.S. and foreign shores, and in inland rivers, bays, and lakes in which piloting techniques apply. Although the orientation of the book is toward surface navigation in these waters, much of the basic information set forth is applicable to all other types of navigation as well.

Throughout seafaring history, the mariner who knew how to navigate was always held in high esteem by his fellows, as without his expertise disaster would surely befall the ship and crew. Until comparatively recent times, the practice of navigation was an art based largely on mathematics and interpretation of written sailing directions, often drafted in Latin or some other foreign language. Only an educated ship's officer or captain could usually master the subject, and his knowledge was often closely guarded, both as a means of enhancing his personal prestige under normal circumstances, and of ensuring his survival and his control over an unruly crew in times of peril. Thus, an aura of mystique became associated with the art of navigation, which has persisted even to the present day to some extent.

Modern navigation, however, is no longer an art that only an educated few can master. Today's mass-produced charts and easily understood reference publications and sailing directions have made it possible for just about anyone with basic reading and math skills and some measure of manual dexterity to navigate effectively, once he or she has learned the basic principles involved. Professional standards of accuracy and effectiveness, however, as in any other field of human endeavor, can usually be achieved only by those who are willing to spend the extra time, energy, and constant attention to detail required for above-average success. And the evident skill with which a professional navigator uses the sextant and the other navigational instruments, and the accuracy and neatness of his position plot, still command the respect of his fellow mariners at sea.

2

The Shipboard Navigation Department Organization

While the basic techniques of safe navigation of a seagoing surface vessel are essentially independent of the size of the vessel, there is, of course, a wide variety of navigation department organizations on board different types of ships. The entire crew of a small boat may consist of only one or two people, who carry out navigational responsibilities in addition to all other responsibilities of operating the vessel. Aboard most merchant vessels, the navigation department usually consists of a single deck officer, typically a second mate, who performs the assignment as ship's navigator as a collateral duty, and who is assisted when the ship is under way by the ship's master and the deck officer on watch. On board Navy ships, the ship's navigator may be a collateral duty assignment on destroyer-type ships and smaller, and a full-time billet on larger ships and staffs. Since the duties of navigator on a warship are generally considerably more complex than on a merchant ship, Navy navigators are assisted by a division of from two to twenty enlisted personnel of the quartermaster rating, with the exact number dependent on the size and mission of the ship. On larger Navy ships such as a cruiser or carrier, the division may even constitute a separate ship's department, with a junior officer assigned as assistant navigator and N-division officer.

Since this text is oriented primarily toward the practice of marine surface navigation in the U.S. Navy, the remainder of this chapter will discuss the duties of the navigator of a Navy ship in some detail. Students of navigation not affiliated with the Navy should nevertheless find the following discussions of some interest, because even though the functional relationships may be unique to Navy warships, the basic responsibilities of navigator on board almost every seagoing vessel are similar.

Duties of the Navigator

Regardless of the size of his ship and his rank, the responsibilities assigned the U.S. Navy navigator are the same. They are set forth in paragraph 323 of *OPNAVINST 3120.32, Standard Organization and Regulations of the U.S. Navy*, reproduced in Figure 2–1.

Relationship of the Navigator to the Command Structure

At this point, it would be well to consider the relationship of the Navy navigator to the commanding officer (CO), the executive officer (XO), and the officer of the deck (OOD). Note that OPNAVINST 3120.32 specifies that the navigator reports directly to the CO on all matters pertaining to the navigation of the ship. Although this seems like a break in the chain of command, it is not, in reality. *Navy Regulations, 1973*, assigns ultimate responsibility for the safe navigation of the ship to the commanding officer:

323 NAVIGATOR

a. GENERAL DUTIES. The head of the navigation department of a ship will be designated the navigator. The navigator normally will be senior to all watch and division officers. The Chief of Naval Personnel will order an officer as navigator aboard large combatant ships. Aboard other ships, the commanding officer will assign such duties to any qualified officer serving under his command. In addition to those duties prescribed by regulation for the head of a department, he will be responsible, under the commanding officer, for the safe navigation and piloting of the ship. He will receive all orders relating to his navigational duties directly from the commanding officer and will make all reports in connection therewith directly to the commanding officer.

b. SPECIFIC DUTIES. The duties of the navigator will include:

1. Advising the commanding officer and officer of the deck as to the ship's movements; and if the ship is running into danger, as to a safe course to be steered. To this end he will:

    (a) Maintain an accurate plot of the ship's position by astronomical, visual, electronic, or other appropriate means.

    (b) Prior to entering pilot waters, study all available sources of information concerning the navigation of the ship therein.

    (c) Give careful attention to the course of the ship and depth of water when approaching land or shoals.

    (d) Maintain record books of all observations and computations made for the purpose of navigating the ship, with results and dates involved. Such books shall form a part of the ship's official records.

    (e) Report in writing to the commanding officer, when underway, the ship's position at 0800, 1200, and 2000 each day and at such other times as the commanding officer may require.

    (f) Procure and maintain all hydrographic and navigational charts, sailing directions, light lists, and other publications and devices for navigation as may be required. Maintain records of corrections affecting such charts and publications. Correct navigational charts and publications as directed by the commanding officer and, in any event, prior to any use for navigational purposes. Corrections will be made in accordance with such reliable information as may be supplied to the ship or as the navigator is able to obtain.

2. The operation, care, and maintenance of the ship's navigational equipment.To this end he will:

    (a) When the ship is underway and weather permits, determine daily the error of the master gyro and standard magnetic compasses and report the result to the commanding officer in writing. He will cause frequent comparisons of the gyro and magnetic compasses to be made and recorded. He will adjust and compensate the magnetic compasses when necessary, subject to the approval of the commanding officer. He will prepare tables of deviations and keep correct copies posted at the appropriate compass stations.

    (b) Ensure that the chronometers are wound daily, that comparisons are made to determine their rates and error, and that the ship's clocks are properly set in accordance with the standard zone time of the locality or in accordance with the orders of the senior officer present.

    (c) Ensure that the electronic navigational equipment assigned to him is kept in proper adjustment and, if appropriate, that calibration curves or tables are maintained and checked at prescribed intervals.

3. Advise the engineer officer and the commanding officer of any deficiencies observed in the steering system, and monitor the progress of corrective actions.

4. The preparation and care of the deck log. He will daily, and more often when necessary, inspect the deck log and will take such corrective action as may be necessary and within his authority to ensure that it is properly kept.

5. The preparation of such reports and records as are required in connection with his navigational duties, including those pertaining to the compasses, hydrography, oceanography, and meteorology.

6. The required navigational training of all personnel such as junior officers, boat coxswains, and boat officers; the training of all quarterdeck personnel in the procedures for honors and ceremonies and of all junior officers in Navy etiquette.

7. Normally, assignment as the officer of the deck for honor and ceremonies and other special occasions.

8. The relieving of the officer of the deck as authorized or directed by the commanding officer (in writing).

c. DUTIES WHEN PILOT IS ON BOARD. The duties prescribed for a navigator in these regulations will be performed by him whether or not a pilot is on board.

d. ORGANIZATIONAL RELATIONSHIPS. The navigator reports to the commanding officer concerning navigation and to the executive officer for the routine administration of the navigation department. The following officers report to the navigator:

1. The engineer officer concerning the steering engine and steering motors.

2. The assistant navigator, when assigned.

Figure 2-1. Duties of the navigator, OPNAVINST 3120.32.

The commanding officer is responsible for the safe navigation of his ship or aircraft, except as prescribed otherwise for ships at a naval shipyard or station, in drydock, or in the Panama Canal. . . .

The CO delegates his responsibility for safe navigation to the navigator, and the navigator reports directly to the CO only on those matters pertaining to the navigation of the ship. He reports to the XO in all other matters having to do with his administrative functions as a department head, such as the administration of his department and the training of junior officers in navigation. The navigator assists the senior watch officer in training watch officers, especially the OODs. He may be empowered to relieve the OOD if he deems it necessary in a dangerous situation and if he is so authorized in writing by the CO; this authorization must be in the form of a letter inserted in the service record of the navigator. When so authorized, he can relieve the OOD when, in his opinion, such action is necessary for the safety of the ship, and the CO or XO are not on deck. Normally, however, the navigator advises the OOD of a safe course to steer, and the OOD may regard this advice as authority to change the course; the OOD should then report this change to the CO.

The Navigator's Staff

The navigator does not personally have to perform all of the tasks indicated in *OPNAVINST 3120.32*, but he is still responsible for seeing that they are carried out. The leading petty officer (LPO) in the navigation division (N-division) assists the navigator and his assistant, if assigned, in carrying out many of the navigator's duties. On smaller ships the LPO may be a first class quartermaster (QM1). On larger ships a chief quartermaster (QMC) is assigned. With his wealth of practical experience, the LPO can be a great help to the navigator.

The navigator and the LPO will usually assign certain specific areas of responsibility to the more junior of the quartermasters, such as chart petty officer, clock petty officer, and training petty officer. The responsibilities associated with these functions will be discussed later. One of the most important collateral duties of all of the junior petty officers and strikers is standing the quartermaster of the watch (QMOW) duty while the ship is under way. They draft the *Ship's Deck Log* on the bridge. This running chronology of all events of significance occurring during their watch is specified in *Navy Regulations* to be one of the legal records of the events occurring on the ship (Figure 2–2).

On smaller ships, the QMOW has several other important functions, in addition to keeping this log. When the navigator is not on deck, he assists the OOD in all matters pertaining to safe navigation,

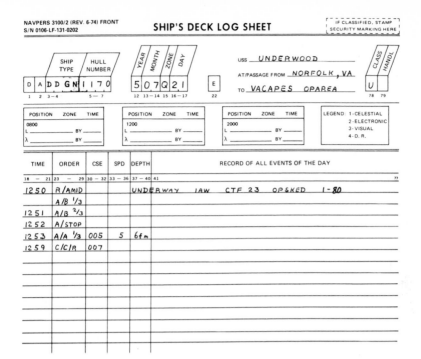

SHIP'S DECK LOG SHEET

IF CLASSIFIED, STAMP
SECURITY MARKING HERE

	SHIP TYPE	HULL NUMBER	YEAR	MONTH	ZONE	DAY			CLASS	HANDL

D A DD GN 1 1 7 0   5 0 7 Q 2 1   E

uss __UNDERWOOD__

AT/PASSAGE FROM __NORFOLK, VA__

TO __VACAPES OPAREA__   U

POSITION	ZONE	TIME	POSITION	ZONE	TIME	POSITION	ZONE	TIME	LEGEND: 1-CELESTIAL

0800   1200   2000

L _____ BY _____   L _____ BY _____   L _____ BY _____   2-ELECTRONIC
λ _____ BY _____   λ _____ BY _____   λ _____ BY _____   3-VISUAL   4-D.R.

TIME	ORDER	CSE	SPD	DEPTH	RECORD OF ALL EVENTS OF THE DAY
18 — 21	23 — 29	30 — 32	33 — 36	37 — 40	41
1250	R/AMID				UNDERWAY IAW CTF 23 OPSKED 1-80
	A/B 1/3				
1251	A/B 2/3				
1252	A/STOP				
1253	A/A 1/3	005	5	6fm	
1259	C/C/R	007			

Figure 2–2. *A portion of the* Ship's Deck Log.

including the maintenance of the ship's position plot. On ships which do not have a separate meteorological organization, the QMOW is also responsible for filling out an hourly weather observation sheet, illustrated in Figure 2–3 next page. A report of these observations is transmitted by radio four times daily to the Naval Oceanography Command Center in whose area of responsibility the ship is operating.

In addition to their functional duties, there are several compartments or spaces normally assigned to the navigation division or department personnel for maintenance. On small ships, these spaces are limited in number to the bridge and pilothouse area, the chartroom, and perhaps a storage area or void. On larger ships, in addition to these spaces, the navigation department may be assigned its own berthing area, head, and secondary conn for maintenance. Of these, the upkeep of the bridge and pilothouse area is usually the most time-consuming, as it is one of the most "visible" and heavily used areas of the ship when under way.

It is not enough, therefore, that the Navy navigator be familiar with the technical aspects of his duties. He is also a division officer, and he must be ready to assume the many and varied responsibilities associated with that office if he would be completely successful.

The Shipboard Navigation Department Organization   **7**

OPNAV FORM 3144-1 (1-68)
Previous editions obsolete
S/N-0107-714-4101

DEPARTMENT OF THE NAVY

SHIP WEATHER OBSERVATION SHEET

USS **Passage (LPA 2)**

AT/PASSAGE FROM **Naples, Italy**   TO **Norfolk, Va.**

DATE (GMT) **15 November** 19 **80**

SECTION I

TIME (GMT)	WINDS DIRECTION (True)	FORCE (Knots)	VISI-BILITY (Miles)	WEATHER (Symbols)	BAROMETER (Inches)	TEMPERATURE DRY BULB	WET BULB	CLOUDS AMOUNT (Tenths)	HEIGHT	TYPE	SEA WATER TEMP. (Degrees and tenths)	SEA WAVES PERIOD (Seconds)	HEIGHT (Feet)	SWELL WAVES DIRECTION (True)	PERIOD (Seconds)	HEIGHT (Feet)
00	010	8	7	BKN	29.96	50	46	8	7000	AC	44	1	2	010	3	1
01	010	8	7	BKN	29.96	50	46	8	7000	AC	44	1	1	010	3	1
02	014	9	7	SCT	29.97	51	46	4	7000	AC	44	1	2	010	3	1
03	017	8	8	SCT	29.97	51	46	4	7000	AC	44	1	2	010	3	1
04	023	10	8	SCT	29.97	54	47	4	7000	AC	45	1	2	020	3	1
05	025	12	9	CLR	29.97	56	48	0		CLEAR	45	1	2	020	3	2
06	030	11	11	CLR	29.98	57	49	0		CLEAR	45	1	2	020	3	2
07																
08																
09																
10																
11																
12																
13																
14																
15																
16																
17																
18																
19																
20																
21																
22																
23																

Figure 2–3.   A portion of a Ship Weather Observation Sheet.

3

The Piloting Team

Unless the navigator is directing the movements of a very small vessel, he cannot effectively navigate in confined waters without some assistance from other personnel. The direction of movements of a vessel in these circumstances by reference to land and seamarks is called *piloting*. Piloting may be done by visual methods or by the use of electronics, or by a combination of the two.

Merchant vessels almost always will bring a local pilot on board to take responsibility for conning the ship in piloting waters and in making a berth. The second or third mate deck watch officer in these circumstances will assist the pilot by taking occasional fixes, and by ensuring that the helmsman interprets and carries out all orders to the helm correctly. Naval vessels, on the other hand, tend to rely much less on pilots than do merchant ships, since by their nature warships are intended to proceed in harms way on their own resources, especially in wartime. Accordingly, the Navy piloting organization is much more extensive than that found on board most merchant ships.

Aboard Navy ships, the persons who assist the navigator in the piloting environment are customarily referred to as the *piloting team*. Its members are normally N-division personnel, if the division is large enough; on small ships having only a few quartermasters on board, recruits are drawn from related rates, such as operations specialists or signalmen. There are usually five functional positions on the team: the plotter, the bearing recorder, the bearing taker, the radar operator, and the echo sounder operator. The number of personnel assigned to each one of these positions varies with the size of the ship; on smaller ships, one person may carry out the functions of two or more positions. The duties of each member of the piloting team on a typical Navy ship are described on the following pages. Again, even though the organization described is unique to the Navy, the functions carried out by the various positions are common to all vessels, so the following descriptions should be of interest to all students of marine surface navigation.

Figure 3–1A. The navigator (right) is in charge of the piloting team. He coordinates the team effort through the bearing recorder (left) who communicates with the other team members over the 1JW sound-powered phone circuit. As each round of bearings is received, he logs them in the Bearing Book.

The Navigator and His Plotter

The navigator is in overall charge of the piloting team on board a Navy ship. In the piloting environment, he usually directs the team from the bridge in close proximity to the captain and officer of the deck, so that he may make his recommendations of course and speed verbally directly to them. The navigator may or may not be plotting on the chart himself. Most navigators find it easier to move about and keep the total picture if they assign one of their more senior petty officers to do the actual mechanical plotting on the chart. This allows the navigator to monitor the navigation plot and at the same time to be cognizant of oncoming shipping, weather, and the many other variables that might affect the ship and its movements in this type of environment.

The Bearing Recorder

Although the navigator is the overall director of the piloting team, the real coordinator of the team is the bearing recorder. He is stationed on the bridge, alongside the navigator and his plotter, and is in constant communication with the other positions on the team via the navigation sound-powered circuit, the 1JW. The navigator keeps

him informed as to which objects he wishes to be used for obtaining lines of position (LOPs) as the ship proceeds along its track, as well as how often to ask for these LOPs. The bearing recorder keeps track of the passage of time and apportions the objects among the bearing takers and radar operator as applicable. When LOPs are received back from the various positions, he then simultaneously reports them to the navigator and plotter and records them in the *Bearing Book* (see Figure 3–4 on page 15). A depth sounding taken at this time is also recorded in the book, for comparison with the charted depth after the ship's position, or fix, is produced from a plot of the LOPs.

The Bearing Taker

On small ships, there may be only one bearing taker stationed at a central position from which he can observe objects on all sides of the ship. On larger ships, the configuration of the superstructure and placement of the gyro repeaters requires the stationing of several bearing takers. Usually, one man is required on the port side or

Figure 3–1B. *A bearing taker on the bridge wing of a Navy training ship passes visual bearings to the bearing recorder via the 1JW sound-powered phone circuit.*

"wing" of the bridge, and one on the starboard side; some ships station a third bearing taker astern as well. The bearing recorder assigns objects to these men according to the side of the ship on which the objects are located—port, starboard, or astern—and the bearing takers then "shoot" the bearings to their assigned objects and report them to the bearing recorder whenever he calls for a round of bearings.

It often becomes necessary to assign a single bearing taker more than one object to shoot for a given round of bearings. In such cases the bearing to the object that is closest to the ship's beam should be obtained first, as it has the greatest angular velocity relative to the ship, and its bearing is drifting most rapidly. Those objects appearing toward the ship's bow or stern should be shot last, because their bearings change quite slowly if at all in the time it takes to shoot a round. In Figure 3–2, the bearing drifts of several objects off the port side of a ship steering course 110°T are depicted, with the lengths of the arrows representing the drift rates. As can be seen from the figure, all bearings except the one directly astern will drift to the left, with the bearing to the object directly abeam at 020°T having the greatest rate of change. Because failure to obtain the bearings in the proper order could result in an erroneous fix position, the bearing takers should be reminded of the proper procedure each time the piloting team is stationed.

The Radar Operator

These visual bearings may be supplemented as desired by radar ranges, and to a lesser extent radar bearings, supplied by the radar operator. Because of space limitations on the bridge, this individual is usually stationed at a radar repeater located in the ship's combat information center (CIC), or in the case of some larger ships so equipped, in the chart house; he is connected to the bearing recorder

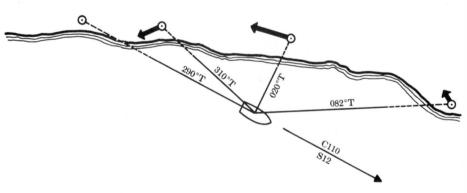

Figure 3–2. Bearing drifts of objects relative to a moving ship.

via the 1JW circuit. When the ship proceeds beyond visual range to land, the radar ranges and bearings then become the primary source of lines of position. Prior to sailing, various prominent landmarks are identified and labeled with letters of the alphabet on the navigator's chart; the radar operator is then briefed and given a small copy of the lettered chart. The bearing recorder requests radar information from the radar operator by referring to point A, B, C, etc., as the ship progresses along its track.

The Echo Sounder Operator

The echo sounder operator is the remaining member of the piloting team. As the console for this instrument is usually located near the chartroom at some distance from the bridge, this man is also on the 1JW sound-powered circuit. When the operator hears a request being

Figure 3–3. A radar operator in CIC provides radar ranges and bearings for the navigator on the bridge and for the CIC radar navigation team, a valuable back-up for the navigator of a Navy ship.

given to the other team members for LOPs, he notes the depth reading at that time, and passes this information to the bearing recorder. As was mentioned earlier, the recorder then logs this reading into the *Bearing Book* along with the other information obtained for that time, so that the charted depth may be compared with the actual depth of water at the ship's fix position.

The Piloting Team Routine

The frequency of obtaining a set or round of LOPs depends upon the judgment of the navigator. He bases his requirements on how constrained the ship is by navigational hazards in proximity to it. In relatively open water, the navigator may obtain a round of bearings only once every fifteen or thirty minutes; in very restricted waters, he may require a round every minute. Most Navy navigators, in all but extreme situations, take a round every three minutes. This not only allows him to use the "three-minute rule" (described in Chapter 7) in plotting his track, but also gives the bearing recorder time to issue new instructions to the team as the ship moves. To let the team know that the time for a round is approaching, the bearing recorder will usually pass the word, "Stand by for a round of bearings," about ten seconds prior to the time of the desired observation. As the time arrives, he gives "Stand by—Mark!" The instant the word "Mark" is given, the bearing takers should note the bearings to their objects, the radar operator should mark his radar range or bearing, and the echo sounder operator should read the depth recorded. All these readings are then recorded in the *Bearing Book*, in the format that appears in Figure 3–4. Note that the gyro error is figured and recorded on the top of the page. Normally, the navigator or his plotter takes this into account when adjusting his plotting instrument. Bearings recorded are those read by the bearing taker or radar operator from their gyro repeaters and radarscopes. The *Bearing Book*, in conjunction with the *Ship's Deck Log*, forms a legal record of a vessel's track. If corrections to an entry are required in either record, an erasure must not be made; a single line is drawn through the entry, initialed by the author, and the correct information is inserted above it.

Although not usually considered part of the piloting team *per se*, two additional billets on a ship's Special Sea Detail bill can provide useful information to the navigator in a piloting situation. On most Navy ships, there is a provision for a leadsman on the bow as part of the deck division's area of responsibility. This individual is little used in today's Navy, but he can provide valuable backup to determine whether or not the echo sounder is operating correctly. He is equipped with a marked line fitted with a weight on the end called

	RECORD GYRO BEARINGS					
Date: *23 June 1969*				Gyro Error *1°W*		
Place: *Entering Norfolk, Va.*						

Time	Cape Henry Light	Cape Charles Light	Thimble Shoals Light	Lynn-Haven Bridge Tower	Checkered Tank	Echo Sounder Reading
0946	205.5	010.0	287.0			4
0949	201.0	012.5	287.0			
0952	197.0	016.0	287.0			3
			10870			
0955	192.5		287.0			3
0958	186.0		287.0			
1001	178.0		287.0	222.5		
1004	171.0		286.5	217.0		3
1007	164.0		287.0	211.0		
1010	155.0		287.0	206.5		3

Figure 3–4. The Bearing Book.

a *leadline* (pronounced lĕd), which can be lowered into the water until it strikes bottom, thereby making a depth reading possible. Communication with this station can be established via the 1JV circuit on the bridge.

On Navy ships having a gunfire control director, the navigator should request that this equipment be manned whenever the special sea detail is set. Its operator can be patched into the 1JW circuit, and its optical and radar systems can be used to provide very precise ranges and bearings to prominent landmarks. On occasion its radar can even be made to lock onto a suitable navigation aid in order to provide continuous range and bearing LOPs from this object to the navigator—a most valuable backup to supplement his conventional radar information.

In conditions of low visibility such as fog, snow, or rain squalls, the normal piloting team routine may be disrupted somewhat because of inability to record visual bearings. In such situations, especially when in constricted waters or ship channels, the navigator may shift primary responsibility for maintaining the navigation plot to a radar piloting team set up in CIC, thus taking advantage of the expertise of the CIC personnel in radar navigation. Even in these circumstances, however, the navigator is still responsible for the ship's safe navigation. The navigator with his piloting team, therefore, acts as a backup to the CIC plot; he attempts to verify all CIC recommendations by using his own team's radar and depth information, and any fixes of oppor-

tunity. If doubt arises as to the ship's position, the navigator should immediately recommend taking all way off the ship, and perhaps even dropping anchor, until the doubt can be resolved. To ensure that the CIC radar navigation team will be ready should it be needed, it has become standard procedure, whenever the special sea detail is set, for the CIC team to act as a backup for the piloting team in the piloting environment. The navigation information is usually passed between the bridge and CIC via phone talkers on the 1JA circuit. The navigator should always make sure, therefore, that the cognizant CIC personnel are briefed prior to getting under way or entering port as to the ship's track and speed of advance that he intends to make through the piloting waters.

Conclusion

In conclusion, it must be remembered that the piloting team, like any other team, needs practice to operate smoothly and efficiently together. The navigator should make it a routine procedure to assemble his team prior to the time the ship enters or leaves port to brief them on the ship's route, the appearance of landmarks and lights, the expected visibility conditions, and any other unusual or pertinent circumstances associated with that particular piloting environment. It must be borne in mind too that it is one thing to try to visualize a navigation aid or landmark from its appearance on a chart or description in a book, and quite another task to pick out the actual object hidden among its natural surroundings. Even an experienced bearing taker occasionally confuses one desired landmark with another when shooting a bearing. The problem is compounded many times for a relatively inexperienced observer operating in poor visibility conditions. For this reason, the bearing taker is considered the "weak link" in the team. With practice and a good idea of what to expect, however, erroneous input from this source can be minimized if not completely eliminated.

4

The Nautical Chart

The nautical chart is historically the most important and certainly the most frequently used tool employed by the navigator in the execution of his functional responsibilities. Maps, charts, and written sailing directions were probably in use by Egyptian and Greek mariners in the Mediterranean Sea well before the birth of Christ. Ptolemy, a Greek astronomer and mathematician, constructed many maps in the second century A.D., among which was a world map based upon an earlier calculation of the earth's circumference as 18,000 miles. His works remained a standard until the Middle Ages; Columbus believed he had reached the East Indies in 1492 in part because he used the Ptolemaic chart as a basis for his calculations of position. In the Pacific, the natives of the South Sea islands constructed and used crude yet effective charts from palm leaves and sea shells, representing islands, ocean currents, and angles of intersection of ocean swells. Gerhardus Mercator, a Flemish cartographer who produced a world chart in 1556 by a type of projection bearing his name, is considered to be the father of modern cartography. As more and more mariners recorded extended voyages throughout the world, the accuracy of charts continued to improve. Until the invention of the printing press, however, they were done entirely by hand, and the mariner considered them much too scarce and valuable to be used for plotting. This led to wide use of mathematical techniques for calculating position known as deduced reckoning or sometimes simply as "the sailings." These methods of determining approximate position continued in use until the late nineteenth century, when charts came to be mass produced and the system of geometric "dead reckoning," as it is practiced today, came into widespread use.

The difference between a map and a nautical chart should be understood. A map is a representation of a land area on the earth's surface, showing political subdivisions, physical topography, cities and towns, and other geographic information. A nautical chart, on the other hand, is primarily concerned with depicting navigable water areas; it includes information on the location of coastlines and harbors, channels and obstructions, currents, depths of water, and aids to navigation.

Aeronautical charts, designed for use by the aviator, show elevations, obstructions, prominent landmarks, airports, and aids to air navigation. Like maps, they usually depict land areas, but they differ in that they emphasize landmarks, restricted areas, and other features of special importance to the air navigator.

This chapter will discuss the nautical chart in some detail, including the terrestrial coordinate system, the major types of chart projections used by the surface navigator, chart interpretation, determination of position, distance and direction, and chart production, numbering, and correction systems.

The Terrestrial Coordinate System

Prior to any discussion of charts or their methods of projection, it is first necessary to understand the nature of the terrestrial sphere and its coordinate system. Our earth is basically round, but it is not quite a perfect sphere, being somewhat flattened at the poles and bulged at the equator. The polar diameter has been calculated to be 6,864.57 miles, while the equatorial diameter is about 6,887.91 miles; the earth is therefore often referred to as a *spheroid*, a close approximation to a sphere. For most navigational purposes, it is considered to be a perfect sphere, with a circumference of exactly 21,600 nautical miles.

On a sphere at rest, any point on its surface is similar to every other point, and all points on the surface are defined as being equidistant from the center. To make measurements on the surface, there must be some point or set of points designated as a reference or references to which all other points can be related. As soon as rotation is introduced, two such reference points are immediately defined—the points at which the spin axis pierces the surface of the sphere. On the earth these points are called the north and south poles; the axis of the earth, together with its poles, constitutes the basic references on which the terrestrial coordinate system is based.

If a straight line is drawn connecting two points on the surface of a sphere, the line drawn actually represents a locus of points formed by the intersection of a plane with the surface of the sphere. Moreover, if the plane passes through the center of the sphere, as well as through the two points of interest on its surface, it can be shown by spherical trigonometry that the resulting line drawn between the two points represents the shortest possible distance between them, as measured across the surface of the sphere. Any line of this type on the surface of a sphere, formed by the intersection of a plane passing through its center, is termed a *great circle*, so named because it is the largest circle that can be formed on the earth's surface. The

shortest distance between any two points on the earth lies along the shorter arc of the great circle passing through them. Figure 4–1 illustrates three great circles, of which two, the equator and a meridian, have special significance in the terrestrial coordinate system.

Any other circle formed on the surface of a sphere by the intersection of a plane *not* passing through the center of the sphere is termed a *small circle*. Figure 4–2 depicts three small circles, of which one, the parallel of latitude, has major importance in the earth's coordinate system.

The great circle formed by passing a plane perpendicular to the earth's axis halfway between its poles is known as the *equator*. The equator, which divides the earth into the *northern* and *southern hemispheres*, is of major importance because it is one of the two great circles from which all locations on the earth's surface are referenced. Any great circle formed by passing a plane through the center of the earth at right angles to the equator is called a *meridian*. There are an infinite number of meridians that could be so formed, but the one that constitutes the second reference line for the terrestrial coordinate system is termed the *prime meridian*. The prime meridian, to which all other meridians are referenced, has been defined by all nations

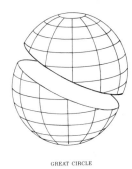

GREAT CIRCLE
(Equator)

GREAT CIRCLE
(Meridian)

GREAT CIRCLE

Figure 4–1. Examples of great circles.

SMALL CIRCLE

SMALL CIRCLE
(Parallel)

SMALL CIRCLE

Figure 4–2. Examples of small circles.

of the world to be the meridian that passes through the original position of the Royal Greenwich Observatory near London, England; although the original structure no longer exists, the spot is marked by a monument visited by thousands of tourists each year. The prime meridian divides the earth in an east-west direction into the *eastern* and the *western hemispheres.*

All meridians, because of their construction, lie in a true north and south direction, and are bisected by the earth's axis. That half of a meridian extending from the north to the south pole on the same side of the earth as an observer is called the *upper branch* of the meridian, and the other half, on the other side of the earth from the observer, is referred to as the *lower branch.* The upper branch of the prime meridian is frequently called the *Greenwich meridian,* while its lower branch is the 180th meridian. In common usage, the word meridian always denotes the upper branch, unless otherwise specified.

Since there are an infinite number of meridians, all points on the earth's surface have a meridian passing through them. The angular distance between the Greenwich meridian and the meridian of a particular point on the earth's surface is the *longitude* of the point; longitude is measured in degrees of arc from 0° to 180°, either in an easterly or westerly direction from the Greenwich meridian. If a point lies from 0° to 180° east of Greenwich, it is described as being in the eastern hemisphere and having *east longitude;* if it is from 0° to 180° west of Greenwich, it is in the western hemisphere and it has *west longitude.* Two meridians are separated by only an infinitely small distance; for the sake of clarity, therefore, meridians of longitude are generally drawn on globes and world maps at intervals of fifteen degrees or so. Figure 4–3 shows the prime meridian of the earth and some other meridians equally spaced around the earth from it.

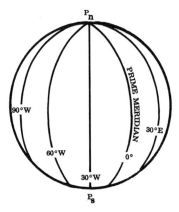

Figure 4–3. The meridians of the earth.

Any small circle perpendicular to the earth's axis formed by passing a plane parallel to the plane of the equator is termed a *parallel of latitude*. As was the case with the meridian, there are an infinite number of parallels of latitude that can be formed in this manner; therefore, every point on the earth's surface has a parallel of latitude passing through it. The angular distance between the equator and the parallel of latitude passing through a particular point is referred to as the *latitude* of that point. Latitude is measured in degrees of arc from 0° to 90°, either in a northerly or southerly direction from the equator; 90° north latitude is the location of the north geographic pole, and 90° south latitude is the location of the south pole. Parallels of latitude are always at right angles with all meridians that they cross, and thus they lie in an east-west direction. Several parallels of latitude, superimposed on the globe of Figure 4–3, are shown in Figure 4–4.

It should be apparent by now that any point on the earth's surface can be exactly located by specifying its latitude and longitude. When the coordinates of a given location are specified, it has become standard procedure to list the latitude, usually abbreviated by the letter *L*, first, and the longitude, abbreviated either by the Greek letter lambda (λ) or the abbreviation *Lo*, second. In the measurement of arc, one degree is made up of 60 minutes of arc, and one minute of arc is made up of 60 seconds. Both latitude and longitude are normally measured to the nearest tenth of a minute of arc, but on some larger scale charts, the two quantities may be read accurately to the nearest second. After the amount of the latitude or longitude has been measured, each must be suffixed by the proper letter indicating the hemisphere in which the given point is located, either N (north) or S (south) for latitude, and E (east) or W (west) for longitude. If these

Figure 4–4. Parallels of latitude.

letters were omitted, a given set of latitude and longitude numbers could refer to any one of four different locations on the earth's surface.

Since latitude is measured in a north-south direction, a convenient location to place a latitude scale is alongside a meridian. In practice, the latitude scale is normally printed on a chart alongside the meridians making up its left and right side boundaries. Since latitude is measured along a meridian, which is a great circle, the length of one degree of latitude is the same everywhere on the earth. It is equal to the earth's circumference, 21,600 miles, divided by 360°, or 60 nautical miles. One minute of latitude, therefore, is equal to one nautical mile of distance, which by strict definition is equal to 1,852 meters or 6,076.1 feet. In the U.S. Navy, however, for distances under 20 miles or so, the nautical mile is usually considered as equivalent to 2,000 yards for most practical purposes.

Longitude is measured in an east-west direction; a logical place for its scale, then, is along a parallel of latitude. On a Mercator chart, the longitude scale is always found alongside the two parallels of latitude constituting the upper and lower boundaries of the chart. There is one very important difference between the latitude and longitude scales. Remember that, with the exception of the equator, parallels of latitude are small circles. One degree of longitude, therefore, will not equal 60 nautical miles except when measured along the equator. At all other locations, as the distance from the equator increases, the length in miles of one degree of longitude decreases. This fact is illustrated in Figure 4–5, where one degree of longitude is shown to be 60 miles wide at the equator, 52 miles at 30° north or

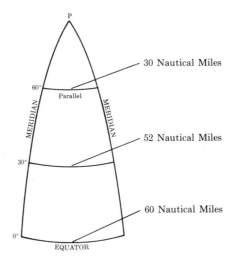

Figure 4–5. Contraction of the longitude scale.

south, 30 miles at 60° north or south, and zero at the two poles. For this reason, distance can never be directly measured along the longitude scale of the Mercator chart.

Chart Projections

In order for a globe of the world to be of practical use to the navigator when operating in restricted waters, it would have to be thousands of feet in diameter to be large enough to show all the necessary details of importance. Since no vessel could carry a globe of this size, even if one could be made, the navigator must rely instead on flat representations of areas of interest on the globe. Unfortunately, this leads to some problems. Experiment would prove that no considerable portion of a rubber ball can be spread out flat without some stretching or tearing. Likewise, because the earth is spherical in shape, its surface cannot be represented on a flat piece of paper without some distortion. The smaller the portion of the globe to be mapped, however, the less the distortion will be.

The surface of a sphere or spheroid is termed *nondevelopable* because of this fact—no part of it can be spread flat without some distortion. Through centuries of experimentation, however, the cartographer has learned to get around this problem by projecting the surface features of the terrestrial sphere onto other surfaces that are developable, in that they can be readily unrolled to form a plane. Two such surfaces are those of a cone and a cylinder. It is also true that a limited portion of the earth's surface can be projected directly onto a plane surface while keeping distortion within acceptable limits, if the area is small in relation to the overall size of the globe. Projections are termed *geometric* or *perspective* if points on the sphere are projected from a single point that may be located at the center of the earth, at infinity, or at some other location. Most modern chart projections are not projected from a single point, but rather they are derived mathematically.

The desirable properties for any projection include the following:

- True shape of physical features
- Correct angular relationships
- Representation of areas in their correct proportions relative to one another
- True scale, permitting accurate measurement of distance
- Rhumb lines (lines on the surface of the earth that cross all meridians at the same angle) represented as straight lines
- Great circles represented as straight lines

It is possible to preserve any one or even several of these desirable

properties in a given projection, but it is impossible to preserve them all in any one type. Although there are several hundred kinds of projections possible, only about half a dozen have ever been used for nautical charts. Of these six, only two have come into general use by the seagoing surface navigator—the Mercator and the gnomonic projections.

The Mercator Projection

As was mentioned in the introductory material of this chapter, the Mercator projection gets its name from a Flemish cartographer, Gerhardus Mercator, who developed it some four hundred years ago. It is the most widely used projection in marine navigation. Position, distance, and direction can all be easily determined, and rhumb lines plot as straight lines; it is also *conformal*, meaning that all angles are presented correctly, and, for small areas, true shape of features is maintained.

The Mercator is a cylindrical projection. To envision the principles involved, imagine a cylinder rolled around the earth, tangent at the equator, and therefore parallel to the earth's axis, as shown in Figure 4–6. Meridians, because they are formed by planes containing the earth's axis, appear as straight vertical lines when projected outward onto the cylinder from within the earth. As has been discussed earlier, however, the distance between successive meridians on the terrestrial sphere lessens as the distance from the equator increases, and finally becomes zero at the poles. On the cylinder, then, as the distance from the equator increases, the amount of lateral distortion steadily

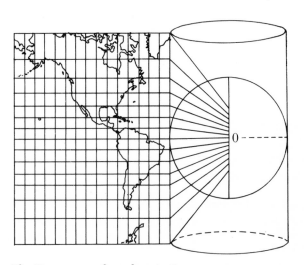

Figure 4–6. The Mercator conformal projection.

increases, and would approach infinity near the poles, projected to the ends of the cylinder. To maintain conformality—true shape—as distance increases from the equator on the cylinder, the latitude scale must be expanded as well. In the Mercator conformal projection, this is done mathematically; the expansion of the latitude scale approximates the secant of the latitude. It should be evident that the greater the distance from the equator, the greater is the distortion of this projection. An example often cited is Greenland, which when completely shown on a Mercator projection, appears to be larger than South America, although in actuality it is only about one-ninth as large. For this reason, most Mercator projections of the world are cut off at about 80° north and south latitudes.

The distortion of the true size of surface features, increasing continually as distance increases from the equator, constitutes the major disadvantage of this type of projection. Although there are Mercator charts that depict the polar regions by using a meridian as the circle of tangency of the cylinder, the gnomonic projection, described below, is generally preferred for this purpose. Another disadvantage of a Mercator chart depicting a large area is that great circles, other than a meridian or the equator, appear as curved lines. For conventional methods of navigation in the mid-latitudes of the world, however, the advantages of easy measurement of position, distance, and direction on the Mercator projection far outweigh its disadvantages, especially when relatively small areas are depicted.

The Gnomonic Projection

The gnomonic projection, in contrast to the mathematically derived Mercator conformal projection, is a geometrical projection in which the surface features and the reference lines of the sphere are projected outward from the center of the earth onto a tangent plane, as illustrated in Figure 4–7 on the following page.

There are three general types of gnomonic charts, based on the location of the point of tangency. It may be on the equator (equatorial gnomonic), at either pole (polar gnomonic), or at any other latitude (oblique gnomonic). An oblique gnomonic using a point of tangency located in the central North Atlantic is illustrated in Figure 4–8.

The gnomonic projection has been adapted to a number of different applications, but in surface navigation it is chiefly used because it shows every great circle as a straight line. Rhumb lines on the gnomonic projection appear as curved lines. Figure 4–9 on page 28 contrasts the appearance of a rhumb line and a great circle line on a Mercator chart with their appearance on a gnomonic chart.

In all three types of gnomonic projection, distortion of shape and

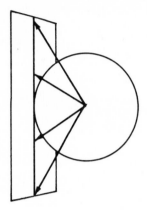

Figure 4–7. A gnomonic projection.

scale increases as the distance from the point of tangency increases. Within about 1,000 miles of the point of tangency the amount of distortion is not greatly objectionable, but beyond that, it increases rapidly. It is not possible to include as much as a hemisphere in a single gnomonic chart, as at 90° away from the point of tangency a point would be projected parallel to the plane of the projection.

Distance and direction cannot be measured directly from a gnomonic projection, although it is possible to determine great circle distances by means of a nomogram printed on the chart. This type of chart projection is used mainly to plot the optimum great circle route as a straight line from one place to another. Coordinates of points along the route are then picked off and transferred to a Mercator projection for further use, as described in Chapter 15 of this text, which deals with voyage planning. The gnomonic chart is useless as a working chart for normal plotting of navigational data.

There is a third type of projection, called a conic projection, which is based on the projection of a portion of the earth's surface onto a cone. With the exception of a series of charts depicting the Great Lakes, this type of projection finds its most extensive use in aeronautical charts; because the marine surface navigator rarely if ever uses this type of chart, it will not be discussed in detail herein.

Chart Interpretation

Before he can effectively use any chart, the navigator must first be able to interpret the chart, in regard to both its scale and also its symbols. The scale of a chart refers simply to the ratio between the actual dimensions of an area that the chart depicts and the size of the area as it appears on the chart. A scale of 1:80,000, for example, would mean that one unit of distance measured on the chart would represent

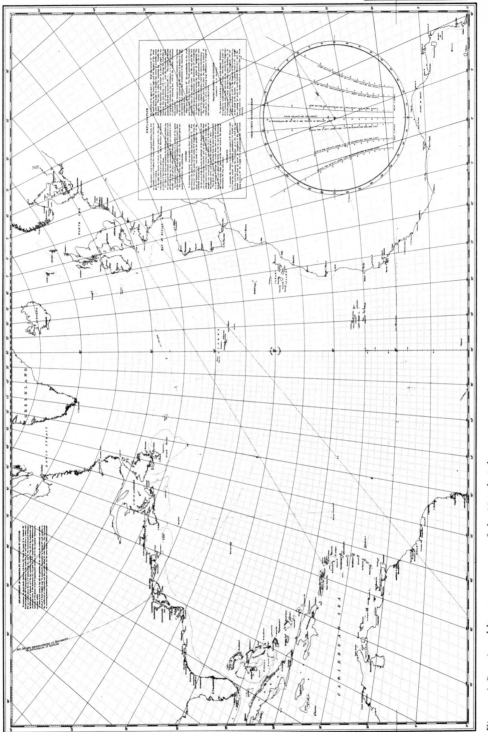

Figure 4–8. An oblique gnomonic of the North Atlantic.

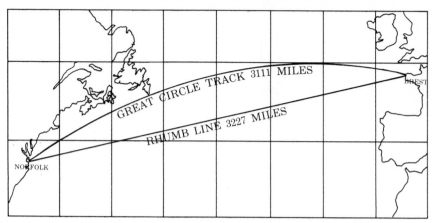

Mercator projection

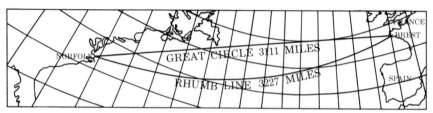

Gnomonic projection

Figure 4-9. *Appearance of a rhumb line and a great circle on a Mercator versus a gnomonic projection.*

80,000 of the same units in the real world. One inch on the chart would represent 80,000 inches, or one foot would represent 80,000 feet.

The terms *large scale* and *small scale* often cause much confusion to those who are not accustomed to working with charts. A ratio can be written as a fraction: the larger the value of a chart scale ratio, the larger is the scale of the chart. A chart having a large scale such as $\frac{1}{80,000}$ can represent only a relatively small area of the earth's surface without becoming prohibitively large. But a chart having a small scale like $\frac{1}{800,000}$ could represent an area 10 times as large on the same size chart. There is no firm definition of the terms large scale and small scale; the two terms are relative. Thus a chart of scale 1:800,000 would be a small-scale chart compared with one of scale 1:80,000; but the same chart of scale 1:800,000 might be called a large-scale chart in comparison with one having a scale of 1:8,000,000. The phrases

Large scale—small area
Small scale—large area

are often used as memory aids when discussing the subject of chart scale.

When a navigator gets out a nautical chart, he should take time to examine it in detail. All explanatory and cautionary notes appearing on the chart should be read and understood. He should check the scale and determine the date of issue of the chart as well as the date of the survey on which it is based. It should be mentioned here that there are some charts still in common use that are based on survey data collected in the last century, especially in the Indian, South Atlantic, and South Pacific oceans. The units in which water depths are recorded should be checked, and areas of the chart in which depth information is either completely missing or widely spaced should be regarded with extreme caution, particularly in coastal regions.

Many chart symbols and abbreviations are used on a chart to describe features of interest and possible use to the navigator. These constitute a kind of shorthand, and make it possible to insert a great deal of information in a small space on a chart. The symbols used are standardized, but some variations do exist, especially in the shading of shallow water, depending on the scale of the particular chart or chart series. It should be noted that it is not possible to provide the same amount of detailed information on all aids to navigation, whether natural or artificial, on a small-scale chart depicting a given area as would be possible on a large-scale chart. The navigator, therefore, should always keep his master navigation plot on the largest-scale chart practical in a given area. Chart symbols and abbreviations employed to present data on a modern chart produced by the United States are contained in a publication entitled *Chart No. 1* published by the Defense Mapping Agency and National Ocean Service, which is included as Appendix A in the back of this volume. The symbols therein, especially the ones in Section O pertaining to dangers to navigation, should be studied until complete familiarity is attained. Figure 4–10 is a portion of a mock chart depicting chart symbols for some of the more commonly encountered dangers to navigation.

As a final note on chart interpretation, it should be mentioned that a changeover is charted depths and heights from the customary (English) to the International (SI) metric system has been in progress for some time , in order to conform to bilateral chart reproduction agreements with other nations. On many new charts produced, water depths and heights of lights are shown in meters; land contours are also shown in meters, except where the source data is expressed in feet. As a consequence of this changeover, many shipboard echo sounders are now equipped with dual scales for use with either the customary English or International metric system units.

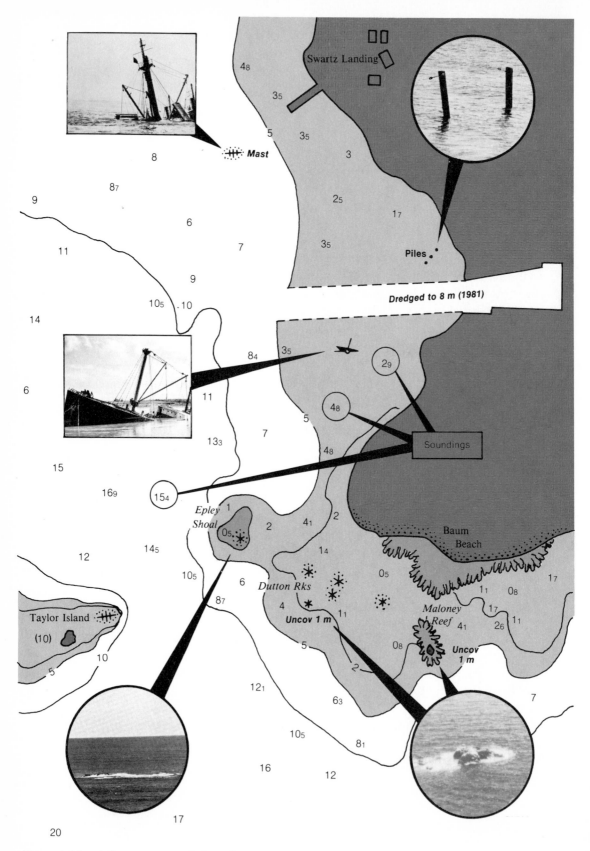

Figure 4–10.   A fictitious nautical chart illustrating symbology for various common hazards to navigation.

Determination of Position, Distance, and Direction on a Mercator Chart

As was stated earlier, the easy determination of position, distance, and direction is one of the chief advantages of the Mercator projection. In most cases, the tools needed to plot on this type of chart are few, consisting of a plotting compass, a pair of dividers, a parallel rule or some other type of plotter, and a sharp pencil.

A position of known latitude and longitude can be quickly plotted on a Mercator projection using only a compass. Since latitude is usually the first coordinate given in a position, it is natural to plot this coordinate first. To accomplish this, the latitude scale on the most convenient side of the chart is referred to, and the two parallels of latitude that bracket the given latitude are located. The pivot point of the compass is placed on the closest line and the compass is spread across the scale until the leaded point rests at the given latitude. Without the spread thus established being changed, the pivot point is then moved horizontally across the printed parallel of latitude until approximately at the correct longitude; the inexperienced navigator will often lay a straightedge or parallel rule vertically on the chart to assist in this endeavor. An arc is then swung, with the crest representing the proper latitude. Figure 4–11A illustrates this process.

To complete the plot, the same procedure is followed to plot the longitude, except that the compass spread is taken off the top or bottom scale of the chart.

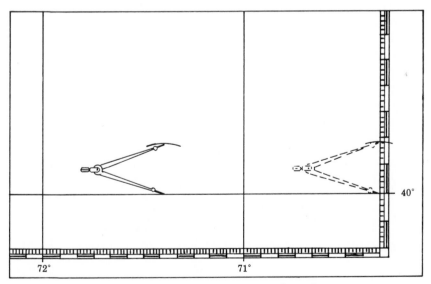

Figure 4–11A. *Swinging an arc to represent a given latitude.*

The desired position is located where the crests of the two arcs intersect, as illustrated in Figure 4–11B.

The reverse problem, that of determining the latitude and longitude of a position on a chart, is also easily accomplished. The compass is also a good instrument to use for this purpose, although the dividers can be used as well. Once again, inasmuch as the latitude is written first in a position, it is picked off first from the chart. The pivot point of the compass or dividers is placed on the nearest printed parallel of latitude directly above or below the given position, and the instrument is spread until the other point rests on the position. Without changing the spread thus established, the instrument is then shifted to the most convenient side of the chart, with the pivot point still on the chosen printed parallel of latitude. If the compass is used, a small arc is swung across the scale, as pictured in Figure 4–11C, and the latitude is read off.

In similar fashion, the longitude is picked off using the compass or dividers set for the distance between the given position and the nearest meridian printed on the chart, as shown in Figure 4–11D.

If dividers are used for this application, chance of error in reading the correct value of latitude and longitude from the chart scale is increased, because there is no definite line drawn across the scale to use as a reference mark.

It has already been explained that one minute of arc on the latitude scale is considered to be equivalent to one nautical mile. Because of

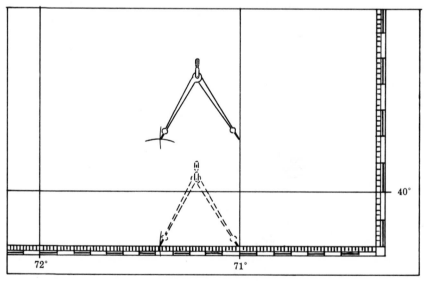

Figure 4–11B. Completing the plot of a given position on a chart.

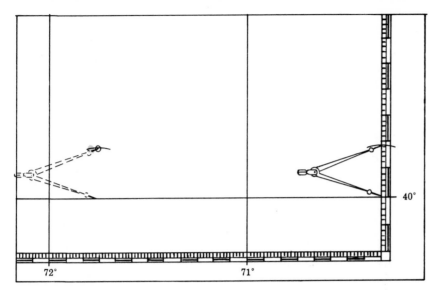

Figure 4–11C. Swinging a latitude arc across the latitude scale.

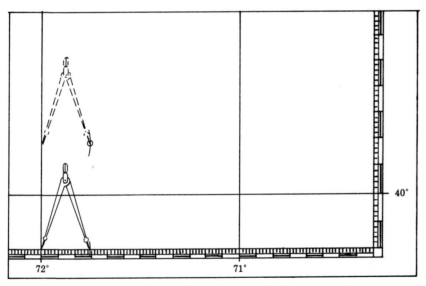

Figure 4–11D. Swinging a longitude mark across the longitude scale.

this fact, distance measurement on a Mercator chart is very simple, because the latitude scale on either the left or right border can be used as a distance scale. There is only one caution that must be observed. Because the latitude scale of a Mercator projection expands in length with increased distance from the equator, the length of a mile on the latitude scale of a Mercator chart is not constant. If the

chart depicts an area in the northern hemisphere, there are more miles per inch in the southerly portion (the bottom) of the chart. Conversely, if the charted area lies in the southern hemisphere, distortion of the scale will increase from north to south, and there will be more miles per inch of scale near the top of the chart. For this reason, that part of the latitude scale that is at the mean latitude of the distance to be measured should always be used for distance measurement. Consider the example in Figure 4–12.

A pair of dividers is the best instrument to use for this type of measurement. If the distance between points A and B is small enough, the span of the dividers may be increased until its points are over A and B, and the total distance can be measured on the side of the chart at the mid-latitude. If the points are so far apart that the dividers cannot reach all the way from A to B in one step, the dividers are spread to some convenient setting, such as 5 or 10 miles, and the distance along a line connecting A and B is then stepped off. After the last full step, there will usually be a remainder, which can be measured as described above and added to the distance stepped off, to obtain the total distance between points A and B. The procedure is illustrated in Figure 4–12.

On Mercator charts having a relatively large scale, such as those showing a harbor or river mouth, the distortion over the small amount of latitude covered by the chart is negligible. If this is the case, the chart will usually have separate miles and yards and kilometers and meters bar scales printed on it, which are very useful for precise navigation and piloting. The scales, which are based on the mid-

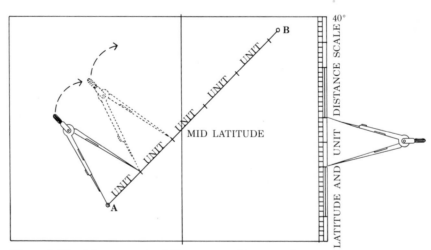

Figure 4–12. Measurement of distance on a Mercator chart.

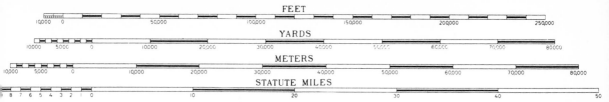

Figure 4–13. Typical chart bar scales.

latitude of the chart, will usually appear over a land area near the chart title block (Figure 4–13).

Measurement of direction on a Mercator projection is, like distance, quite simple because of the conformality of the Mercator. All straight line directions measured on a Mercator chart are *rhumb line* directions; a rhumb line is a line making a constant angle with all meridians it crosses. Since true directions are given with respect to a meridian, it follows that all straight rhumb lines drawn on a Mercator chart are true directions. As soon as a ship is steadied on a given true direction or heading, her course can be represented by a rhumb line drawn across the Mercator chart. Measurement of a rhumb line direction can be made with reference to any convenient meridian or parallel of latitude using any one of several instruments used for measuring direction that incorporates a protractor. A description of the more common instruments of this type is given in Chapter 7. Measurement of direction on a Mercator projection can also be accomplished by using a parallel rule or universal drafting machine to transfer the direction of a rhumb line to a nearby *compass rose*. A Mercator chart compass rose is nothing more than a 360° directional scale referenced to true north by means of a meridian; most roses also contain an inner magnetic direction scale for use with a magnetic compass. Whenever direction is measured on a Mercator chart, care must be taken to read the proper direction off the protractor or true scale of the compass rose; it is very easy to read a direction 180° away from the one desired.

Production of Nautical Charts

In the United States, almost all nautical charts, except those covering some inland rivers and lakes, are produced by one of two government activities—the Defense Mapping Agency (DMA) and the National Ocean Service (NOS). The former agency is concerned mainly with the production and upkeep of charts and related navigational publications covering all ocean areas of the world outside U.S. territorial waters, while NOS produces charts covering inland and coastal waters of the United States and its possessions. This latter organization is also charged with survey responsibilities in support

The Nautical Chart   **35**

of DMA. In cases where U.S. chart coverage of various ocean and foreign coastal areas is only superficial or lacking entirely, the DMA will often obtain and reproduce applicable charts of various foreign chart-production agencies, especially the British Admiralty.

Charts of some inland waterways, most notably the Mississippi, Ohio, and Missouri rivers and their tributaries, are prepared by the U.S. Army Corps of Engineers. They are referred to as "navigational maps," and are available for purchase from district offices of the Corps of Engineers. Because these charts are of relatively minor importance to seagoing navigators, they will not be discussed further in this text.

The Defense Mapping Agency

Until 1972, there were a number of independent organizations funded by the U.S. government to produce and supply charts and maps to the armed services and other government and private users. In 1972, several of these organizations were combined into the *Defense Mapping Agency* (DMA), which prints and distributes maps, charts, and supporting publications for all users within the Department of Defense and also for private sale. Between 1972 and late 1978, there were three production centers within DMA, each of which issued its own products: the Hydrographic Center (DMAHC) issued nautical charts, the Topographical Center (DMATC) issued land maps, and the Aerospace Center (DMAAC) issued aeronautical charts. In September 1978, in order to increase efficiency by sharing the use of similar production equipment and personnel, the Hydrographic and Topographic Centers were consolidated into a joint Defense Mapping Agency Hydrographic/Topographic Center (DMAHTC). Simultaneously with this consolidation, the distribution functions of all three previous centers were combined into a single agency named the Office of Distribution Services (DMAODS). The mission of DMAHTC is to provide topographic, hydrographic, navigational, and geodetic data, maps, charts, and related products to all armed services, other federal agencies, the Merchant Marine, and private and commercial mariners.

Since the 1978 consolidation, DMAHTC has been making rapid strides toward the integration of large-scale data processing equipment to automate all phases of chart production and supporting publication and correction services. Concurrently, a navigational data base is being built that eventually will be made available worldwide via telephone and communication-satellite links to users on a real-time basis, both to supply new navigational data to the base, and to obtain from the data base the latest corrections that apply to the charts and publications covering users' current areas of operation.

The National Ocean Service

The National Ocean Service* (NOS), formerly the U.S. Coast and Geodetic Survey (C&GS), is an independent activity within the National Oceanic and Atmospheric Administration (NOAA) in the Department of Commerce. The Coast and Geodetic Survey was established by Congress in 1807 and charged with survey responsibilities for all U.S. coastal waters, harbors, and off-lying island possessions. Since its formation in 1973, NOS has retained the old C&GS responsibility for maintaining accurate surveys of all U.S. coastal waters, and it performs surveys of other areas upon request in support of DMA. As previously mentioned, the NOS also produces a series of large-scale charts covering coastal and certain intracoastal waters of the United States, including the Great Lakes. NOS charts are cataloged in both the DMA chart catalog and a separate series of NOS catalogs, and are distributed by NOS sales offices and representatives.

The Chart Numbering System

Because of the tremendous quantity and variety of modern nautical charts in existence, the need for a logical chart numbering system became apparent to both producers and users of charts. At one time all charts produced by the various U.S. agencies engaged in chart production were simply numbered by series in the order in which they were printed, but in 1971 the former U.S. Navy Hydrographic Office (now the DMAHTC) began to convert the numbers of all charts depicting mid-ocean areas, foreign coastal waters, and certain U.S. harbor and coastal regions to the system now in effect. This conversion process was continued by the DMAHC upon its formation from the Hydrographic Office in 1972, and was finally completed in 1974. In July of that same year, conversion of all large-scale U.S. coastal, harbor, and river charts produced by the National Ocean Service to the DMA standard numbering system was completed.

The U.S. Chart Numbering System

In the U.S. chart numbering system, all commonly used nautical charts produced by both DMAHTC and NOS are assigned a number consisting of from one to five digits, according to their scale and the area they depict. The relationship between the number of digits appearing in the chart number and the chart scale is shown below:

Number of Digits	Scale
1	No scale involved
2	1:9,000,001 and smaller

*Prior to 1 December 1982, called the National Ocean Survey.

3	1:2,000,001 to 1:9,000,000
4	Miscellaneous and special nonnavigational charts
5	1:2,000,000 and larger

Because of the chart scale required to represent a given area, it happens that only charts with two- or three-digit identification numbers are of suitable scale to depict large ocean basins or their subdivisions. Likewise, only charts having a five-digit number (a scale larger than 1:2,000,000) are suitable for charting coastal regions with the great detail necessary for piloting applications. A chart number, therefore, not only classifies a chart as to its scale, but it also indicates the size of the geographic region it represents.

Charts bearing a single digit are in reality not nautical charts at all, but rather they are various supporting publications that have no scale. The booklet of nautical chart symbols (*Chart No. 1*) previously mentioned in this chapter is one such publication. Others include chart symbol sheets for other nations, two sheets illustrating national flags and symbols, and a sheet showing the international signal flags and their meanings.

Charts labeled with two- or three-digit numbers are relatively small-scale charts, which for the most part depict either ocean basins or their subdivisions. For the purposes of this type of chart, all ocean areas of the world have been included in one of nine designated basins, numbered as shown in Figure 4–14. The first digit of a two- or three-digit chart number, with three exceptions, denotes the ocean basin in which the area represented by the chart is located. Because of the small size of the Mediterranean (basin 3), the Caribbean (basin 4), and the Indian Ocean (basin 7), there can be no useful two-digit charts of scale smaller than 1:9,000,000 covering these areas. The two-digit numbers 30 through 49 and 70 through 79 are therefore available for other purposes; they are used for large charts which, because of their nature, would not refer to a single ocean basin. Charts of this type include the magnetic dip chart (No. 30), the magnetic variation chart of the world (No. 42), and the standard time zone chart of the world (No. 76).

Charts identified by four-digit numbers are so-called nonnavigational plotting charts and sheets. Examples of this type of chart include large wall and planning charts with scales ranging from 1:1,096,000 to 1:12,000,000, Omega and Loran-C plotting charts, and special gnomonic or azimuthal equidistant charts produced for communications planning purposes. There are over 4,000 different special-purpose charts in existence.

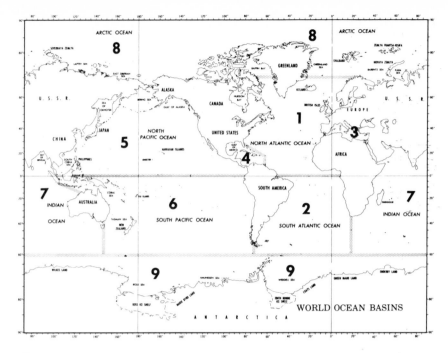

Figure 4–14.   Ocean basins of the world.

Five-digit charts represent coastal areas; these charts range in scale from 1:2,000,000, which might be used for a chart depicting an entire coastline, to scales as large as 1:8,000, which could depict a river entrance or small harbor. All coastal areas in the world are divided into nine coastal regions, as illustrated in Figure 4–15 next page. Note that these coastal regions are independently numbered in comparison to the ocean basin numbering concept discussed earlier, which forms the basis of the numbering system for two- and three-digit charts. The coastal regions do not have the same numbers as the ocean basins of which they are a part. Where possible, all coastal waters surrounding a major continent are located in the same-numbered coastal region. Each of the nine coastal regions is further subdivided into a number of subregions; altogether there are 52 different coastal subregions throughout the world. The first two digits of a five-digit coastal chart identify the coastal region and subregion in which the charted area is located. The last three digits place the chart in geographic sequence through the subregion.

Figure 4–16 is an expanded view of coastal region 5 taken from the DMAHTC *Catalog of Nautical Charts*, described in detail in the next chapter. Shown in the figure are the intermediate-scale charts presently available that portray the coastal areas within region 5 and its

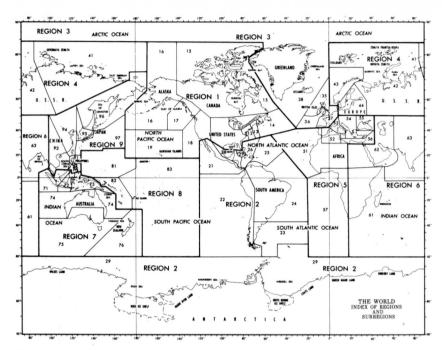

Figure 4–15. Coastal regions and subregions of the world.

seven subregions. Notice that the charts are not numbered consec-utively. The reason for this is that there are many large-scale charts in existence that depict small portions of the coastal waters within the areas of coverage of most of these intermediate-scale coastal charts. Each of these are also assigned a five-digit number such that they will fall into the proper geographic order. Moreover, even among the larger-scale charts many numbers have been left unused, so that future charts not yet in existence may be placed in their proper geographic order as they are produced, without having to change any numbers on existing charts.

Most of the charts produced by DMAHTC and NOS and used by the navigator in the normal course of his duties will be marked as described above. This numbering system is also applied to nautical charts produced by foreign countries that the DMAHTC maintains within its distribution system. Standard five-digit numbers are as-signed to these charts so that they may be filed in a logical sequence with charts produced by the United States. The numbering system provides many benefits to the navigator; not only does it indicate a chart's scale and the area it portrays, but it also facilitates the ar-rangement of charts into *portfolios*, which are aggregations of charts grouped for the most part by coastal subregion for ease of indexing

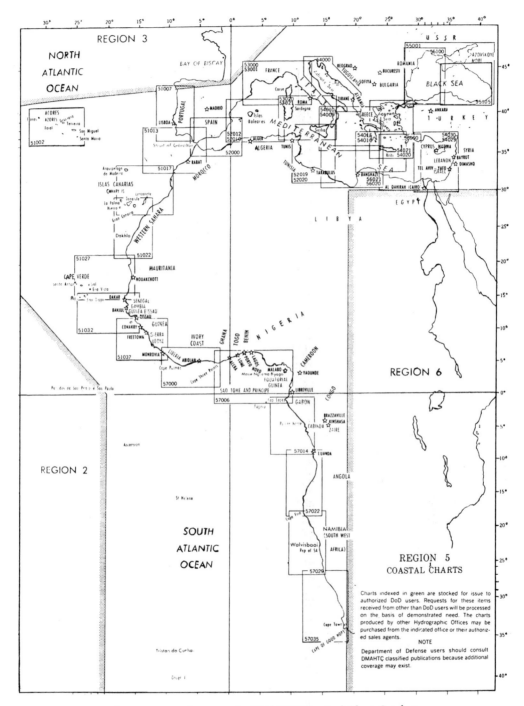

Figure 4–16. Coastal region 5, as shown in the DMAHTC Nautical Chart Catalog.

and storage. Altogether there are 55 chart portfolios, each containing anywhere from about 30 to over 250 charts. Fifty-two of these correspond to the 52 coastal subregions of the world (see Figure 4–15), and three contain general charts of the Atlantic, Pacific, and the world. Complete listings of the charts comprising each portfolio are contained in *Part 1–N–L* of the DMAHTC *Catalog of Nautical Charts* described in Chapter 5. Another part of this same publication contains allowance listings of all portfolios and publications that are to be carried on board each Navy ship, staff, and shore activity. Navy destroyer-type and larger ships might carry as many as twenty to thirty portfolios of charts on board, depending on the fleet assignment, in order that they might be prepared to operate anywhere within their assigned fleet area on short notice. Most merchant ships of the larger steamship companies, on the other hand, carry far fewer portfolios, as they are usually assigned to specific runs for fairly long periods of time. When necessary, new portfolios of charts can be ordered from the DMA Office of Distribution Services.

Most of the *Catalog of Nautical Charts* is composed of booklets containing graphic drawings of each region and subregion, designed to enable the navigator to quickly find the numbers of charts covering areas of interest to him. Having found the chart numbers he requires, the navigator then goes to the appropriate portfolio storage locations and pulls out the charts he needs. An example of the use of the *Catalog* to find a desired chart number is given in the next chapter.

DMAHTC Charts Outside the Five-Digit Numbering System

There are three types of DMAHTC charts with which the surface navigator will occasionally come into contact that are not covered by the standard five-digit numbering system. These are combat charts, a new series of naval warfare planning charts, and general bathymetric charts. Because of their nature, many of these charts bear a security classification, and must be kept separately locked up in a secure storage area.

Combat charts are printed in special grid patterns for use in offshore coastal bombardment. Although they are almost all drawn to a scale of 1:50,000 and describe portions of a coastal region, it is nevertheless desirable to distinguish them from normal coastal region navigational charts because of their special military use. They are identified by a five-character group, consisting of two digits, a central alphabetic letter, and a final two digits. The initial two digits identify the coastal subregion in which the area depicted is located, and the last two digits place the chart in geographic sequence with respect to other combat charts of the area. The middle letter has no significance other

than to identify the chart as a combat chart; two adjacent combat charts in the same subregion can have completely different letters.

The *naval warfare planning chart* series is a relatively new series of charts now being produced; among the series are a number of ASW environmental prediction charts and specialized grid reference charts. These charts are identified by an alphanumeric code similar to the following: NAR 7–0408–2401. For the purposes of this series, all oceans of the world have been divided into seven index areas; each index area has been subdivided into rectangular index regions, and each region has been further subdivided into rectangular subregions. The letters in the first group of characters identify the index area, and the number(s) the index region. In the example given above, NAR would refer to the North Atlantic area, and 7 would refer to the index region 7 therein. The second group of four digits identifies the subregion, and the third group, the type of operation and operational data portrayed.

General bathymetric charts and *plotting sheets* are also relatively new, and consist of compilations of sounding data presented on Mercator charts at a scale of 1:1,000,000. The soundings are obtained from random track and survey data from naval, merchant, and research vessels of the United States and foreign nations. These charts carry a five-character designation beginning with the letter G, followed by four numbers that identify the subregion covered. (The subregion codes are the same as those used for the naval warfare planning chart series described above.) The general bathymetric charts currently available covering the Atlantic are shown in Figure 4–17 next page.

The Chart Correction System

All charts published by both the Defense Mapping Agency and the National Ocean Service are edited and corrected to reflect the latest information available at the time the chart was printed. Unfortunately, a considerable amount of time often elapses between successive printings of a given chart or other navigational publication. As a result, provisions must be made for keeping mariners apprised of changes in hydrographic conditions that will affect the accuracy of their charts and publications as soon as possible after the changes occur.

The principal means by which necessary periodic corrections to DMAHTC and NOS charts and publications are disseminated are the *Notice to Mariners* and *Local Notice to Mariners*. These are bulletins in pamphlet form distributed by mail each week that contain all corrections, additions, or deletions to all DMAHTC and NOS charts and most DMAHTC publications reported during the week preceding issue of the *Notices*. The *Notice to Mariners* published by DMAHTC

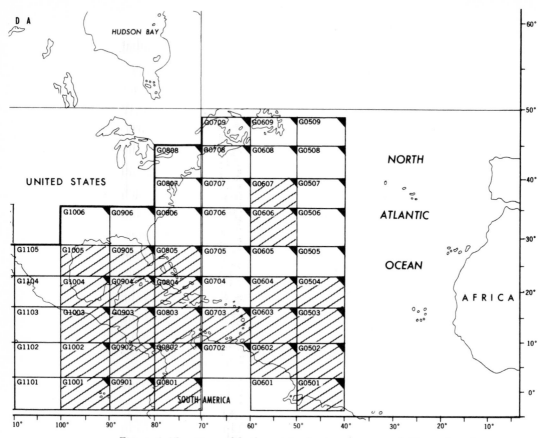

Figure 4–17. *General bathymetric chart coverage available in the Atlantic, 1980.*

contains all changes relating to oceanic and coastal areas worldwide, while the *Local Notice to Mariners* issued by each of twelve U.S. Coast Guard districts contains changes pertaining to U.S. inland waters within each district. Navigators of seagoing vessels that also operate in the inland waterways of one or more Coast Guard districts will need to obtain and use the *Local Notices* issued for each district, as well as the weekly *Notice to Mariners*. Corrections reported in the *Notices* are generated both within and without DMAHTC and NOS; all mariners everywhere are urged to report to the applicable agency any recommended corrections whenever changes are observed. A form for this purpose is included inside the back cover of each *Notice*.

Format and Use of *Notice to Mariners*

Each *Notice to Mariners* is made up of three sections, as shown in Figure 4–18, which depicts a title page of a *Notice*. Within each section, corrections are listed sequentially by chart/publication/item

Figure 4–18. *Cover of a* Notice to Mariners.

number in the format shown in Figure 4–19A on page 46. As noted in the figure, those corrections preceded by a star(★) are based on original U.S. source information; a T indicates a temporary correction; and a P indicates a preliminary correction. Figure 4–19B illustrates a typical correction page.

The navigator and his staff are not obliged to enter every correction into every chart and publication to which it applies week by week. To conserve the use of nautical charts and publications, and to reduce the amount of chart correction work aboard ship, the chart/publication correction card system was established.

Under this sytem, every chart and publication kept on board a ship should have a correction card on file (Figure 4–20 page 48). As each weekly *Notice to Mariners* is received, the navigator, or more often in the Navy one of his quartermasters designated as the chart petty officer, scans the *Notice* and enters its number together with the page number bearing the correction onto the cards of any charts or publications that he carries on board. He actually corrects only those frequently used charts and publications that cover the area in which the ship is operating, plus any additional charts and publications that the ship's commanding officer might specify. After the required changes

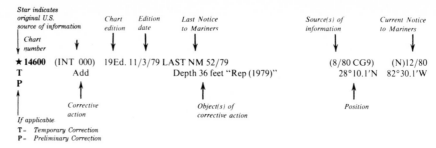

International No.

Star indicates original U.S. source of information		Chart edition	Edition date	Last Notice to Mariners	Source(s) of information	Current Notice to Mariners
Chart number		↓	↓	↓	↓	↓
★ **14600**	(INT 000)	19Ed. 11/3/79 LAST NM 52/79			(8/80 CG9)	(N)12/80
T	Add			Depth 36 feet "Rep (1979)"	28°10.1′N	82°30.1′W
P						
↑	↑			↑	↑	
If applicable	Corrective action			Object(s) of corrective action	Position	

T = Temporary Correction
P = Preliminary Correction

Figure 4–19A. Format of a Notice to Mariners *correction.*

	SECTION I	NM 45/80
★ 11344	21Ed. 8/9/80 LAST NM 44/80	(39/80 CG8) 45/80
	Delete Platform	29°34′28″N 93°02′17″W
★ 11349	24Ed. 3/15/80 LAST NM 44/80	(41/80 CG8) 45/80
	Add Paltform (lighted) HORN	29°30′13″N 92°19′42″W
★ 11356	20Ed. 1/26/80 LAST NM 38/80	(40/80 CG8) 45/80
	Substitute Platform (lighted) HORN for buoy and submerged well (covered 53 feet) (See 18/80-11356)	28°38′46″N 91°01′11″W
★ 11357	20Ed. 1/19/80 LAST NM 44/80	(40, 41/80 CG8) 45/80
	Add Platforms (lighted) HORN	28°54′29″N 90°27′57″W
		28°56′29″N 90°41′21″W
★ 11358	29Ed. 8/9/80 LAST NM 44/80	(40/80 CG8) 45/80
	Add Platform (lighted) HORN	29°09′15″N 90°01′08″W
11360	24Ed. 2/9/80 LAST NM 42/80	(40/80 CG8) 45/80
	Delete Stranded wreck (PA) (See 38/80-11360)	28°58.3′N 89°08.3′W
★ 11361	43Ed. 8/9/80 LAST NM 42/80	(40, 41/80 CG8) 45/80
	Add Platform (lighted) HORN	28°49′20″N 89°23′42″W
	(And Inset South Pass)	
	Delete Stranded wreck (PA) (See 42/80-11361)	28°58.3′N 89°08.3′W
★ 11364	23Ed. 2/16/80 LAST NM 44/80	(NOS) 45/80
	Add Light flare symbol to light dot 91	29°48′54″N 89°35′12″W
★ 11370	7Ed. 9/8/79 LAST NM 44/80	(39, 41/80 CG8) 45/80
	(Side A)	
	Delete Passing light (Fl R 2.5sec) from Range "A" front light	29°58′20″N 90°16′54″W
	Change Range "A" rear light to Occ R 4sec Light (105.1) to 10 Fl R 2½sec HORN (Priv maintd)	29°58′26″N 90°16′48″W
		29°57′11″N 90°09′46″W
	Range lights, front to Qk Fl R 17ft	30°01′06″N 90°46′02″W
	rear to Occ R 4sec 34ft	30°01′08″N 90°45′59″W
★ 11372	14Ed. 7/21/79 LAST NM 41/80	(CL1208/80) 45/80
	(Side B)	
	Add Tabulation of controlling depths from back of Section I	

Figure 4–19B. Excerpt from a correction page of the Notice to Mariners.

are made, the *Notice to Mariners* is then inserted into a file for possible future reference.

When a ship is scheduled to begin operating in a new area, the cards of all the charts and publications that will be used therein are pulled from the card file. Corrections are made as necessary using the file of old *Notice to Mariners* to bring the charts and publications up to date.

Summary of Corrections

For convenience in cases where older editions of charts and publications have been newly obtained, such as when a new ship is initially outfitted or when a ship is to proceed to a new area of operations, or where charts and publications stowed on board have not been updated for some time, the DMAHTC issues semiannually a set of five *Summary of Corrections* volumes, with areas of coverage as follows:

Volume Number	Area Covered
1	East Coast of North and South America
2	Eastern Atlantic and Arctic Oceans, including the Mediterranean Sea
3	West Coast of North and South America, including Antarctica
4	Western Pacific and Indian Oceans
5	World and Ocean Basin Charts, U.S. *Coast Pilots*, *Sailing Directions*, and Miscellaneous Publications

Each of the *Summary* volumes is cumulative, and contains corrections from previous volumes as well as all applicable items for the most recent edition of each chart or publication that appeared in the last six months' *Notice to Mariners*. If a new edition of a chart or publication appears during the period of coverage of a particular *Summary*, that volume will contain only those cumulative corrections to be applied to the new edition, and will drop all items pertaining to the previous edition. Because the *Summary* does not list corrections affecting many of the DMAHTC and NOS navigational publications described in the next chapter, it cannot be used as a substitute for a complete *Notice to Mariners* file.

Radio Broadcast Warnings

Occasionally it becomes necessary to promulgate changes affecting the safe navigation of a body of water more rapidly than can be done

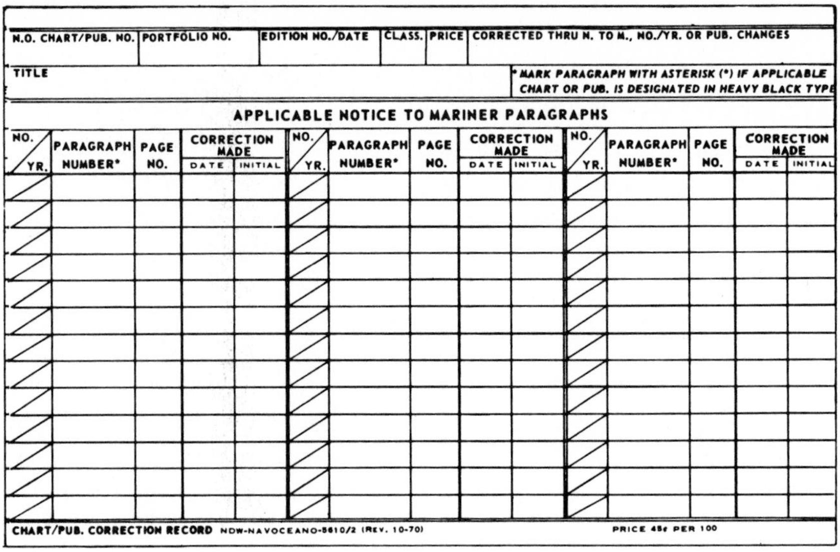

N.O. CHART/PUB. NO.	PORTFOLIO NO.	EDITION NO./DATE	CLASS.	PRICE	CORRECTED THRU N. TO M., NO./YR. OR PUB. CHANGES		
TITLE					* MARK PARAGRAPH WITH ASTERISK (*) IF APPLICABLE CHART OR PUB. IS DESIGNATED IN HEAVY BLACK TYPE		

APPLICABLE NOTICE TO MARINER PARAGRAPHS

NO. /YR.	PARAGRAPH NUMBER*	PAGE NO.	CORRECTION MADE		NO. /YR.	PARAGRAPH NUMBER*	PAGE NO.	CORRECTION MADE		NO. /YR.	PARAGRAPH NUMBER*	PAGE NO.	CORRECTION MADE	
			DATE	INITIAL				DATE	INITIAL				DATE	INITIAL

CHART/PUB. CORRECTION RECORD NDW-NAVOCEANO-8610/2 (REV. 10-70)   PRICE 48¢ PER 100

Figure 4–20. A chart correction card.

by means of the weekly DMAHTC *Notice to Mariners* or Coast Guard *Local Notice to Mariners*. Radio broadcasts are used for this purpose. Within each of the twelve Coast Guard districts covering U.S. waters, *Broadcast Notices to Mariners* are transmitted as required by various Coast Guard, Navy, and commercial radio stations. The information disseminated in these broadcasts is included in the next *Local Notice to Mariners* issued, if still valid. Information pertaining to littoral and mid-ocean areas plied by ocean-going vessels is broadcast by the Worldwide Navigational Warning System operated by member nations of the International Hydrographic Organization. For purposes of the system, the oceanic areas of the world have been divided into 16 so-called *NAVAREAs*, as shown in Figure 4–21. A different nation or nations are assigned responsibility for coordinating the long-range broadcast of all urgent navigational warnings pertaining to their assigned NAVAREAs; the United States is Area Coordinator for NAVAREAs IV and XII.

In addition to the NAVAREA region IV and XII broadcasts, DMAHTC also transmits as required special messages called *HYDROLANTS* and *HYDROPACS* that provide somewhat redundant coverage to that offered by the NAVAREA system. HYDROLANTS cover the Atlantic Ocean, Gulf of Mexico, and Caribbean Sea, and HYDROPACS cover the Pacific and Indian oceans. As in the case of the *Broadcast Notices* covering U.S. waters, if of continuing interest the information disseminated in the NAVAREA and HYDROLANT

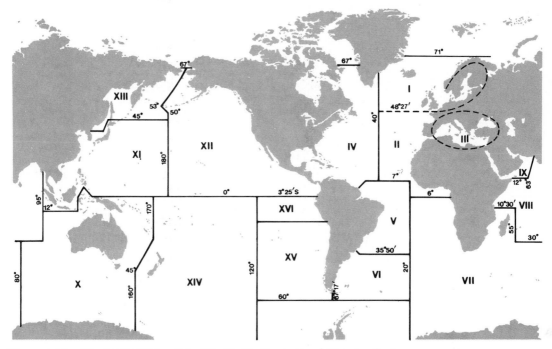

Figure 4–21. The NAVAREAs *of the* Worldwide Navigational Warning System.

and HYDROPAC broadcasts is included in the subsequent issue of the DMAHTC *Notice to Mariners*, as well as in the DMAHTC *Daily Memoranda* described in the following section.

Full details of all navigational warning broadcasts by both U.S. and foreign stations are contained in the DMAHTC publications *Radio Navigational Aids, Nos. 117A* and *117B*.

Other Printed and Broadcast Warnings

The DMAHTC publishes a set of *Daily Memoranda* each working day in two editions, one for the Atlantic, and one for the Pacific. Each *Daily Memorandum* contains the text of appropriate NAVAREA IV and XII warnings, as well as any HYDROLANT or HYDROPAC messages broadcast in the past 24 hours or since the previous working day. The *Daily Memoranda* are sent to naval operating bases, naval stations, customs houses, and major shipping company offices, where they may be picked up for use by navigation personnel of both naval and merchant vessels in port.

Special Warnings are occasionally broadcast by U.S. Navy and Coast Guard radio stations to disseminate official U.S. governmental proclamations affecting shipping. Each warning is consecutively numbered, and is included in the next edition of the appropriate *Daily*

Memorandum and *Notice to Mariners.* All *Special Warnings* in effect are published in the first *Notice to Mariners* of each year, which is issued in January.

Summary

This chapter has described the fundamental characteristics of nautical charts and the methods by which they are kept up to date. The modern chart is probably the single most important aid to the marine navigator, because without it piloting in coastal waters and navigating at sea would be virtually impossible. The navigator should be thoroughly familiar with this most fundamental of all nautical tools, in order that he may use it correctly and effectively. Having selected and obtained the most appropriate chart for his intended area of operations, the navigator must remember always to ensure that it is brought up to date prior to use. Numerous cases of grounding have been at least partially attributed to failure of the navigator to have an up-to-date chart of the area at his disposal.

In addition to nautical charts, the navigator has other sources to which he can refer to obtain information about any area of the world in which he may be operating. This information is contained in a variety of publications, which may be broadly classified by type as chart supplemental publications, manuals, navigation tables, and almanacs. Most of these publications are produced by the Defense Mapping Agency Hydrographic/Topographic Center (DMAHTC) and distributed by the Office of Distribution Services in the same manner as charts, with the exception of certain manuals and almanacs obtained from the U.S. Coast Guard, the National Ocean Service (NOS), and the Naval Observatory. On board ship, the publications are usually stowed in the chartroom, for easy access by the navigator, his assistants, and the ship's deck watch officers.

This chapter will describe the contents and use of the more important navigational publications usually carried on board almost all seagoing vessels, and particularly those of the U.S. Navy. Some of the information given herein is reiterated in the chapters of this text dealing with applications of the various publications described.

Catalog of Nautical Charts, Publication No. 1–N

The DMAHTC *Catalog of Nautical Charts*, more familiarly known as *Publication 1–N*, is an illustrated catalog of all unclassified nautical charts produced by the DMAHTC. The *1–N* contains serialized graphic drawings of the world that allow the navigator to locate visually the numbers of all charts and applicable *Sailing Directions* that describe an area of interest to him. The basic *Catalog* is a set of eleven large booklets divided into two parts: an introductory booklet containing general product and ordering information and listings of special-purpose charts and tables, and a second part made up of a booklet containing numerical listings of all charts and publications by portfolios, plus nine coastal region booklets. A summary of the contents of the booklets in the *Catalog* follows:

Part 1

PUB. 1–N–A Contains information on miscellaneous and special-purpose navigational charts, sheets, and tables, including:
- Ordering procedures and agents
- Lists of DMAHTC publications
- Lists of world charts
- Lists of Loran and Omega charts and lattice tables
- Lists of unclassified bathymetric charts

Part 2

PUB. 1–N–L Contains complete numerical listings of all unclassified charts and publications by portfolio number

PUB 1–N, REGION 1 United States and Canada

PUB 1–N, REGION 2 Central and South America and Antarctica

PUB 1–N, REGION 3 Western Europe Iceland, Greenland and the Arctic

PUB 1–N, REGION 4 Scandinavia, Baltic and U.S.S.R.

PUB 1–N, REGION 5 West Africa and the Mediterranean

PUB 1–N, REGION 6 Indian Ocean

PUB 1–N, REGION 7 Australia, Indonesia and New Zealand

PUB 1–N, REGION 8 Oceania

PUB 1–N, REGION 9 East Asia

In addition to the basic publication, there are also two other parts not available for general use. These are *Pub. 1–N–S*, which contains listings of classified charts and publications available only to U.S. Navy and other U.S. governmental activities, and *Pub 1–N–P*, which contains allowance listings of unclassified nautical charts and publications for commands and ships of the U.S. Navy, Military Sealift Command (MSC), Naval Control of Shipping (NCOS), and U.S. Coast Guard. The use of this latter part is described below.

Arranging the *Catalog* in booklet form facilitates the making of changes in its contents. New booklets are produced as required by DMAHTC, and interim corrections are disseminated via change sheets and the *Notice to Mariners*.

Use of the Basic *Catalog of Nautical Charts*

Each coastal region booklet contains a series of interrelated graphics that indicate the numbers of all available DMAHTC or NOS charts

and *Sailing Directions* volumes covering that region (the latter publications are described later in this chapter). Page 1 of each booklet shows the geographic area of the region divided into a number of color-coded subdivisions (see Figure 5–1A). Blue-bounded subdivi-

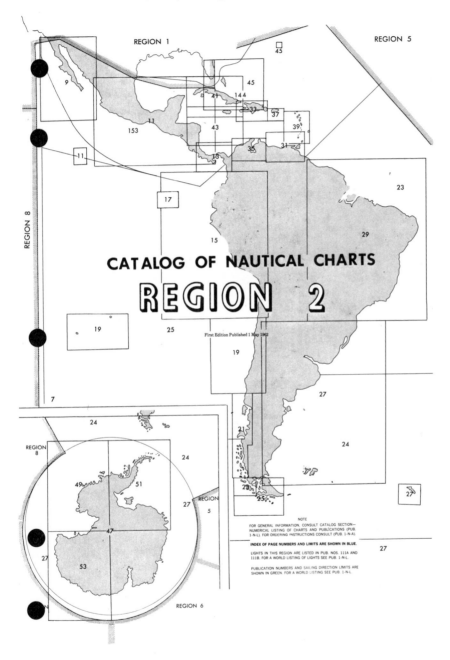

Figure 5–1A. Page 1 of Coastal Region 2 *of the* Catalog of Nautical Charts.

sions are marked with the page number within the booklet on which the large-scale chart coverage for that area is depicted. Green-bounded subdivisions indicate the *Sailing Direction Enroute* and *Planning Guide* volume numbers (see pages 60–63) applicable within the subdivision; the two types of volumes are differentiated by a rectangle printed around *Planning Guide* numbers. The next few pages consist of an index of world charts, immediately followed by a graphic depicting the intermediate-scale chart coverage throughout the subregion. The remaining pages depict all other large-scale charts available within each of the blue subdivisions of page 1.

As an example of the use of the *Catalog*, suppose that a navigator wished to find the numbers of all charts he would need in entering

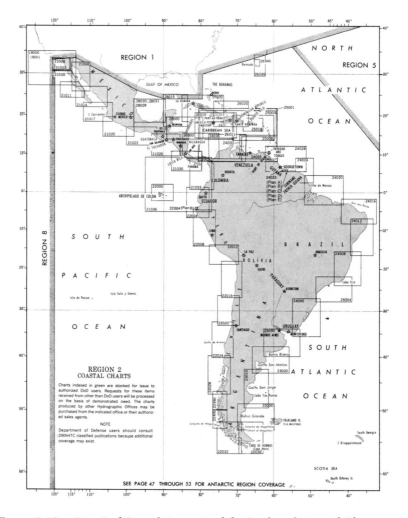

Figure 5–1B.   Page 7 of Coastal Region 2 *of the* Catalog of Nautical Charts.

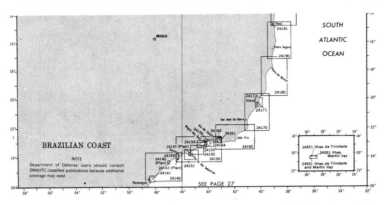

Figure 5–1C. *A portion of page 29 of* Coastal Region 2 *of the* Catalog of Nautical Charts.

Rio de Janeiro, Brazil. First, he ascertains from a diagram of the coastal regions of the world that the eastern coast of South America lies in coastal region 2, so he refers to the Coastal Region 2 booklet. Opening the booklet to page 1, reproduced in Figure 5–1A, he notes that Rio de Janeiro lies within a blue subdivision marked with page number 29, and a green subdivision marked with the *Sailing Directions* volume number 23, which is the volume describing the coastal area of South America within which Rio is located. Turning next to page 7, shown in Figure 5–1B, he obtains and records the numbers of the intermediate-scale charts covering the approaches to Rio, Nos. 24004 and 24008. Finally, turning to page 29, reproduced in part in Figure 5–1C, he obtains and records the number of the large-scale charts covering the approaches and harbor of Rio de Janeiro. They are chart Nos. 24160, 24161, and 24162, in order of increasing scale.

After having thus determined the numbers of the charts and the *Sailing Directions* volume he needs, the navigator then goes to the appropriate portfolio storage location, pulls the needed charts, and corrects them up to date by use of the chart correction card system explained in the preceding chapter. In this case, all the charts would be located in Portfolio 24. He would obtain the *Sailing Directions* volume from the location where they are stored, normally the charthouse.

The chart and *Sailing Directions* numbers applying to any location throughout the world can be found by using *Publication 1–N* in the manner described above.

Use of *Pub. 1–N–P* by Navy Ships

As previously mentioned, *Pub. 1–N–P* contains allowance listings that specify those unclassified chart portfolios and publication num-

bers that are to be carried on board activities and ships of the U.S. Navy Atlantic and Pacific Fleets, MSC, NCOS, and the U.S. Coast Guard. In the case of Navy ships, each fleet and submarine force commander also promulgates in the form of supplemental instructions allowance lists for classified charts and publications to be carried by each staff and ship type. The various allowance listings represent the minimum chart and publication requirements for all active and reserve ships and staffs. All Navy ships except those specifically authorized by higher commands to carry a reduced or modified allowance must carry on board all chart portfolios and publications listed as being in the ship's allowance by the *1–N–P* and the supplemental fleet commanders' instructions. Unless otherwise specified, two copies of each chart and publication listed are supplied to most ships. The DMA Office of Distribution Services (DMAODS) will automatically service each of these as new editions or change pages are issued.

When a ship is scheduled for a yard overhaul exceeding one year in length, or if a lack of storage facilities during a shorter overhaul period precludes proper maintenance, DMAODS will normally suspend automatic distribution of charts. Generally, publications are kept on board throughout the overhaul, as they are less subject to change.

Naval ships under construction or ships reactivated for duty are automatically provided a commissioning allowance of charts and publications approximately 30 days prior to the scheduled commissioning date, unless requested at an earlier date. Navigators attached to ships in this category should check with DMAODS well in advance of the desired distribution date to ensure that the agency is aware of the impending commissioning.

As an example of the use of *Pub. 1–N–P*, if a Navy navigator's ship were attached to the Second Fleet home-ported in Norfolk, Virginia, he might determine from the Allowance Listing section that all ships of that type in the Atlantic Fleet are required to carry on board among other portfolios all charts of Portfolio 14. He would then consult part *1–N–L* of the *Catalog* to ascertain the numbers and latest edition dates of all charts in Portfolio 14. Any charts in Portfolio 14 presently on board could be checked against this list to determine whether they were the latest editions available. Any charts superseded by new editions and any missing charts could then be ordered from DMAODS to bring the ship's supply of charts into conformance with the allowance requirements for that ship.

NOS *Nautical Chart Catalogs*

The National Ocean Service *Nautical Chart Catalogs* consist of five separate folded sheets printed with graphic drawings similar to those

in *Publication 1–N*. Each sheet contains the numbers of all NOS charts and related publications that pertain to the area covered, as well as descriptions of other NOS publications of general interest and ordering information. The areas of coverage of each of the five *Catalog* sheets are indicated below:

Nautical Chart Catalog Number	Area Covered
1	Atlantic and Gulf Coasts
2	Pacific Coast and Pacific Islands
3	Alaska and the Aleutian Islands
4	Great Lakes
5	Bathymetric Maps and Special Purpose Charts

A portion of *Nautical Chart Catalog 1* covering the Chesapeake Bay appears in Figure 5–2 on the following page.

As mentioned in the preceding chapter, all NOS coastal charts are assigned a standard five-digit chart number and most are also cataloged in Region 1 of *Publication 1–N*.

Coast Pilots

The U.S. *Coast Pilots* are a series of nine bound volumes containing a wide variety of supplemental information concerning navigation and piloting in the coastal and intracoastal waters of the United States and its possessions, including the Great Lakes. They are published by the National Ocean Service and distributed to Navy users by DMA. Civil users may purchase them at any NOS distribution agency.

The nine volumes of the *Coastal Pilots* are arranged as follows:

Volume Number	Area Covered
	Atlantic Coast
1	Eastport, Maine to Cape Cod
2	Cape Cod to Sandy Hook
3	Sandy Hook to Cape Henry
4	Cape Henry to Key West
5	Gulf of Mexico, Puerto Rico, and Virgin Islands
	Great Lakes
6	Great Lakes and St. Lawrence River

Pacific Coast
7	California, Oregon, Washington, and Hawaii

Alaska
8	Dixon Entrance to Cape Spencer
9	Cape Spencer to Beaufort Sea

The *Coast Pilots* are of great value to the navigator during the planning stages of any voyage through the coastal waters of the United

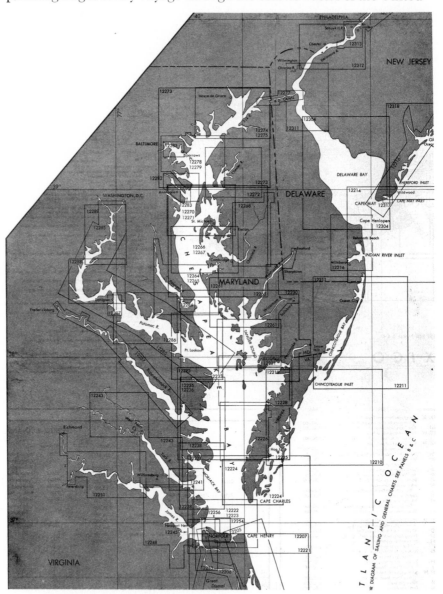

Figure 5–2.   Chesapeake Bay area, NOS Nautical Chart Catalog 1.

13. CHESAPEAKE BAY, PATUXENT AND SEVERN RIVERS

This chapter describes the western shore of Chesapeake Bay from Point Lookout, on the north side of the entrance to Potomac River, to Mountain Point, the northern entrance point to Magothy River. Also described are Patuxent River, Herring Bay, West River, South River, Severn River, and Magothy River, the bay's principal tributaries; the ports of Solomons Island, Benedict, Chesapeake Beach, Shady Side, Galesville, and Annapolis; and several of the smaller ports and landings on these waterways.

COLREGS Demarcation Lines.–The lines established for Chesapeake Bay are described in **82.510**, chapter 2.

Charts 12230, 12263, 12273.– From Potomac River to Patuxent River, the western shore of Chesapeake Bay is mostly low, although the 100-foot elevation does come within 1 mile of the water midway between the two rivers. Above Patuxent River, the ground rises and 100-foot elevations are found close back of the shore along the unbroken stretch northward to Herring Bay. Above Herring Bay, the 100-foot contour is pushed back by the tributaries. Except for the developed areas, the shore is mostly wooded.

The bay channel has depths of 42 feet or more, and is well marked by lights and buoys.

The **fishtrap areas** that extend along this entire section of the western shore are marked at their outer limits and are shown on the charts.

Ice is encountered in the tributaries, particularly during severe winters. When threatened by icing conditions, certain lighted buoys may be replaced by lighted ice buoys having reduced candlepower or by unlighted buoys, and certain unlighted buoys may be discontinued. (See Light List.)

Tidal Current Charts, Upper Chesapeake Bay, present a comprehensive view of the hourly speed and direction of the current northward of Cedar Point, at the south entrance to Patuxent River. The series of 12 charts may be obtained from NOS sales agents and from the National Ocean Survey, Distribution Division (C44), 6501 Lafayette Avenue, Riverdale, Md. 20840.

Weather.–Storm warning display locations are listed on the NOS charts and shown on the Marine Weather Services Charts, published by the National Weather Service.

Chart 12230.–The **danger zone** of an aerial gunnery range and target area begins off Point Lookout and extends northward to Cedar Point. (See **204.42**, chapter 2, for limits and regulations.)

A middle ground with depths of 10 to 18 feet is about 8 miles eastward of Point Lookout; the area is about 7 miles long in a north-south direction and 2 miles wide. The stranded wreck near the middle of the shoal is marked by lighted buoys.

Chart 12233.–St. Jerome Creek, 5 miles north of Point Lookout, is entered by a marked channel. In 1966-1971, the controlling depth was 7 feet. There are general depths of 8 to 4 feet above the marked channel. The creek is used principally as an anchorage for oyster and fishing boats.

There are several small wharves along St. Jerome Creek. The landing at **Airedele**, on the south side just above the entrance, has depths of about 5 feet at the channel face; gasoline is available.

Point No Point, on the west side of Chesapeake Bay 6 miles north of Point Lookout, has no prominent natural marks. **Point No Point Light** (38°07.7'N., 76°17.4'W.), 52 feet above the water, is shown from a white octagonal brick dwelling on a brown cylinder, in depths of 22 feet, 1.6 miles southeastward of the point; a seasonal fog signal is sounded at the light. The light is 1.7 miles due west of a point on the bay ship channel 76.4 miles above the Capes.

An aerial bombardment **prohibited area** is 5.5 miles north-by-west of Point No Point Light. (See **204.42**, chapter 2, for limits and regulations of the prohibited area.) The 200-yard-square target area has rock and concrete piers at the corners and in the center, all in depths of 37 feet. Each pier is 50 feet in diameter and 12 feet high; lighted buoys are moored east and west of the target. The steel piling of a Navy radar target (38°14'15"N., 76°20'25"W.) is about 1.6 miles northwestward of the center of the aerial target. The piling is marked by a light; mariners are advised to exercise caution when transiting the area.

Hooper Island Light (38°15.4'N., 76°15.0' W.), 63 feet above the water, is shown from a white conical tower on a brown cylindrical base, in depths of 18 feet near the outer edge of the shoals, 3 miles westward from Hooper Islands; a seasonal fog signal is sounded at the light. The light is 2.8 miles due east of a point on the bay ship channel 84.4 miles above the Capes.

Charts 12264, 12265.–The enclosed Navy seaplane basin 8.5 miles north-northwestward of Point No Point and 2 miles southwestward of Cedar Point has depths of about 10 feet. The entrance to the basin is between two breakwaters, each marked at their outer ends by a light.

Cedar Point (38°17.9'N., 76°22.5'W.) is 10 miles north-by-west of Point No Point. The ruins of an abandoned lighthouse are on the tiny islet 0.3 mile off the point. The shoal extending 0.5 mile eastward from the islet is marked at its outer end by a lighted buoy.

Figure 5–3. A sample page from Coast Pilot 3.

States. Typical information in this series includes the appearance of coastlines, topographical features, navigation aids, normal local weather conditions, tides and currents, local rules of the road, descriptions of ports and harbors, pilot information, general harbor regulations, and many other miscellaneous items of interest to the navigator. (A sample page from Volume 3 of the *Coast Pilots* appears in Figure 5–3.) Some of the more important areas of concern in which the *Coast Pilots* are especially valuable references are the following:

- Recommended tracks
- Physical descriptions of navigation aids and prominent landmarks
- Description and boundary of navigation hazards
- Procedures for obtaining pilots
- Local rules of the road and local speed limits
- Normal weather, tide, and current information
- Berthing information

New editions of *Coast Pilots 1* through *7* are published annually, with each edition containing all changes reported during the preceding year. *Coast Pilots 8* and *9* are published every two years. Interim corrections to all volumes are contained in the *Notice to Mariners*.

Sailing Directions

The *Sailing Directions* consist of twelve geographic groups of loose-leaf volumes that provide information about foreign coasts and coastal waters similar to that found in the *Coast Pilots* for U.S. coastal areas, as well as information on adjacent mid-ocean areas. The *Sailing Directions* are published and kept up to date by DMAHTC.

At one time, the *Sailing Directions* were composed of a series of 70 independent volumes, whose format of often verbose, overly detailed descriptions of approaches, harbors, and channels had not changed since the institution of written compilations of navigational information in the fifteenth century. Moreover, these volumes contained almost no data on open-ocean areas for use by navigators planning an offshore voyage or ocean transit.

To correct these and other deficiencies of the *Sailing Directions*, in 1971 DMAHC (now DMAHTC) began a ten-year program to completely revise and modernize the entire series of publications. Under the new concept, the 70 volumes of the original *Sailing Directions* will ultimately be replaced by 43 new publications; eight of these will be *"Planning Guides"* for ocean basin transits, and 35 will be *"Enroute"* directions for piloting in coastal waters throughout the world. In addition to these volumes, one other publication, the *World Port Index, Publication No. 150*, has been established, to provide information on the facilities available in some 7,000 seaports throughout the world. This data was previously scattered randomly among the pages of the old *Sailing Directions*.

For the purposes of the *Planning Guides*, the world's large land-sea areas have been arbitrarily divided into eight "ocean basins," as shown in Figure 5–4; each of these eight basins will ultimately be covered by a separate *Planning Guide* volume. Note that this division into ocean basins is not the same as the division for the purposes of the chart numbering system, discussed in the previous chapter.

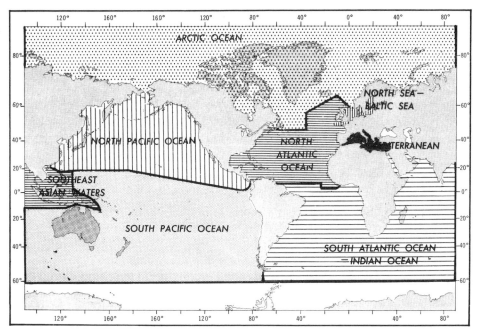

Figure 5–4. The ocean basin concept for Sailing Directions Planning Guides.

Each of the *Planning Guides* contains five chapters, their contents being as follows:

1. COUNTRIES	Governments Regulations Search & Rescue Communications	
2. OCEAN BASIN ENVIRONMENT	Oceanography Climatology Magnetic Disturbances	
3. DANGER AREAS	Operating Areas, Firing Areas Reference Guide to Warnings — NM No. 1 — NEMEDRI, DAPAC — Charts	
4. OCEAN ROUTES	Route Chart & Text Traffic Separation Schemes	
5. NAVAID SYSTEMS	Electronic Navigation Systems Systems of Lights & Buoyage	

By using the appropriate *Planning Guide* volume in conjunction with the applicable small-scale ocean charts, the navigator should find most of the mid-ocean navigational information he needs to plan intelligently for a prospective ocean voyage in the area covered.

The 35 *Enroute* publications are designed to be used in conjunction with applicable large-scale DMAHTC coastal charts to provide all information required for piloting in foreign coastal and intracoastal waters. The currently available *Enroute* volumes, plus the remaining older format *Sailing Directions* volumes not yet superseded, are divided into twelve geographic groups. The number of volumes in each group varies with the size of the geographic area covered. The Mediterranean area, for example, is covered by two *Enroute* volumes, while the Northern Pacific area requires five.

Each *Enroute* publication is divided into a number of geographic subdivisions called sectors; each sector contains the following information:

CHART INFORMATION—Index: Gazetteer
COASTAL WINDS & CURRENTS
OUTER DANGERS
COASTAL FEATURES
ANCHORAGES (COASTAL)
MAJOR PORTS
    —Directions; Landmarks; Navaids; Depths;
    —Limitations; Restrictions; Pilotage; Regulations;
    —Winds; Tides; Currents; Anchorages

The *Enroute* volumes incorporate several new features never before included in the old format *Sailing Directions*. Of these, three of the more valuable are panoramic photographs of coastal features of the more heavily traveled coasts and harbors, graphic keys to charts within sectors, as illustrated in Figure 5–5, and "graphic direction" plates that combine an annotated chartlet of an area with an orientation photograph and line drawings of prominent navigational features, as in the example shown in Figure 5–6 on page 64. After he has transited an ocean basin using the *Planning Guide*, the navigator should then be able to brief all personnel concerned as to the characteristics of the piloting waters to be entered at the end of the voyage, using the appropriate volume of the *Enroute* publications, together with the complementary large-scale coastal charts of the area.

All *Sailing Directions* volumes, once published, are kept current by the periodic issuance of sets of changed pages, supplemented by the *Notice to Mariners*. Because its content is subject to continuous

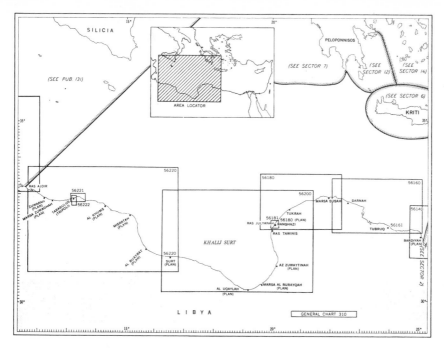

Figure 5–5. Graphic key to charts within Sector 1, "Coast of Libya," from the Enroute *publication "Eastern Mediterranean Sea," of the* Sailing Directions.

change, the *World Port Index* volume is updated and reissued annually, with interim changes promulgated via the *Notice to Mariners*.

Fleet Guides

The *Fleet Guides* are a set of two volumes published in looseleaf form by DMAHTC, which contain information primarily designed to acquaint incoming naval ships with pertinent command, navigational, operational, repair, and logistical information on frequently visited ports in both the United States and foreign countries. One volume is intended for use by ships of the Atlantic Fleet, the other for use by ships of the Pacific Fleet. Both volumes are divided into chapters, each of which deals with a separate port. Following is a list of ports described in the current edition of the *Fleet Guides*:

Atlantic Area	Pacific Area
1. Portsmouth, N.H.	1. Puget Sound
2. New London	2. San Francisco
3. New York	3. Port Hueneme, Cal.
4. Philadelphia	4. Long Beach
5. Hampton Roads	5. San Diego

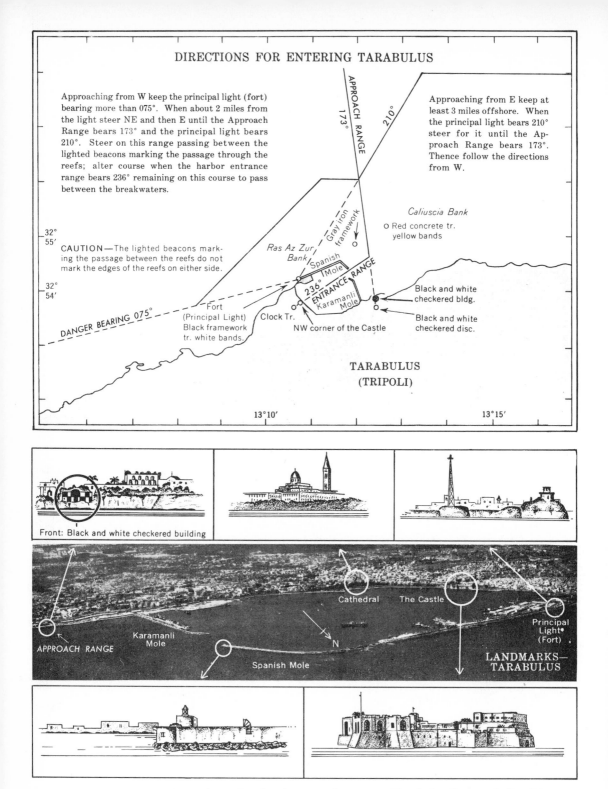

DIRECTIONS FOR ENTERING TARABULUS

Approaching from W keep the principal light (fort) bearing more than 075°. When about 2 miles from the light steer NE and then E until the Approach Range bears 173° and the principal light bears 210°. Steer on this range passing between the lighted beacons marking the passage through the reefs; alter course when the harbor entrance range bears 236° remaining on this course to pass between the breakwaters.

CAUTION—The lighted beacons marking the passage between the reefs do not mark the edges of the reefs on either side.

Approaching from E keep at least 3 miles offshore. When the principal light bears 210° steer for it until the Approach Range bears 173°. Thence follow the directions from W.

APPROACH RANGE 173°

210°

Caliuscia Bank
O Red concrete tr. yellow bands

Ras Az Zur Bank

Gray iron framework

Spanish Mole

ENTRANCE RANGE
236°

Karamanli Mole

Black and white checkered bldg.

Black and white checkered disc.

DANGER BEARING 075°

Fort (Principal Light) Black framework tr. white bands.

Clock Tr.

NW corner of the Castle

TARABULUS
(TRIPOLI)

32° 55'

32° 54'

13°10'

13°15'

Front: Black and white checkered building

LANDMARKS— TARABULUS

APPROACH RANGE

Karamanli Mole

Spanish Mole

Cathedral

The Castle

Principal Light (Fort)

N

Figure 5–6.   Graphic directions for entering Tarabulus, Enroute Sailing Directions.

Atlantic Area (cont.)	Pacific Area (cont.)
6. Canal Zone	6. Canal Zone
7. Charleston, S.C.	7. Pearl Harbor
8. Mayport, Fla.	8. Midway
9. Port Everglades, Fla.	9. Guam
10. New Orleans, La.	10. Subic Bay
11. Bermuda	11. Tokyo Wan
12. Guantanamo Bay, Cuba	12. Adak
13. Roosevelt Roads, Puerto Rico	
14. Morehead City, N.C.	
15. Kings Bay, Ga.	

All *Fleet Guides* are restricted for use only by agencies of the U.S. government, and are not issued to private users or the Merchant Marine.

Much of the information contained in the *Fleet Guides* is similar to that found in applicable volumes of the *Coast Pilots*, *Sailing Directions*, and the *World Port Index*, but the *Fleet Guides* emphasize areas of special interest and concern to U.S. Navy ships, such as command relationships, operational responsibilities, and munitions support capabilities. A sample page from the logistics section of a Hampton Roads *Fleet Guide* appears in Figure 5–7 on the next page.

New editions of the various *Fleet Guide* chapters are published at intervals as conditions warrant, while interim changes are promulgated via the *Notice to Mariners*.

Light List

The *Light List* is a series of five bound volumes describing lighted aids to navigation, and unlighted buoys, daybeacons, fog signals, radiobeacons, and Loran-C coverage in the coastal and intracoastal waters of the continental United States and the islands of Hawaii. (A radiobeacon is a navigation aid incorporating a radio transmitter, to be used by vessels equipped with radio direction finders; see *Marine Navigation 2* for further information). The *Light List* is published annually and distributed by the U.S. Coast Guard; its five volumes are arranged as indicated:

Volume Number	Contents
I	Atlantic Coast from St. Croix River, Maine, to Little River, South Carolina
II	Atlantic and Gulf Coasts from Little River, South Carolina, to Rio Grande, Texas

LOGISTICS

5-8 LOCAL SUPPLY ACTIVITIES

NAVY REGIONAL FINANCE CENTER (NAV REGFINCEN), NORFOLK

Mission

Pay dealers' bills; examine and consolidate disbursing returns; review and consolidate property returns; perform accounting functions; maintain accounts of, and make payment to, personnel of the regular Navy and Naval Reserve on active duty and training duty and personnel of assigned Organized Naval Reserve Units; prepare and pay military and civilian travel claims; receive miscellaneous cash collections; issue transportation requests and meal tickets; settle accounts of disestablished activities as directed; act as field representative of the Comptroller of the Navy; and perform such other functions as may be assigned by the Controller of the Navy.

Command and Support

NAVREGFINCEN Norfolk, administered by a Commanding Officer, is under the command of, and receives primary support from, the Comptroller of the Navy.

Commander, Naval Base, Norfolk, exercises area coordination authority.

Location

The NAVREGFINCEN and the Military Pay Department are in Building E (Figure 3), NAVSTA Norfolk.

Finance Services

Ships without disbursing officer, and desiring to be paid, should submit requests to the Navy Regional Finance Center, Norfolk, immediately upon arrival. Ships in the Little Creek area should contact the Disbursing Office, Building 3015, which is a Branch Office of the NAVREGFINCEN Norfolk. Services include payments for paydays, travel and reenlistments. (See NAV PHIBASE "Logistical Services Bulletin.")

Finance services for ships in the Norfolk Naval Shipyard are provided by the Finance Office, Building 11 (Figure 15).

Official funds required by ship's disbursing officers can be obtained at the office of the Virginia National Bank in Building Z-133 (Figure 3). In an emergency, funds may be arranged for by telephoning the Navy Regional Finance Center, Norfolk.

NAVY PUBLICATIONS AND PRINTING SERVICE OFFICE (NAVPUBPRINTSERVO), FIFTH NAVAL DISTRICT, NORFOLK

Mission

Serve as the publications and printing service for the Fifth Naval District in accordance with the Navy Industrial Fund Charter and assure the economical and efficient provision of the publications and printing requirements for the area.

Tasks

See NAVPUB INST 5450.5 series.

Command and Support

NAVPUBPRINTSERVO Norfolk (Building K-BB, Figure 3), administered by a civilian Director, is under the command of, and receiving primary support from, the Chief of Naval Material (NAVSUPSYSCOM).

5-9 FUEL AND WATER

FUEL

Fueling Arrangements

Requests for fuel to be delivered at piers or in the stream should be placed with the Bulk Fuel Scheduler, Fuel Department, Naval Supply Center, Norfolk, by message, requisition, or telephone, giving as much advance notice as practicable and informing him of complete accounting information, quantity, desired time and place of delivery. Have requisitions prepared and ready to be presented to the Fuel Supervisor at time of fueling.

Navy Special Fuel Oil (NSFO) is not available at the Naval Amphibious Base; however, NSFO can be delivered to ships by barge (YOG). Arrangements may be made with SOPA(ADMIN) Little Creek Sub-Area, at least 72 hours in advance, specifying requirements. In an emergency, NSFO can be obtained at the Craney Island Fuel Terminal at any time with arrangements made through SOPA(ADMIN) and NSC, Norfolk Duty Officer.

Ships in the Norfolk Naval Shipyard should request fuel oil from the Fuel Division, NSC Norfolk, by requisition, message, or telephone, giving as much advance notice as practicable. Deliveries must be reduced to a minimum and must be limited to actual emergencies. Ships should notify the Shipyard Operations Officer of any fueling arrangements proposed to take place in the Shipyard.

Figure 5–7. A sample page from the "Hampton Roads" chapter of the Atlantic Fleet Guide.

Information for each aid to navigation described is arranged in seven columns as illustrated in Figure 5–8 next page. Data in the *Light List* is arranged sequentially in the order in which the mariner would encounter the lights and buoys when approaching from seaward. For the purposes of the *Light List*, "from seaward" is defined as proceeding in a clockwise direction around the continental United States and in a northerly direction in the Chesapeake Bay and the Mississippi River.

The range given for lighted aids in column 4 is the nominal range of the light, defined as the distance the light should be seen in "clear" meteorological visibility conditions, disregarding the height of eye of the observer (the nominal range of a navigation light and other related topics are covered in detail in Chapter 6). In column 1, italics are used for the names of floating lighted navigation aids, while roman type is used for unlighted buoys and light structures fixed to the bottom or on land. Further differentiation of the various kinds of aids to navigation covered is done by the use of different type sizes, as indicated below:

PRIMARY SEACOAST LIGHTS AND LIGHTSHIPS
Secondary Lights.
Radiobeacons.
RIVER, HARBOR, AND OTHER LIGHTS.
Lighted Buoys.
Unlighted Fixed Aids and Unlighted Buoys.

Place names are indexed alphabetically in the back of each volume for ease of reference if the name of the location of a particular navigation aid is known. Each volume also contains a luminous range diagram, described in the next chapter, for use in light visibility computations. Much of the data in the *Light List* also appears in abbreviated form near the symbol for the nav aid on a chart, but the *Light List* is much more complete. Information in the *Light List* is updated as changes occur by means of the *Notice to Mariners*.

List of Lights

The List of Lights is a series of seven bound publications containing detailed information on the location and characteristics of lighted navigational aids, fog signals, and radiobeacons located in all foreign and selected U.S. coastal areas. The *List of Lights* volumes are pro-

(1) No.	(2) Name Characteristic	(3) Location Lat. N. Long. W.	(4) Nominal Range	(5) Ht. above water	(6) Structure Ht. above ground Daymark	(7) Remarks Year
	Severn River ANNAPOLIS HARBOR (MAIN CHANNEL) (Chart 12282)					
	—Channel Buoy 1	In 31 feet 38 56.5 76 25.6	. . . .	. . . .	Black can	Green reflector.
3500	—*Channel Lighted Bell Buoy 2* . . . **Fl. W., 2.5ˢ**	In 36 feet	5	. . . .	Red	Ra ref. Replaced by nun if endangered by ice.
	—Channel Buoy 3	In 21 feet	. . . .	. . . .	Black can	Green reflector.
	—Channel Buoy 4	In 24 feet	. . . .	. . . .	Red nun	Red reflector.
	—Channel Buoy 5	In 22 feet	. . . .	. . . .	Black can	Green reflector.
	—Channel Buoy 6	In 17 feet	. . . .	. . . .	Red nun	Red reflector.
3501	GREENBURY POINT SHOAL LIGHT. **Fl. W., 4 ˢ**	In 13 feet 38 58.1 76 27.3	8	40	NR on red skeleton tower on piles.	Higher intensity beam toward entrance to Annapolis Harbor and toward south approach to Bay Bridge Channel. BELL: 1 stroke ev 15 ˢ continuously from Sept. 15 to June 1. 1891–1934
3502	—*Channel Lighted Buoy 8* . . . . **Fl. R., 2.5ˢ**	In 25 feet	3	. . . .	Red	Replaced by nun if endan- gered by ice.
	—Channel Buoy 10	In 22 feet	. . . .	. . . .	Red nun	Red reflector.
3503	—*Channel Lighted Buoy 11* . . . . . . **Fl. W., 4ˢ**	In 23 feet	5	. . . .	Black	Replaced by can if endan- gered by ice.
	Horn Point Obstruction Daybeacon.	In 8 feet 38 58.4 76 28.2	. . . .	. . . .	NW worded DANGER SHOAL, on pile.	
	—Channel Buoy 13	In 21 feet	. . . .	. . . .	Black can	Green reflector.
	—Channel Buoy 14	In 22 feet	. . . .	. . . .	Red nun	Red reflector.
3504	—*Channel Lighted Bell Buoy 15* . . **Qk. Fl. W.**	In 23 feet	5	. . . .	Black	Ra ref. Replaced by can if endangered by ice.
	— Anchorage Buoy A	In 18 feet			White nun	Green reflector.
	Severn River Restricted Area Buoy A	In 26 feet	. . . .	. . . .	Orange and white horizontal bands; can.	White reflector. Marks southeast corner of restricted area.
	—Channel Buoy 17	In 20 feet	. . . .	. . . .	Black can	Green reflector.
3506	TRITON LIGHT . . . . **Gp. Fl. G. (4+5)30ˢ** 0.3ˢfl., 1.3ˢec. 0.3ˢfl., 1.3ˢec. 0.3ˢfl. 1.3ˢec. 0.3ˢfl. 3.4ˢec. 0.3ˢfl., 1.3ˢec. 0.3ˢfl., 1.3ˢec. 0.3ˢfl., 1.3ˢec. 0.3ˢfl., 14.8ˢec.	On shore, about 950 yards 091° from Naval Academy Dome. 38 58.9 76 28.6	8	25	Bronze structure	Maintained by U. S. Naval Academy. 1959
3507	NAVAL ACADEMY LIGHT . . . . . . **Qk. Fl. W., R. sector.**	On shore	7W	10	. . . .	Red from 220 to 259.2° white from 259.2° to 273.2°. Maintained by U. S. Naval Academy. 1962
	Severn River Restricted Area Buoy B	In 31 feet	. . . .	. . . .	Orange and white horizontal bands; can.	White reflector. Marks southwest corner of restricted area.
	Severn River Restricted Area Buoy C.	In 18 feet 38 59.0 76 28.7	. . . .	. . . .	Orange and white horizontal bands, can.	
	Santee Basin Entrance Buoy J	In 17 feet	. . . .	. . . .	Black and white vertical stripes, can.	Private aid.
	Severn River Restricted Area Buoy D.	In 20 feet	. . . .	. . . .	Orange and white horizontal bands, can.	
	—Channel Buoy 18	In 24 feet	. . . .	. . . .	Red nun	Red reflector.

Figure 5–8. A sample page of Volume 1 of the Light List.

duced, distributed, and supported by DMA. A major point of difference between the *Light List* and *List of Lights* is that the latter does not contain descriptions of unlighted buoys. Only those nav aids incorporating either a light, fog signal, or radiobeacon are described in the *List of Lights*.

Arrangement of the data for each light is basically similar to the *Light List*, except that the visibility given for a light is always expressed as the distance that a light can be seen in clear weather with the eye of the observer 15 feet above water. Column 1 contains numbers assigned to each light by DMAHTC, with any international numbers beneath. Names of lights having a geographic range of 15 miles or more, intended for making a landfall, are printed in bold-faced type; italics are used for all floating aids, and ordinary roman type for all other lights. Heights of lights are given in feet, followed by meters in bold-faced type. A sample page from the *List of Lights* appears in Figure 5–9 on the following page.

An updated set of volumes is published annually, with interim changes being promulgated as they occur via weekly *Notice to Mariners*.

Tide and *Tidal Current Tables*

Tide Tables are published annually by the National Ocean Service in four volumes covering the following areas:

Europe and the West Coast of Africa, including the Mediterranean Sea;
East Coast, North and South America, including Greenland;
West Coast, North and South America, including the Hawaiian Islands;
Central and Western Pacific Ocean and the Indian Ocean.

Together they contain daily tide time predictions for some 190 reference ports and listings of time and height difference data for an additional 5,000 locations, referred to as substations. The navigator can construct a daily tide table for any one of the 5,000 substations by applying the time difference data for that substation to the tabulated daily data for its reference port. The *Tide Tables* and their use are described in greater detail in Chapter 11.

Tidal Current Tables are published annually by the National Ocean Service in two volumes, one for the Atlantic Coast of North America, and the other for the Pacific Coast of North America and Asia. They are arranged somewhat like the *Tide Tables*, with one part of each volume containing daily tidal current predictions for a number of reference stations, and a second part containing time and maximum

(1) No.	(2) Name and location	(3) Position lat. long.	(4) Characteristic and power	(5) Height (feet)	(6) Range (miles)	(7) Structure, height (feet)	(8) Sectors. Remarks. Fog signals
			SCOTLAND—WEST COAST				
9080 A 4254	Loch Indail, Rudha an Duin, 0.5 mile N. of Port Charlotte.	N. W. 55 45 6 22	Gp. Fl. W. R. (2)..... period 7ˢ fl. 0.5ˢ, ec. 1.5ˢ fl. 0.5ˢ, ec. 4.5ˢ Cp. W. 6,000 R. 3,000	50	12	White brick tower; 42......	W. 218°–249°, R.–350°, W.–36°. obscured elsewhere.
9090 A 4252	— Bruichladdich, head of pier.	55 46 6 21	F. W. R .... Cp. under 100	20	5	Column; 13	R. 251°–15°, W.–251°.
10010 A 4272	Mull of Kintyre, SW. head-land of Kintyre.	55 19 5 48	Gp. Fl. W. (2) .... period 30ˢ fl. 0.5ˢ, ec. 4.0ˢ fl. 0.5ˢ, ec. 25.0ˢ Cp. 546,000	297	24	Yellow tower on white building; 38.	Visible 347°–178°. Siren: (0.6 mile SSE. of light-house): 2 quick bl. ev. 90ˢ. Radiobeacon.
10020 A 4274	The Ship, S. side of Sanda Island.	55 16 5 35	Fl. W. R .... period 24ˢ fl. 8ˢ, ec. 16ˢ Cp. W. 61,000 R. 20,000	W. 18 R. 16	165	White tower; 48..........	R. 245°–267°, W.–land to West. Distress signals. Siren: 1 bl. ev. 60ˢ.
10030 A 4276	Da Voar, NE. part of island..	55 26 5 32	Fl. W.............. period 20ˢ fl. 0.6ˢ, ec. 19.4ˢ Cp. 254,000	120	17	White tower; 65..........	Visible 73°–330°. Siren: 1 bl. ev. 20ˢ. Distress signals.
10040 A 4278	CAMPBELTOWN: — Millmore .... U.	55 25 5 34	Fl. W.............. period 6ˢ fl. 0.5ˢ, ec. 5.5ˢ Cp. 100	26	9	White beacon surmounted by a black tank; 30.	
10045 A 4279	— Range, front .... U.	55 25 5 35	F. Or.............. Cp. 130	24	6	Beacon with yellow tri-angle, point up; 20.	
10045.1 A 4279.1	— — Rear, 280 yards 240° 30′ from front. U.	....	F. Or.............. Cp. 130	91	6	Beacon with yellow tri-angle, point down; 23.	
10050 A 4284	— Head of old pier .... U.	55 26 5 36	F. W. R Cp. under 100	22	4	Column; 17	R. eastward: W. toward harbor. Storm signals.
10060 A 4282	— Head of new pier .... U.	55 25 5 36	F. W. G Cp. under 100	17	4	Mast; 12..................	G. eastward, W. toward harbor.
10070 A 4286	— Head of Dalintober Pier . U.	....	F. W Cp. under 100	18	5	Metal column; 14	
10072 A 4290	Port Grannaich Breakwater, Head. U.	55 36 5 28	Fl. G.............. period 10ˢ Cp. 200	15	6	Iron column; 10..........	Obscured 279°–99°.
10080 A 4340	Loch Ranza, pier head	55 42 5 18	F. R .... Cp. under 100	20	4	White post; 12...........	
10090 A 4296	LOCH FYNE: — Sgat More, on S. end .... U.	55 51 5 18	Fl. W period 3ˢ fl. 0.7ˢ, ec. 2.3ˢ Cp. 4,000	30	10	White circular tower on concrete base; 28.	
10100 A 4298	— East Loch Tarbert, Madadh Maol, S. side of entrance. U.	55 52 5 24	Fl. R.............. period 2.5ˢ fl. 0.5ˢ, ec. 2ˢ	14	.....	Perch	

Figure 5–9.   A sample page from the List of Lights.

strength difference data for a large number of subordinate locations. The *Tidal Current Tables* and their use are thoroughly described in Chapter 12.

Pilot Charts

Pilot charts are special charts of portions of the major ocean basins designed to assist the navigator during voyage planning. Their name is misleading, because they are small-scale charts of ocean areas, of little use in the actual practice of piloting as defined in this text. Nevertheless, they are invaluable to the navigator. They present, in color-coded graphic form, a complete forecast of the hydrographic, navigational, and meteorological conditions to be expected in a given ocean area during a given time of year. Included is information concerning average tides, currents, and barometer readings; frequency of storms, calms, or fogs; possibility of the presence of ice, including the normal limits of iceberg migration; and a great variety of other meteorological and oceanographic data. Lines of equal magnetic variation—*isogonic* lines—are given for each full degree of variation. The shortest and safest routes between principal ports are also indicated.

Pilot charts of the North Atlantic and North Pacific are issued quarterly by the DMAHTC, with each edition consisting of three monthly charts, plus an interesting and informative article on a topic of current interest to navigators. Pilot charts are also published by DMAHTC in an annual atlas format for certain years for the following regions: Central American waters and the South Atlantic; South Pacific and Indian oceans; the northern North Atlantic; and the Indian Ocean separately.

A portion of a pilot chart of the North Atlantic is shown in Figure 5–10 on the next page.

Distances Between Ports, Publication No. 151

The publication *Distances Between Ports, Publication No. 151*, is a useful reference publication made available to the navigator by DMAHTC for use in the preliminary planning stages of a voyage. It is simply a tabulated compendium of great-circle distances calculated along the most frequently traveled sea routes between U.S. and foreign ports, and between foreign ports. A sample page, showing distances from Norfolk, Virginia, to various locations throughout the western hemisphere is shown in Figure 5–11.

This publication finds its greatest use during the preplanning phase of voyage planning to determine estimated total distances between the points of departure and arrival, measured along the most common routes between them. An example of its use to find the initial estimate

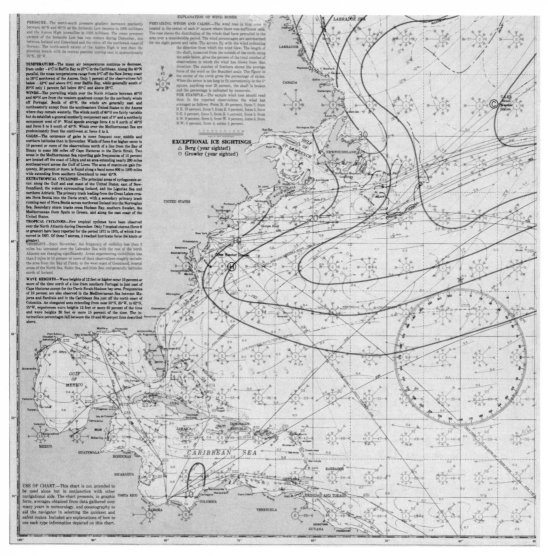

Figure 5–10. A portion of a pilot chart of the North Atlantic.

of the distance for a typical voyage from Norfolk, Virginia, to Naples, Italy, is given in Chapter 15.

A similiar publication listing distances between U.S. ports is published and sold by NOS.

Almanacs

Books giving the positions of the various celestial bodies used for celestial navigation, times of sunrise and sunset, moonrise and moonset, and other astronomical information of interest to the navigator

Pago Pago, American Samoa, 4,748
Petropavlovsk, U.S.S.R., 1,317
Pevek, U.S.S.R., 700
Prince Rupert, Canada, 1,846
San Francisco, California, U.S.A., 2,636
Sasebo, Japan (via Osumi Kaikyo), 3,373
Seattle, Washington, U.S.A. (via Juan de Fuca Strait), 2,288
Sydney, Australia, 6,280
Valparaiso, Chile, 7,728
Vancouver, Canada, 2,304
Yokohama, Japan, 2,696

NONOUTI, GILBERT ISLANDS
(0°43'00'' S., 174°17'00'' E.) to:

Junction Points*
Panama, Panama, 6,439

Ports
Apia, Samoa, 1,136
Apra, Guam, 1,981
Brisbane, Australia, 2,032
Funafuti, Ellice Islands, 544
Honolulu, Hawaii, U.S.A., 2,100
Levuka, Fiji Islands, 1,162
Ponape, Caroline Islands, 1,082
Yokohama, Japan, 2,931

NONUTI, GILBERT ISLANDS. See Nonouti.

NORDENHAM, GERMANY
(53°29'00'' N., 8°29'40'' E.)

Add 5½ miles to Bremerhaven distances.

NORD-OSTSEE-KANAL. See Kiel Canal.

NORFOLK, VIRGINIA, U.S.A.
(36°51'00'' N., 76°18'00'' W.) to:

Junction Points*
Bishop Rock, England (track A), 3,202
Bishop Rock, England (track B), 3,168
Bishop Rock, England (track C), 3,107
Cape of Good Hope, Republic of South Africa, 6,802
Fastnet, Ireland (track A), 3,109
Fastnet, Ireland (track B), 3,072
Fastnet, Ireland (track C), 3,002

Inishtrahull, Ireland (track A), 3,235
Inishtrahull, Ireland (track B), 3,191
Inishtrahull, Ireland (track C), 3,100
Montreal, Canada (via Cabot Strait), 1,700
Panama, Panama, 1,822
Pentland Firth, Scotland (track A, eastbound), 3,449
Pentland Firth, Scotland (track B, eastbound), 3,395
Pentland Firth, Scotland (track C, eastbound), 3,280
Punta Arenas, Chile, 6,900
Strait of Gibraltar (track A), 3,349
Strait of Gibraltar (track B), 3,340
Strait of Gibraltar (direct), 3,335
Straits of Florida (via outer route), 980
Yucatan Channel, 1,165

Ports
Angmagssalik, Greenland (direct), 2,458
Angmagssalik, Greenland (via 75 miles off Kap Farvel), 2,472
Argentia, Canada, 1,189
Arkhangelsk, U.S.S.R., 4,399
Banes, Cuba (via Crooked Island Passage), 1,018
Belem, Brazil, 2,832
Belize, British Honduras, 1,503
Bocas del Toro, Panama (via Crooked Island Passage and Windward Passage), 1,853
Bordeaux, France (track A), 3,495
Bordeaux, France (track B), 3,466
Bordeaux, France (track C), 3,423
Buenos Aires, Argentina, 5,824
Cadiz, Spain (track A), 3,315
Cadiz, Spain (track B), 3,306
Cadiz, Spain (direct), 3,302
Cape Town, Republic of South Africa, 6,790
Cartagena, Colombia (via Crooked Island Passage and Windward Passage), 1,658
Castries, St. Lucia, 1,620
Charlotte Amalie, Virgin Islands, 1,296
Cienfuegnos, Cuba (Windward Passage), 1,482
Colon, Panama (via Crooked Island Passage and Windward Passage), 1,779
Dakar, Senegal, 3,408
Elizabeth Harbor, Bahamas, 848
Fort de France, Martinique, 1,597
Freetown, Sierra Leone, 3,821
Funchail, Madeira Island, 2,907
Georgetown, British Guiana, 2,090
Guantanamo, Cuba, 1,117

Habana, Cuba (southbound, outside), 985
Halifax, Canada, 790
Hamilton, Bermuda, 683
Ivigtut, Greenland, 2,116
Kingston, Jamaica (via Crooked Island Passage and Windward Passage), 1,279
La Guaira, Venezuela, 1,687
Lagos, Nigeria, 4,941
Las Palmas de Gran Canaria, Canary Islands, 3,130
Limon, Costa Rica (via Crooked Island Passage and Windward Passage), 1,852
Lisboa, Portugal (track A), 3,149
Lisboa, Portugal (track B), 3,137
Lisboa, Portugal (tracks C and D, eastbound), 3,128
Liverpool, Canada, 724
Livingston, Guatemala, 1,595
Maracaibo, Venezuela, 1,682
Montevideo, Uruguay, 5,710
Nassau, Bahamas, 758
Nuevitas, Cuba, 1,076
Parrsboro, Canada, 784
Pointe a Pitre, Guadeloupe, 1,527
Ponta Delgada, Azores (track A), 2,402
Ponta Delgada, Azores (direct), 2,401
Port Antonio, Jamaica (via Crooked Island Passage and Windward Passage), 1,228
Port au Prince, Haiti, 1,178
Port of Spain, Trinidad, 1,799
Porto Grande, Cape Verde Islands, 2,971
Puerto Barrios, Guatemala, 1,603
Puerto Cortes, Honduras (via Straits of Florida), 1,568
Puerto Mexico, Mexico, 1,736
Recife, Brazil, 3,651
Reykjavik, Iceland, 2,677
Rio de Janeiro, Brazil, 4,723
St. John, Canada, 731
St. John's, Canada, 1,277
Salvador, Brazil, 4,042
San Juan, Puerto Rico, 1,252
San Juan del Norte, Nicaragua (via Straits of Florida), 1,846
San Juan del Norte, Nicaragua (via Windward Passage), 1,837
Santa Cruz de Tenerife, Canary Islands, 3,057
Santa Marta, Colombia (via Crooked Island Passage and Windward Passage), 1,588
Santiago de Cuba, Cuba, 1,167
Santo Domingo, Dominican Republic, 1,329
Santos, Brazil, 4,910

*JUNCTION POINT.—See Preface for use of junction points.

Figure 5–11. A sample page from DMAHTC Publication No. 151, Distances Between Ports.

Navigational Publications   **73**

are called *almanacs*. There are several of these published in the United States both by commercial presses and by the government. The two most commonly used by American marine navigators are the *Nautical Almanac* and the *Air Almanac*, both of which are prepared by the U.S. Naval Observatory and published in the United States by the Government Printing Office. The *Nautical Almanac* is published annually in a single volume, and the *Air Almanac* twice a year, with each edition covering six months. Both contain data pertaining to celestial navigation from either surface vessels or aircraft, with the *Air Almanac* specially designed to facilitate air navigation. These publications, used in conjunction with several different types of standard navigation celestial sight reduction tables, are discussed in detail in *Marine Navigation 2: Celestial and Electronic*.

In addition to the conventional *Nautical* and *Air Almanacs*, in 1978 the Government Printing Office began annual publication of a third almanac designed for the growing number of mariners who are using hand-held calculators and in some cases small computers to perform some or all of their celestial navigation computations. This almanac, called the *Almanac for Computers*, contains data tables, formulas, and procedures for the precise calculation of all the chronological data presented in the daily pages of the *Nautical* and *Air Almanacs*, and for the calculation of all necessary corrections to observed sextant altitudes of celestial bodies. Celestial navigation by electronic calculators is discussed in Appendix A of *Marine Navigation 2*.

Reference Texts and Manuals

There are two reference texts found on board most Navy and other seagoing ships that deserve mention. The first of these is considered the bible for marine navigation—*The American Practical Navigator*, usually called *Bowditch* after its original author. This book was originally published in 1802, and has been popular with mariners ever since. It is not only an encyclopedic compendium of navigational information and technique, but it also contains many tables that are useful in solving navigational problems; one of these, the table of horizon distances for various heights above sea level, is used in the chapter of this book dealing with light visibility computation. *Bowditch* is published by the DMAHTC in two volumes as *Publication No. 9*.

The second text is *Dutton's Navigation and Piloting*, published by the Naval Institute Press, Annapolis, Maryland. It has long been regarded as the standard textbook for the professional mariner on all phases of marine navigation.

In addition to the texts mentioned above, DMAHTC and NOS

publish a number of other manuals and reference publications of interest to the navigator. Some of the more important of these are described below.

The DMAHTC *Handbook of Magnetic Compass Adjustment and Compensation,* more familiarly known as *Publication No. 226,* is a simply written and well-organized technical manual that provides step-by-step instruction for magnetic compass adjustment and compensation and the procedures for swinging ship. It has been considered the standard reference on its subject by three generations of Navy, Merchant Marine, and private navigators.

Radio Navigational Aids, Publication Nos. 117A and *117B,* are a set of two bound volumes arranged in tabular format that provide information on a wide variety of radio navigational aids and broadcasts throughout the world. They are produced in annual editions by DMAHTC, and are discussed in greater detail in *Marine Navigation 2: Celestial and Electronic.*

The DMAHTC *Radar Navigation Manual, Publication No. 1310,* is an excellent treatment of the use of radar at sea for both navigation and collision avoidance. Although oriented primarily toward the merchant mariner, much of the material is also applicable to Navy and civilian navigators whose vessels are equipped with radar.

Publication Correction System

The *Notice to Mariners* system of correcting charts and publications has already been described in some detail in the preceding chapter of this text. The system as it applies to publications is briefly reviewed below.

Each navigational publication supported by DMAHTC and NOS, like every chart, should have a correction card made up for it and entered into the central chart/publication correction card file when the publication is initially received on board. Each time a *Notice to Mariners* is received and found to contain a correction affecting an on-board publication, the navigator or an assistant enters the number of the *Notice* containing the correction onto the correction card of the affected publication. If the publication is currently being used, the correction is then immediately entered into the publication. If, as happens in a great many cases, the publication is not in current use, nor anticipated to be used in the near future, the correction is not entered into the publication. In the latter case, after the *Notice* number has been written on the correction card, the card is returned to the central file until such time as the publication is needed. At that time, the card is pulled out of the file, along with the file copies of old *Notices to Mariners,* and each correction annotated on the card

is entered into the publication. In this way, only a few of the many publications on board need be corrected week by week; the rest are not corrected until an occasion for their use arises.

Summary

This chapter has discussed and described the major navigational publications usually found on board almost all seagoing naval, commercial, and private vessels. The prudent navigator will make extensive use of these and all other applicable publications both in the piloting environment and while at sea.

The navigator should always remember that he is not only a user of the various publications described herein, but he is also expected to be a contributor. Many observers contribute valuable data from time to time concerning currents, aids and dangers to navigation, port facilities, and related suggestions that help the Defense Mapping Agency and the National Ocean Service materially in keeping all publications described in this chapter in agreement with actual conditions. Much of this type of data could be otherwise obtained only by vast expenditures of time and money, if at all. The DMA and NOS solicit such cooperation and greatly appreciate the receipt of any data that may increase the accuracy and completeness of their publications.

6

Visual Navigation Aids

In the earliest days of sail the need for lights and buoys to aid the mariner in the navigation of coastal waters and rivers was recognized. It has been established that a lighthouse was built at Sigeum, near Troy, in the seventh century B.C., and in Alexandria in Egypt in the third century B.C. Wood fires furnished their illumination; wood and sometimes coal continued to be used for this purpose until the invention of the electric light in the nineteenth century. The first lighthouse in the United States was built at Boston in 1716, and logs and empty kegs were used to mark channels in the Delaware River as early as 1767.

An aid to navigation, sometimes abbreviated and called a "nav aid," is defined as any device external to a vessel or aircraft intended to assist a navigator in determining his position and safe course, or to warn him of dangers or obstructions to navigation. The modern visual aids to be discussed in this chapter are those local to the area in which the navigator is operating and which appear on his chart. These include permanent structures attached to the shore or bottom, such as lighthouses, light towers, automated lights and beacons, as well as floating aids, such as lightships and buoys.

Before any aid to navigation can be used, it must first be positively identified. Once having identified the aid, the navigator can then correlate its position on his navigational chart, and proceed to use it to assist in the determination or verification of his vessel's position. By day, the general location, shape, color scheme, auxiliary features, and markings of a navigation aid can all assist in its identification. By night, most of these identifying characteristics become indistinguishable, and the navigator must then obtain all identifying information primarily from the light shown by lighted navigation aids. Fortunately, over the years a rather extensive code-like system of light colors, patterns of flashes, and time cycles has been developed and standardized, which allows a great deal of identifying information to be imparted by the light itself. By carefully analyzing the various visual characteristics of aids to navigation by day, and their lights at night, in conjunction with the nautical chart covering the area and

the applicable volumes of the *Light List* and *Coast Pilots* in U.S. waters, and the *List of Lights* and *Sailing Directions* in foreign waters, even an inexperienced navigator should have no undue difficulty in identifying most aids.

In this chapter, the characteristics, identification, and use of lighted and unlighted aids to navigation will be discussed. Since almost all modern aids of importance are lighted so that they may be used both by day and by night, the chapter will begin with a discussion of the characteristics and functions of navigation lights. The last half of the chapter will describe in some detail the major U.S. and foreign systems of buoyage. As will be seen, the majority of buoys incorporate a light, so a general understanding of navigation light characteristics is also of fundamental value in the study of buoyage systems.

Characteristics of Lighted Navigation Aids

Lights intended to function as navigational aids are broadly classified into two groups—*major* and *minor* lights. Major lights have high intensity and reliability, and are normally placed in lightships, lighthouses, and highly automated light towers and other permanently installed structures. They are intended to indicate key navigational points along seacoasts, channels, and in harbor and river entrances. Major lights are further subdivided into *primary* and *secondary* lights, with the former being very strong lights of long range established for the purpose of making landfalls or coastal passages, and the latter being those lights of somewhat lesser range established at harbor entrances and other locations where high intensity and reliability are required.

Minor lights are automated lights of low to moderate intensity, placed on fixed structures, and intended to serve as navigational aids within harbors, along channels and rivers, and to mark isolated dangers. They usually have the same numbering, coloring, and light and sound characteristics as the buoyage system in the surrounding waters.

Lighthouses and lightships were until the early 1900s the principal types of major navigation lights. Today, however, most lightships and many manned lighthouses have been replaced by highly reliable automatic light towers and large navigational buoys (LNBs). Most of these and other primary and secondary lights incorporate radio monitoring devices that give warning at centrally located manned stations if their light should fail.

Whether manned or not, the essential requirements for any major light structure are as follows: it should be placed at the best possible location dependent on the physical conditions at the site, should be built up to a sufficient height for the location, and should incorporate

Figure 6–1. *(Top) The Ambrose Lightship off New York Harbor was retired in 1967 after having been replaced by a manned light tower. (Bottom) The Portland Lightship off Cape Elizabeth, Maine, was replaced by a large navigation buoy in 1975. (Courtesy U.S. Coast Guard)*

a rugged support for the lantern and a housing for the power source, usually electricity. Most structures are painted in order to make them readily distinguishable from their background by day.

As was mentioned in the introductory section of this chapter, the first objective of the navigator upon sighting a navigation aid is its identification. In the case of a lighted navigation aid at night, the navigator uses three attributes of the light in this endeavor—the light phase characteristic, the duration of its period, and its color. It is important that each of these attributes be understood. A light *phase characteristic* is defined as the light sequence or pattern of light shown within one complete cycle of the light; the light *period* is defined as the length of time required for the light to progress through one

complete cycle of changes; and the light *color* refers to the color of the light during the time it is shining.

Light Phase Characteristics

The following is a summary of the more common light phase characteristics seen for lights in both permanent structures and on buoys. Their standard abbreviations appear in parentheses alongside the characteristic.

Fixed (F.) This is a light that shines with steady, unblinking intensity.

Flashing (Fl.) This light appears as a single flash at regular intervals; the duration of the light is always less than the duration of darkness. Normally, a flashing light should not flash more than 30 times per minute.

Quick Flashing (Qk.Fl.) Basically similar to a flashing light, the quick flashing light shows more frequently to indicate a greater degree of cautionary significance. The duration of the flash is less than the duration of darkness, and the light will flash at least 60 times per minute.

Interrupted Quick Flashing (I.Qk.Fl.) By convention, this is a light that quick flashes six times, followed by a time of darkness, with a standard period of ten seconds.

Group Flashing (Gp.Fl.) This light shows groups of two or more flashes at regular intervals. Its period usually appears on the chart, and may be of any duration.

Morse (Mo.(A)) A light showing a pattern of flashes comprising a Morse Code character, by convention normally the letter "A."

Equal Interval (E.Int.) Sometimes called an *Isophase* (Iso.) light, this is a light having equal durations of light and darkness. Its period may be of any length, and will usually be written on the chart.

Occulting (Occ.) Any light that is on longer than it is off during its period is termed an occulting light.

Group Occulting (Gp.Occ.) This is an occulting light, broken by groups of eclipses into two or more flashes. The pattern of the eclipses is indicated on a chart enclosed by parentheses following the basic abbreviation, as for example, Gp.Occ. (2 + 3), which indicates a light interrupted by a group of two, then three, eclipses.

Composite This refers to a light showing two or more distinct light sequences within its period. There is no standard abbreviation

Illustration	Symbols and meaning		Phase description
	Lights which do not change color	Lights which show color variations	
	F.= Fixed...	Alt.= Alternating.	A continuous steady light.
	F.Fl.=Fixed and flashing	Alt. F.Fl.= Alternating fixed and flashing.	A fixed light varied at regular intervals by a flash of greater brilliance.
	F.Gp.Fl. = Fixed and group flashing.	Alt. F.Gp.Fl = Alternating fixed and group flashing.	A fixed light varied at regular intervals by groups of 2 or more flashes of greater brilliance.
	Fl.=Flashing	Alt.Fl.= Alternating flashing.	Showing a single flash at regular intervals, the duration of light always being less than the duration of darkness.
	Gp. Fl. = Group flashing.	Alt.Gp.Fl.= Alternating group flashing.	Showing at regular intervals groups of 2 or more flashes
	Gp.Fl.(1+2) = Composite group flashing.	....	Light flashes are combined in alternate groups of different numbers.
	Mo.(A) = Morse Code.	....	Light in which flashes of different duration are grouped in such a manner as to produce a Morse character or characters.
	Qk. Fl. = Quick Flashing.	....	Shows not less than 60 flashes per minute.
	I.Qk. Fl. = Interrupted quick flashing.	....	Shows quick flashes for about 5 seconds, followed by a dark period of about 5 seconds.
	E.Int.= Equal interval. *(Isophase)*	....	Light with all durations of light and darkness equal.
	Occ.=Occulting.	Alt.Occ. = Alternating occulting.	A light totally eclipsed at regular intervals, the duration of light always greater than the duration of darkness
	Gp. Occ. = Group Occulting.	....	A light with a group of 2 or more eclipses at regular intervals
	Gp.Occ.(2+3) = Composite group occulting.	....	A light in which the occultations are combined in alternate groups of different numbers.

Light colors used and abbreviations: W = white, R = red, G = green.

Figure 6–2. Light phase characteristics.

for a composite light; the fact that it is composite is indicated on a chart by a set of parentheses placed after the basic light characteristic, which contains the number of flashes or occul-

tations (eclipses) within each period. For example, Gp.Fl. (2 + 3) indicates a light that flashes a total of 5 times during its period, in two groups composed of 2, then 3, flashes.

Occasionally a light combines two or more of the preceding characteristics, so as to be distinctive when several lights are located in the same general vicinity. A light abbreviated F.Gp.Fl. (2 + 3) on a chart would be a composite fixed flashing light in which the flashes appeared in two groups of 2, then 3, flashes. Figure 6–2 illustrates the different light phase characteristics described above, as well as some of the more common light phase combinations. Those phase characteristics which can be further enhanced by changing or alternating the light colors are also indicated; light color alterations will be more fully explained later.

When working with charted composite light abbreviations, care must be taken to distinguish between the meanings of the numbers in the parentheses following a composite flashing light, and similar numbers following a composite occulting light. In the case of the composite flashing light, the numbers refer to the pattern of the *flashes* of the light. Contrariwise, when the light is a composite occulting light, the numbers within the parentheses denote the pattern of the *eclipses* in the light. This possible confusion is avoided in the *Light List* and *List of Lights*, because the parentheses are not used to describe a composite light. In these publications, the lengths of each eclipse and duration of light of an occulting light are given in tabular form, similar to the entry shown in Figure 6–3A below for Brandywine Shoal Light, a group occulting light. As will be explained in the following section, this information allows the light to be diagrammed in bar-chart form, thus eliminating all doubt as to its appearance.

(1) No.	(2) Name Characteristic	(3) Location Lat. N. Long. W.	(4) Nominal Range	(5) Ht. above water	(6) Structure Ht. above ground Daymark	(7) Remarks Year
			NEW JERSEY			THIRD DISTRICT
2102	BRANDYWINE SHOAL LIGHT ..... Gp. Occ. W., R. sector, 12ˢ 2ˢlt., 2ˢec. 6ˢlt., 2ˢec. 2 eclipses.	On shoal, 0.9 miles from south end. 38 59.2 75 06.8	17W 13R	60	Cylindrical concrete structure; adjacent to old screw-pile structure.	Red from 151° to 340°, covers shoal area southwest of Cape May. RADIOBEACON: See p. XVIII for method of operation. HORN, 1 blast ev 15ˢ (2ˢbl). See p.XX for Special Radio Direction Finder Calibration Service. 1823–1914

Figure 6–3A. Brandywine Shoal Light data, Light List.

It should also be mentioned here that on nautical charts, abbreviated phase characteristics and other identifying data are printed in roman type for fixed light structures, and in italics for lighted floating aids, to assist in differentiation between the two types of navigation aids.

Period of a Light

As previously mentioned, the period of a navigational light is defined as the time required for the light to progress through one complete cycle of changes. The periods of all lights except those having either quick flashing or interrupted quick flashing phase characteristics are indicated both on charts depicting them and in the *Light List* and *List of Lights*. For all types of lights, the period is measured from the start of the first flash of one cycle to the start of the first flash of the succeeding cycle.

It is often advantageous to diagram the period of a light in order to aid in its identification. To diagram a light, it is necessary to obtain the length and pattern of its flashes and eclipses from the applicable volume of the *Light List* or *List of Lights*. Once these data are on hand, the light may then be diagrammed in the form of a simple bar-chart, as in the example shown below (Figure 6–3B) for Brandywine Shoal Light. As can be seen from the diagram, the period of this light is 12 seconds in length. It is a group occulting light, broken by two eclipses of two seconds' duration each.

Color of a Light

There are only three light colors in common use on lighted navigation aids in U.S. and most foreign waters—white, green, and red (some European buoys have yellow lights). All lighted navigation aids, regardless of the color of their light, are symbolized on a chart either by a purple-colored ray in the form of an exclamation point or by a $\frac{1}{8}$-inch purple circle, superimposed over a black dot or small open

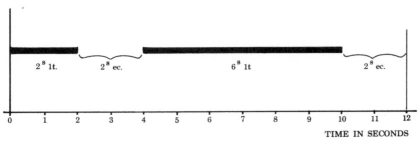

Figure 6–3B. Diagram of the period of Brandywine Shoal Light.

Visual Navigation Aids **83**

circle indicating the location of the light. Sections K and L of *Chart No. 1, Chart Symbols and Abbreviations*, included as Appendix A of this volume, may be referred to for examples. On charts, the color of the light, if other than white, is indicated by the abbreviations "R" for red, and "G" for green, printed near the light symbol. A white light has no abbreviation on a chart. Thus, if a purple light symbol appears on a chart with no color abbreviation nearby, the navigator should assume its color to be white. In the *Light List* and *List of Lights*, however, the color of a white light is indicated by the abbreviation "W."

Alternating Lights

If a light is made to change color in a regular pattern, either by alternately energizing different-colored lights or by passing colored filters around the same light, the light is termed an *alternating* light, abbreviated "Alt." Alternating lights used in conjunction with different phase characteristics, as described in Figure 6–2, show a very distinctive appearance that cannot be easily mistaken. Their use is generally reserved for special applications requiring the exercise of great caution, such as airport beacons, harbor entrance lights, and lighthouses.

Sector Lights

A type of light often confused with alternating lights is the *sector* light. Sector lights are used to warn the navigator of hazards to navigation when approaching the light from certain dangerous arcs or sectors. A sector light may be separated into two or more colored arcs, or rendered invisible in all but one or two narrow arcs, by permanently positioned shields built around the light. If it is multicolored, a color seen in one sector should not be visible in an adjacent sector, except when the observer is located exactly on the border between them. A sector may change the color of a light, as the observer moves around it, but not its phase characteristic or its period. For example, a flashing white light having a red sector will appear flashing red if viewed from within the red sector, but outside of this sector the light will appear flashing white. Sectors may be only a few degrees in width, marking an isolated rock or shoal, or so wide as to mark an entire deep water approach toward shore. Any bearings given to describe the limits of the various sectors are always expressed in degrees true as observed from a hypothetical vessel moving in a clockwise direction around the light. As an example, consider Harbor of Refuge Light in Delaware Bay; its description from the *Light List* appears in Figure 6–4A.

(1)	(2)		(3)		(4)	(5)	(6)		(7)	
	Name		Location		Nominal	Ht.	Structure			
					Range	above	Ht. above			
No.	Characteristic		Lat. N.	Long. W.		water	ground	Daymark	Remarks	Year
				NEW JERSEY					THIRD DISTRICT	
2098	Harbor of Refuge Light.......... Fl. W., 2R sector, 5ˢ		On south end of Harbor of Refuge Break- water 38 48.9 75 05.6		23W 20R	72	White conical tower, on brown cylindrical substructure. 66		Red from 325˚ to 351˚ and 127˚ to 175˚ covers hen and Chicken and Brown Shoals. HORN, 2 blasts ev 30ˢ (2ˢbl- 2ˢsi-2ˢbl-24ˢsi). Oriented on bearings 120˚ and 000˚ 1901 –1926	

Figure 6–4A. Harbor of Refuge Light, Light List.

Figure 6–4B. Harbor of Refuge sector light.

In the Remarks column, this light is described as having red sectors from 325°T to 351°T, and from 127°T to 175°T, as viewed from seaward. An illustration of this light appears in Figure 6-4B.

Range Lights

Two or more lights in the same horizontal direction, so located that one appears over the other when they are sighted in line, are known as *range* lights. They are usually visible in only one direction, and are mainly used to mark straight reaches of a navigable channel between hazards on either side. The light nearest the observer is called the front light, and the one farthest away, the rear light. By steering his ship in such a way as to keep the front and rear lights always in line, one over the other, the navigator will continually remain on the centerline of the channel.

The lights of ranges may be any of the three standard colors. In U.S. waters, their phase characteristics are being standardized, with the front light always quick flashing, and the rear light an equal interval (isophase) light with a period of six seconds. Their structures are always fitted with conspicuously colored daymarks for daytime use. Most range lights have very narrow sectors of visibility, and lose brilliance rapidly as the ship diverges from the range. The range lights marking the Cape May Harbor entrance in New Jersey, for instance,

are visible only a few degrees on each side of the range centerline. When using range lights for navigation, care must be taken to examine the charts beforehand to determine how far a particular set of lights can be safely followed, especially in the vicinity of a bend in the channel.

Identifying a Navigational Light

As previously mentioned, the navigator will take all three attributes of a navigational light into account when seeking to identify it—the phase characteristic, period, and color. Of the three, the light period is the feature usually weighted most heavily, as it is normally the most unambiguous attribute of a light, especially when other lights with which it might be confused are in close proximity. To eliminate all possible error, no light should ever be considered positively identified until its period is timed by a stopwatch and found to be identical with published information.

One note of caution that should be borne in mind when identifying a sector and especially an alternating light is that the colored segments of the light will almost always be of lesser range than the white segment (see Figures 6–3A and 6–4A). This is so because the same light source is normally used for both the white and colored segments, with one or more colored filters used to produce the latter. In addition to changing the color of the basic white light passing through it, such a filter will also absorb some of the energy of the light, thus decreasing its luminous range. Hence, it is possible at extreme range that only the white portion of an alternating light might be visible, which could lead to false interpretation of the light phase characteristic and an invalid measurement of its period. Likewise, when approaching a sector light on the borderline between a white and a colored sector, only the white sector may be visible initially, until the vessel has drawn within the more limited luminous range of the colored segment of the light.

Computing the Visibility of a Light

It frequently happens that the navigator desires to know at what time and position on the ship's track he can expect to sight a given light. This is especially important when the ship is making a landfall, as failure to sight certain lights when expected could mean that a serious error had been made in the determination of the ship's position. Moreover, in some circumstances, the navigator will need to determine when the light will be lost from sight, after it has been acquired and used for some time for position-finding, as when making a coastal transit. The basic quantity required for both these deter-

minations is the *computed visibility*, the maximum distance at which the light can be seen in the meteorological visibility conditions in the immediate vicinity. Once this distance is computed, the navigator can use it to swing a corresponding distance arc centered on the light symbol across his projected DR track on the chart, to ascertain the approximate bearing and time at which he may expect to sight the light initially, and if desired, the bearing and time at which he may expect to lose sight of the light.

The following sections describe how to calculate the computed visibility, and how to use the distance thus determined to derive the time and true bearing at which a light may be expected to be sighted, or once acquired, lost from view.

Meteorological Visibility

The concepts of meteorological and computed visibility are often confused by the inexperienced navigator. Meteorological visibility results primarily from the amount of particulate matter and water vapor present in the atmosphere at the location of an observer. It denotes the range at which the unaided human eye can see an unlighted object by day in a given set of meteorological conditions. Several of the more common terms used to describe different meteorological visibility conditions are listed below:

Term	Meteorological Visibility Range	International Visibility Codes
Dense to moderate fog	0—500 yards	0—2
Light to thin fog	500 yards—1 mile	3, 4
Haze	1—2 miles	5
Light haze	2—5-1/2 miles	6
Clear	5-1/2—11 miles	7
Very clear	11—27 miles	8
Exceptionally clear	Over 27 miles	9

There are several means whereby the navigator may obtain an estimate of the meteorological visibility range. If in an area covered by marine weather broadcasts, he may tune in to one of these to obtain this as well as other useful weather information (stations broadcasting marine weather are listed in DMAHTC *Publications 117A and B, Radio Navigation Aids*). Probably the most common procedure, however, is simply to determine the meteorological visibility empirically through observations of nearby land and seamarks, other ships in the vicinity, and at night, land lights or minor navigation lights.

In calculating computed visibility distances for lights, it is important to realize that the visibility so computed is not limited to the existing meteorological visibility. A very strong light such as that in a lighthouse might "burn through" even a dense fog and be seen several miles away, yet the prevailing meteorological visibility could be less than 500 yards.

Terms Associated with Light Visibility Computations

In addition to meteorological visibility, there are several other terms that come into use when determining the visibility of a light:

Horizon distance This is the distance measured along the line of sight from a position above the surface of the earth to its visible horizon, the line along which earth and sky appear to meet. The higher the position, the further its horizon distance will be.

Luminous range Luminous range is the maximum distance at which a light may be seen under the existing meteorological visibility conditions. It depends only on the intensity of the light itself, and is independent of the elevation of the light, observer's height of eye, or the curvature of the earth.

Nominal range Nominal range is a special case of the luminous range; it is defined as the maximum distance at which a light may be seen in clear weather (considered for this computation to be meteorological visibility of 10 nautical miles). Like luminous range, it takes no account of the elevation, height of eye, or the earth's curvature, and depends only on the light intensity.

Geographic range Geographic range is the maximum distance at which a light may be seen in perfect visibility by an observer whose eye is at sea level. It is analogous to the horizon distance of the light.

Charted range Charted range is the range printed on the chart near the light symbol as part of the data describing the light; in some texts, it is referred to as the charted visibility. On charts edited after June 1973, the charted range is the nominal range rounded to the nearest whole nautical mile. On previously issued charts, the charted range is the lesser of the nominal range or a geographic range that was then defined as the maximum distance at which a light could be seen in perfect visibility by an observer with a height of eye of 15 feet.

Computed range Computed range is the distance at which a light could be seen in perfect visibility, taking its elevation, the ob-

server's actual height of eye, and the curvature of the earth into account.

Computed visibility Computed visibility is the maximum distance at which a light could be seen in the existing visibility conditions, taking the intensity and elevation of the light, the observer's actual height of eye, and the curvature of the earth into account. It is always the lesser of the luminous range or the computed range.

Figure 6–5 may help to visualize the relationships existing among these terms as they apply in a situation in which an observer with a height of eye of 100 feet is located exactly at the computed range to a light having a geographic range of 12.5 miles and a nominal range of 19.6 miles.

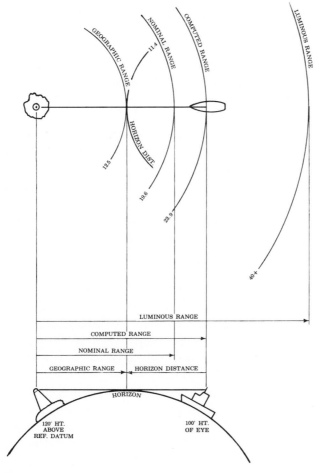

Figure 6–5. Relationships among light visibility terms.

In order to compute the visibility of a light, the navigator must first have some information about it, especially its intensity and elevation above the water. The *Light Lists* provide this and other information for lights and buoys in the coastal areas and rivers of the United States and its possessions, and the *Lists of Lights* provides similar information on lights found in foreign waters. The former series is published by the U.S. Coast Guard, and the latter is published by the Defense Mapping Agency Hydrographic/Topographic Center. Descriptions of these publications are given in Chapter 5 of this text dealing with navigation publications.

In addition to the information contained in the body of the various volumes, the *Light List* and the *List of Lights* contain another valuable aid for the navigator in the form of the Luminous Range Diagram. Figure 6–6 depicts the diagram found in the *Light List* volumes, based on the international visibility code.

By the use of the diagram, the luminous range of any light of given nominal range can be found for any meteorological visibility condition prevailing. The diagram is entered from either the top or bottom, using the nominal range of the light as an entering argument. The vertical line adjacent to the entry point is followed to the appropriate curve corresponding to the estimated visibility at the time of the observation. The luminous range for the given light under these visibility conditions is then read directly across to the left or right. As an example, consider Harbor of Refuge Light, from Figure 6–4A on page 85. The nominal range of its white sector is given as 23 miles. Entering the diagram in Figure 6–6 with this number, and going to the 11-mile visibility curve, a luminous range of about 25 miles is obtained from the scale on the sides. For 5-1/2-mile (clear) visibility, the luminous range is about 14 miles, and for 2-mile visibility, the luminous range is just under 7 miles.

If a 10-mile visibility curve were plotted by interpolation on the diagram, it would be seen that the luminous range of any light in 10-mile meteorological visibility conditions would closely agree with the listed nominal range of the light. This would be as expected, since the nominal range and the luminous range in 10-mile visibility conditions are identical by definition.

In situations in which meteorological visibility conditions are different from the curves on the luminous range diagram, an interpolated curve may be drawn in between the two curves bracketing the given value. In practice, however, the navigator usually estimates the visibility conditions by eye, based on experience, and he then uses the curve on the diagram closest to his estimated value for luminous range computations.

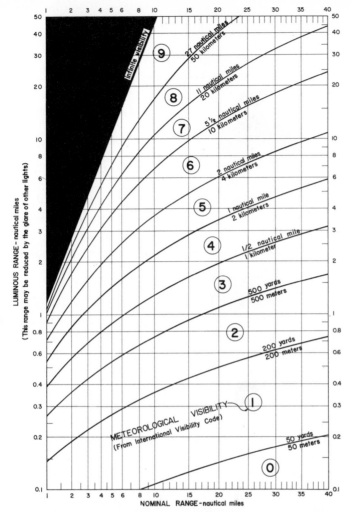

Figure 6–6. The luminous range diagram, Light List.

Procedure for Computing the Visibility of a Given Light

Once the navigator has determined the luminous range of the light through the use of the luminous range diagram, he is ready to proceed with the calculation of the computed visibility of the light. In addition to the luminous range, the other data that the navigator needs are his height of eye, and the elevation of the light, obtained either from the chart or the applicable *Light List* or *List of Lights* volume. After having obtained this information, the visibility of the light is computed in stepwise fashion as described below.

The first step is to find the geographic range or horizon distance that a light ray emanating from the light in question would travel

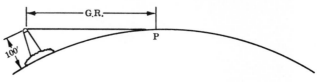

Figure 6–7. Geographic range for a light of 100-foot elevation.

before it is cut off by the curvature of the earth, as illustrated in Figure 6–7.

In Figure 6–7 a light ray originating from a light elevated 100 feet above sea level is cut off by the earth's curvature at point P which lies on the visible horizon. Although geographic range or horizon distances can be calculated by means of the following formula,

$$D \text{ (horizon distance)} = 1.144 \sqrt{h \text{ (elevation)}}$$

the usual practice is to look up the desired range or horizon distance from a table of precomputed values, such as shown in Figure 6–8 opposite. For an elevation of 100 feet, the table yields a geographic range of 11.4 miles.

The second step is to find the horizon distance for the observer's height of eye, as illustrated by Figure 6–9.

For an observer's height of eye of 50 feet, this distance from the table in Figure 6–8 is 8.1 miles.

As the third step, the geographic range of the light and the horizon distance of the observer are added together to form the *computed range* of the light—the distance a light could be seen in perfect visibility, taking the elevation of the light, the observer's height of eye, and the curvature of the earth into account. The computed range to a light having a 100-foot elevation from an observer with a height of eye of 50 feet is 11.4 + 8.1 = 19.5 miles; it is illustrated in Figure 6–10 on page 94.

For the fourth and final step in determining the distance at which he should sight a given light, the navigator figures the *computed visibility* of the light by comparing its luminous range in the existing visibility to the computed range. If the luminous range is greater than the computed range, he will be able to sight the light as soon as his vessel comes within the computed range; under these conditions, the computed visibility is the same as the computed range. Figure 6–11 is an illustration of this situation for a light having a luminous (nominal) range of 23 miles in 10-mile visibility.

If the luminous range is less than the computed range, the navigator will not be able to sight the light until the ship approaches within the

Height feet	Nautical miles	Height feet	Nautical miles
1	1.1	41	7.3
2	1.6	42	7.4
3	2.0	43	7.5
4	2.3	44	7.6
5	2.6	45	7.7
6	2.8	46	7.8
7	3.0	47	7.8
8	3.2	48	7.9
9	3.4	49	8.0
10	3.6	50	8.1
11	3.8	55	8.5
12	4.0	60	8.9
13	4.1	65	9.2
14	4.3	70	9.6
15	4.4	75	9.9
16	4.6	80	10.2
17	4.7	85	10.5
18	4.9	90	10.9
19	5.0	95	11.2
20	5.1	100	11.4
21	5.2	105	11.7
22	5.4	110	12.0
23	5.5	115	12.3
24	5.6	120	12.5
25	5.7	125	12.8
26	5.8	130	13.0
27	5.9	135	13.3
28	6.1	140	13.5
29	6.2	145	13.8
30	6.3	150	14.0
31	6.4	160	14.5
32	6.5	170	14.9
33	6.6	180	15.3
34	6.7	190	15.8
35	6.8	200	16.2
36	6.9	210	16.6
37	7.0	220	17.0
38	7.1	230	17.3
39	7.1	240	17.7
40	7.2	250	18.1

Figure 6–8. Horizon distance table, from Bowditch.

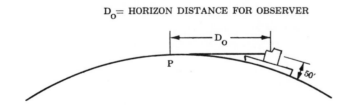

D_O = HORIZON DISTANCE FOR OBSERVER

Figure 6–9. Horizon distance for an observer's height of eye of 50 feet.

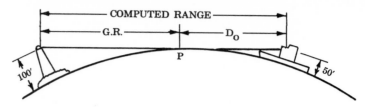

Figure 6–10. *Computed range for a 100-foot light and an observer's height of eye of 50 feet.*

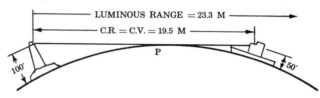

Figure 6–11. *Computed visibility in clear weather for a 100-foot, 23-mile light and an observer's height of eye of 50 feet.*

luminous range, regardless of his height of eye. In this situation, the luminous range becomes the computed visibility of the light, as depicted in Figure 6–12, where meteorological visibility has decreased to 5½ miles.

Thus, the distance at which the navigator should expect to sight a light, the computed visibility, is always the lesser of the computed range or the luminous range. In the absence of a predicted or expected visibility condition in the vicinity of the light, nominal range is normally used as the luminous range for purposes of the light visibility computation. The small boatman not equipped with a *Light List* or *List of Lights* will usually use the charted nominal range as his computed visibility.

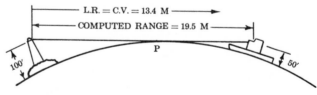

Figure 6–12. *Computed visibility for a 100-foot, 23-mile light in 5½-mile visibility, for an observer at 50 feet.*

Plotting the Computed Visibility of a Light on the Chart

Having determined the computed visibility distance at which he should sight a given light, the navigator can then find the approximate position and time at which the light should be sighted by referring to his dead reckoning plot. To do this, the navigator swings an arc centered on the charted position of the light, with radius equal to the computed visibility, across the projected DR course line, as shown in Figure 6–13.

The intersection of the arc and the projected DR course line is the approximate position at which the ship should be when the light becomes visible on the horizon. The bearing of the light from this position should be the approximate bearing at which the light will appear; it can be expressed in both degrees true and degrees relative, for the benefit of the lookouts. Finally, the approximate time of the sighting can be obtained by computing how long it should take the ship to reach the intercept position.

In practice, the navigator computes the expected distance and time and bearing for each light the ship should sight during a night's steaming by swinging computed visibility arcs for all lights which should be sighted across the ship's intended track for the night. On Navy ships and on most merchant vessels, this information is entered in the night orders for the night-time deck watch officers, and the watch officer on duty is required to notify the commanding officer or master and navigator if the actual sightings differ too greatly from the precomputed times and bearings. Because of unanticipated changes in the local meteorological visibility conditions, a small amount of variance between actual and expected sighting data is normally to be expected. As mentioned earlier, on some occasions the navigator will also swing the visibility arc for each light across the intended exit track to determine when the ship can be expected to pass out of the range of the light.

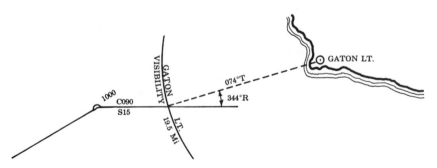

Figure 6–13. Predicting the ship's position, time and bearing of a light sighting.

Buoys and Beacons

The second category of visual aids to navigation to be described in this chapter, of equal importance to the fixed light structures predominantly described up to this point, is buoys and related immobile aids called beacons (more fully described on pages 100–102). The primary function of these aids is to warn the mariner of some danger, obstruction, or change in the contour of the sea bottom, and to delineate the limits of safe channels through relatively shallow water. A secondary function is to assist the navigator to a limited degree in the determination of the ship's position.

Systems of Buoyage

There are two general systems of buoyage in use throughout the world—the *lateral* system and the *cardinal* system. In the lateral system, the location of each buoy or beacon indicates the direction of the danger it marks relative to the course that should normally be followed; it is best suited for well-defined channels. In the cardinal system, the location of each buoy indicates the approximate true bearing of safe water from the danger it marks. The cardinal system is best suited for marking offshore rocks, shoals, islets, and other dangers in and near the open sea.

Over the last two centuries, most of the major maritime nations of the world began to recognize the need to standardize both of these systems of buoyage. Although the cardinal system has been standardized through a number of international accords into what is now the Uniform Cardinal System, the same cannot be said of the lateral system. In 1889, an international marine conference held in Washington, D.C., recommended that under the leadership of the United States, in the lateral system right-hand channel buoys should be painted red, and left-hand buoys black. With the subsequent introduction of lighted aids to navigation about the turn of the century, the logical extension of this pattern led to the use of red or white lights on the right side and green or white lights on the left side of channels. This system became known as the U.S. Lateral System. In 1936, however, in the last attempt to reach an international accord on the subject, a League of Nations subcommittee recommended a system virtually opposite to that of the U.S. Lateral System. The 1936 system has become known as the Uniform Lateral System. In this system, black buoys with green or white lights are used to mark the right side of a channel, and red buoys with red or white lights are used for left-side markings.

For the next forty years until the mid-1970s, almost all foreign maritime countries used the Uniform Lateral System to mark their coastal waters and navigable rivers, and the Uniform Cardinal System

to mark dangers in offshore areas, while the United States and its possessions used the U.S. Lateral System. In 1977, however, most western European nations began adopting a new system called the IALA (International Association of Lighthouse Authorities) Combined Cardinal and Lateral System, which combined features of both the old Uniform Systems. By 1980, this system had been implemented throughout western Europe. Then in the spring of 1982, some eighty of the major maritime nations, including the United States, signed an international agreement to implement the IALA system in one of two forms worldwide. The original European system, designated IALA System "A", was prescribed for Europe, Africa, and Asia, and a new IALA System "B", for North, Central, and South America, Korea, and the Philippines. The most significant new feature of both systems is that green vice black buoys are used as channel markers; in System "A", green buoys mark the starboard side, while in System "B", green buoys mark the port side.

Implementation of IALA System "B" in U.S. waters began in 1983, and should be completed by 1989. The main changes from the current U.S. Lateral System will be the repainting of black channel buoy colors to green, and the use of new vertically striped red-and-white midchannel buoys and new yellow special buoys.

The remainder of this chapter will discuss the U.S. Lateral System, and the IALA Combined Cardinal and Lateral System, in some detail.

The U.S. Lateral System

The lateral system used in the territorial waters of the United States and its possessions, with its own peculiarities, is known as the U.S. Lateral System. This system employs an arrangement of colors, shapes, numbers, letters, and light phase characteristics to indicate the side on which the buoy or beacon should be left when proceeding in a given direction with respect to the sea. As was mentioned in Chapter 5 in connection with the *Light List*, "from seaward" is defined as proceeding in a clockwise direction around the continental United States, in a northerly direction up the Chesapeake Bay, and in a northerly and westerly direction on the Great Lakes (except southerly on Lake Michigan).

Buoys of the U.S. Lateral System

There are four main categories of buoys in the U.S. Lateral System—unlighted, lighted, sound, and combination. There are several configurations of buoys within each of these four categories:

Can Buoys Always unlighted, these buoys are so named because they are built in the cylindrical shape of a can.

Nun Buoys Like can buoys, this kind of buoy is always unlighted.

They have the shape of a truncated cone above water, resembling the old-time habit of a nun.

Lighted Buoys This type of buoy consists of a metal float on which a short skeleton tower of any shape is mounted. The tower supports a lantern powered by electric batteries in the body of the buoy.

Bell Buoys These are steel floats of any shape, surmounted by a structure able to accommodate a single-toned bell. Most bell buoys are sounded by the motion of the sea, with a very few powered by bottled gas or electricity.

Gong Buoys Similar in construction to a bell buoy, these sound buoys are fitted with a series of gongs, each having a distinctive sound, rather than with a bell.

Whistle Buoys Like bell and gong buoys, this type of sound buoy can be of any shape. They are fitted with a whistle or horn, powered by a system of bellows actuated by the motion of the sea.

Horn Buoy Similar to a whistle buoy, a horn buoy is powered by an electric battery or land line rather than the random motion of the sea. Its horn therefore sounds in a regular cadence, rather than in the irregular cadence of most motion-actuated bell, gong, and whistle buoys.

Combination Buoy This is a descriptive term that applies to any buoy in which a light and a sound signal are combined, such as a lighted bell buoy, lighted gong buoy, or lighted whistle buoy.

Radar reflectors that enhance the buoy as a radar target are now being incorporated in all buoys of the U.S. Lateral System to assist in their location and use in radar navigation. All buoys serve as daytime navigation aids; those fitted with lights are available at night, while those equipped with sound signals are more readily located in times of poor visibility, whether in darkness or in fog, rain, or snow.

Chart No. 1 in Appendix A of this volume contains five plates depicting lighted and unlighted aids to navigation found on the navigable coastal waters, intracoastal waterways, western rivers, and state waterways of the United States. The first three of these plates show buoys and daymarks (described on page 100) of the U.S. Lateral System found in U.S. coastal waters, harbors, and intracoastal waterways. The chart symbol and abbreviated descriptive chart data appear by each type of aid (see also Section L for other buoy and beacon symbols). Note that for buoys, the abbreviated chart phase charac-

teristic data is printed in italics, as opposed to the roman type used for fixed light structures. As can be seen by an inspection of the third of these plates, the intracoastal waterway aids are identified as such by the addition of a yellow marking to the basic color scheme. The remaining two plates depict aids found only on western rivers and state waterways of the United States; the seagoing navigator will seldom if ever encounter these.

The left-hand columns of the first four plates contain those buoys to be kept on the port side of the ship when returning from sea. They are painted black (or green) and carry odd numbers; if neither lighted nor fitted with a sound device, port-side buoys are always can buoys. The righthand columns of these plates depict the starboard-side buoys; they are always painted red and carry even numbers. If they are neither lighted nor equipped with a sound device, right-hand channel buoys must be in the truncated cone shape of the nun buoy. If sound buoys are used to mark a channel, gong buoys are placed on the left side, and bell buoys on the right. The color-coding scheme of port and starboard channel buoys can be remembered by means of an old mnemonic aid—*Red Right Returning*.

The central columns of the plates show buoys that are used in or near channels for purposes other than to delineate their boundaries. Junction or obstruction buoys mark either junctions in the channel or obstructions such as wrecks or sand bars. They are painted in red and black (or green) horizontal bands. If the preferred channel, when entering from seaward, requires that the buoy be left to port in order to proceed down the preferred starboard channel, the top band is black or green; conversely, if the preferred channel calls for leaving the buoy to starboard, the top band is red. Fairway or midchannel buoys have vertical black (or red) and white stripes, and serve to mark the center of the channel. Although they may be safely passed to either side, in practice they are usually left to port, inasmuch as the nautical rules of the road require the mariner always to proceed down the right side of the channel.

The preceding types of buoys that either delineate channel boundaries or mark junctions or fairways are often referred to as *channel buoys*. The remaining buoys are those used to denote special areas of caution usually found outside of channels. Collectively, these are referred to as *special purpose buoys*. They are subdivided into four basic categories: white anchorage buoys, black and white horizontally banded fish net buoys, green and white dredging buoys, and miscellaneous orange and white horizontally banded special purpose buoys.

As can be seen, in the U.S. Lateral System the size, shape, coloring, signaling equipment, and numbering on a buoy all help to identify

it. Note especially the various light phase characteristics and colors associated with the different types of buoys. As has been mentioned earlier, the *Light List* contains full data on each buoy in U.S. coastal waters and rivers, regardless of whether it is lighted or unlighted. The location of each buoy is also shown on charts of the area; the position of a buoy is indicated by an appropriate buoy symbol with a small open circle or dot marking the location of its anchor. If the buoy is lighted, the dot is overprinted by a purple circle or ray. Abbreviated italicized descriptive data are given by the buoy symbol, indicating its color scheme, sound signal, markings, and light color and phase characteristic, if any; buoys fitted with a radiobeacon or equipped with some other special feature are also indicated.

Buoy Light Colors and Phase Characteristics

The light color and phase characteristic scheme used on buoys in the U.S. Lateral System is well standardized. Red lights are used only on red starboard-hand channel buoys and on junction buoys with the topmost band red. Green lights are used only on black or green port-hand channel buoys and junction buoys with the top band black or green. White lights can be used on any buoy in the system.* As was the case for immobile light structures, if the light is red or green the abbreviations "R" or "G" will appear on the chart alongside the purple buoy light symbol, printed in italics in the case of buoys. A white light has no abbreviation, so that any lighted buoy symbol without color abbreviation nearby should be considered to be fitted with a white light.

The light phase characteristics associated with each type of buoy in the U.S. Lateral System are illustrated in *Chart No. 1*. Port- and starboard-hand channel buoy lights are either fixed, flashing, occulting, quick flashing, or equal interval. Junction or obstruction buoys are readily identified by their interrupted quick-flashing characteristic. Midchannel or fairway buoys show a distinctive morse code pattern, by convention the letter "A." The light phase characteristic always appears in abbreviated form alongside the charted buoy light symbol, along with the color designation, if any.

Beacons

In addition to buoys, there is another type of aid to navigation in U.S. waters that is extensively used to delineate channels and mark hazards to navigation, especially in the shallower harbors, bays, and inland waterways. Called *beacons*, they are not floating aids, but rather are firmly attached to the bottom or the shore. Older beacons

---

*White lights are not allowable on channel marks and buoys in the new IALA B system now being implemented in U.S. waters. See buoy plate one in *Chart No. 1* in Appendix A.

Figure 6–14. Two kinds of typical beacons found in U.S. waters are pictured off the Coast Guard base at Miami, Florida. The traditional type of wooden structure on the left is gradually being replaced by concrete piles fitted with plastic or metal daymarks. Note the minor light at the top of each beacon. (Courtesy U.S. Coast Guard)

normally consist of a single wooden pile, or a structure made up of several wooden piles fastened together, but newer beacons are constructed of masonry, steel beams, or other suitable materials. If unlighted, they are referred to as *daybeacons*. A geometrically shaped visual indicator called a *daymark* is normally affixed to the beacon to convey identifying information and to denote the kind of marker it is by day. Daymarks marking channels covered by the U.S. Lateral System are colored and numbered in basically the same manner as buoys, with red triangular daymarks indicating a right-side boundary, and green square marks a left-side channel boundary, progressing from seaward. The shapes and color schemes for channel, junction

or obstruction, and midchannel daymarks, along with their chart symbols, are shown on the U.S. Lateral System plates of *Chart No. 1* in Appendix A referred to earlier. Daymarks are often fitted with reflective tape to aid in their identification at close range by searchlight at night.

The basic chart symbols for daymarks are squares and triangles. Left-side channel daymarks are represented by either a green square or an open triangle and the letter "G," and right-side daymarks by a purple triangle. Midchannel and junction/obstruction daymarks are symbolized by an open triangle with the color abbreviations "BW" for a black-and-white midchannel mark, or "RG" (red uppermost) or "GR" (green uppermost) for junction/obstruction marks.

In addition to daymarks, *minor lights* are commonly mounted on beacons to facilitate their use at night. These minor lights have the same color and phase characteristics as similarly placed lighted buoys. On charts, they are differentiated from lighted buoys by the use of the standard symbols for structural navigational lights, and by the use of roman vice italic type to describe their phase characteristics and color.

U.S. Lateral Numbering System

The standard numbering scheme used on buoys and daymarks of the U.S. Lateral System, like light colors and characteristics, greatly facilitates the identification and location of the aid on the chart. Odd numbers are used only on black or green port channel buoys or green daymarks, while even numbers are found only on red starboard channel buoys or daymarks. The numbers on both sides increase sequentially from seaward. Thus, buoys marked with the numbers 1, 3, 5, *etc.* might mark the port side of a channel entrance, while the numbers 2, 4, 6, *etc.* would appear on starboard-hand buoys. If there are more buoys or daymarks on one side of a channel than the other, some numbers are omitted, so that a buoy marked "4" would never appear opposite one marked "9." If a buoy or daymark is added after a particular system is established, it is marked with the same number as the channel marker preceding it, plus a letter suffix, as for example "1A." Letters without numbers are applied in some cases to black or red and white vertically striped midchannel or fairway buoys or daymarks, red and black or green horizontally banded junction or obstruction buoys or daymarks, and other buoys or daymarks which are not all red or all black or green. The markings on a buoy or daymark are indicated on the chart by the placement of quotation marks around the letters or numbers, with italics used for the buoy data. Care must be taken not to confuse an identification mark with the symbol for a

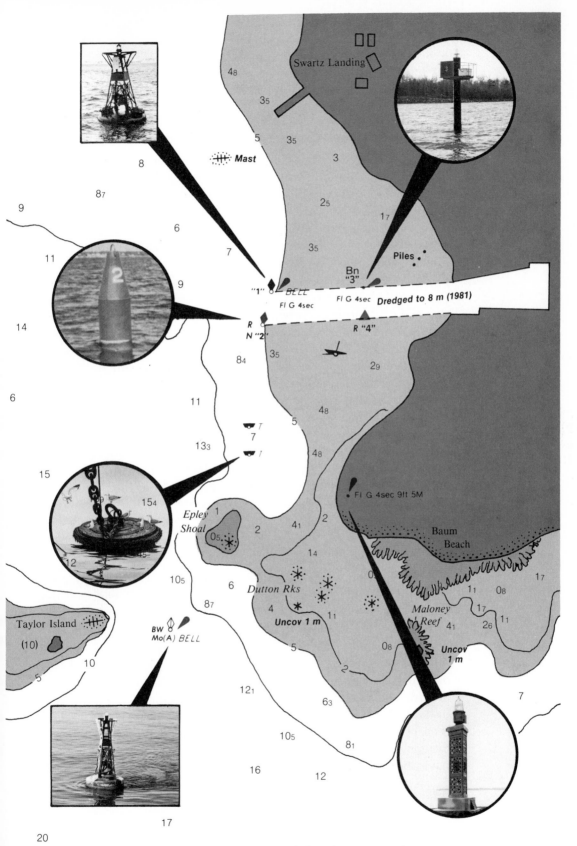

Figure 6–15. A fictitious nautical chart illustrating symbology for various aids to navigation found in U.S. waters.

color or with a nearby sounding. For example, a charted buoy symbol with the abbreviations *RN* "2" printed nearby would indicate a red nun buoy marked with the number 2.

Figure 6–15 is a portion of a mock chart illustrating typical chart symbols for various types of visual aids to navigation that might be encountered in U.S. waters.

The IALA Combined Cardinal and Lateral System

Because of the progressive implementation of the IALA Combined Cardinal and Lateral System in both foreign and U.S. waters during the latter 1980s, it follows that navigators should have some familiarity with both IALA Systems "A" and "B" and the differences between these systems and the Uniform Lateral, Uniform Cardinal, and U.S. Lateral Systems which will all be found at various locations in the interim years. Accordingly, the following sections will briefly describe the IALA Combined Cardinal and Lateral System (hereafter referred to as the IALA Combined System), highlighting, where they exist, the major differences between this system and both the Uniform Cardinal and Uniform Lateral systems, and the U.S. Lateral System. If it is desired, further information on all these systems can be obtained from *The American Practical Navigator* (Bowditch) and from the applicable volumes of the *Enroute Sailing Directions* covering particular foreign areas of interest. In the case of lighted buoys, full descriptive data on them is also contained in the appropriate DMAHTC *List of Lights* volume.

The last five pages of *Chart No. 1* included in this volume as Appendix A consist of five plates depicting the buoys and light colors and characteristics used in the IALA Combined Cardinal and Lateral Systems "A" and "B". These plates will be referred to throughout the following discussion.

Topmarks

A major difference between buoys of the U.S. Lateral System and those of the Uniform Lateral, Cardinal, and IALA Combined Systems—especially System "A"—is that in many cases buoys of the latter systems are fitted with a type of visual indicator called a *topmark*. Somewhat similar in concept to the daymark of the U.S. system, topmarks are intended to convey information as to the type of buoy on which they are placed, primarily by their shape, and in the case of cardinal marks, by their orientation. These topmarks are a necessity in the Uniform Cardinal, Uniform Lateral, and IALA Combined systems because, with a few exceptions in the case of the Uniform Lateral System, the shape of unlighted as well as lighted buoys does not have navigational significance. The different topmarks associated with the

various kinds of buoys in the IALA Combined System are shown in the IALA plates of *Chart No. 1.*

Channel Markers

From the point of view of the American navigator, probably the greatest difference between the U.S. Lateral System and the Uniform Lateral and IALA Combined System "A" is that in the latter systems red channel markers are used to indicate the *left* side of channels, rather than the right side as in the U.S. system. Black markers in the Uniform Lateral and green markers in the Combined System "A" are used to delineate the right side of channels (green is used on the *left* in the IALA System "B"). At night, red or white lights are used on left-hand channel markers of the Uniform Lateral System, and green or white lights on the right side. In both IALA Systems, no white lights may be used on channel markers. The buoys, topmarks, and associated lights adapted from the lateral system for use as channel markers in the IALA Combined Systems are illustrated on the IALA plates of *Chart No. 1.*

Channel markers in the Uniform Lateral and IALA Combined System "A" and "B" may be identified with numbers or letters, as are similar aids in the U.S. Lateral System, except that in the former two systems left-hand channel markers are marked with even numbers, and right-hand markers with odd numbers, commencing from seaward. This convention, like the color scheme, is opposite to the practice in the U.S. Lateral System. In addition to these markings, many buoys in European waters are stenciled with a name related to their locale or the danger they mark, as for example, *"CASTLE"* buoy or *"SPIT REFUGE"* buoy.

Midchannel and Junction Markers

The equivalent of the vertically striped black and white midchannel buoy of the U.S. Lateral System is called a "safe water mark" in the IALA Combined Systems. The equivalent of the horizontally banded black or green and red junction buoy of the U.S. Lateral System is called a "preferred channel mark" in the IALA Combined Systems. Both these markers are shown in the IALA plates of *Chart No. 1.* Note the red topmark of the lighted safe-water mark in both IALA Systems "A" and "B".

The Uniform Lateral System features both midchannel markers very similar to those of the U.S. Lateral and IALA Combined systems, and "middle-ground" markers corresponding to the U.S. junction and IALA preferred channel buoys. The former markers are either red and white or black and white vertically striped, with corresponding red or black topmarks. Middle-ground markers are used to indicate

bifurcations (splits) or junctions of channels. If the main channel is to the right of the buoy, or if the channels are of equal importance, a horizontally banded red and white marker with red topmarks is used. Horizontally banded black and white markers with black topmarks are used if the main channel is to the left. Both types of markers may be lighted with lights distinctive from others in the vicinity.

Special Buoys

There are no exact equivalents in the Uniform Lateral, Cardinal, or IALA Combined systems to any of the special purpose buoys of the U.S. Lateral System. There are, however, a number of special marks in the IALA Combined System that may be used to indicate a special area or feature, illustrated on the IALA plates. Colored yellow, and in IALA System "A" fitted with single yellow "X"-shaped topmarks, they are used as traffic separation marks, and as markers for spoil grounds, military exercise zones, cable or pipe lines, and recreation zones. If lighted, they carry yellow lights having distinctive phase characteristics. Both Uniform systems and the IALA Combined Systems feature a red-and-black horizontally banded isolated-danger mark, fitted with spherical topmarks, and optional distinctive lights (see IALA plate 3). Additionally, the Uniform Lateral and Cardinal systems both provide for green-colored wreck buoys fitted with a green light for marking the position of wrecks. There is no corresponding marker in the IALA Combined Systems.

Cardinal Marks

The other type of buoy for which there is no equivalent in the U.S. or Uniform Lateral systems is, of course, the cardinal mark of the Uniform Cardinal and IALA Combined systems. As indicated in IALA plate 1, there is a different cardinal mark for each of the four cardinal points of the compass, i.e., north, south, east, and west. The buoy indicates the direction of safe water from the point of interest marked by the buoy. A cardinal mark may be used to indicate the direction of deepest water, the safe side on which to pass a danger, or to draw attention to a feature in a channel such as a bend, junction, bifurcation, or end of a shoal.

An important feature of a cardinal mark is the two black double-cone topmarks always associated with it, which by their arrangement, coupled with the color scheme of the buoy, identify by day the quadrant indicated by the mark. The shape of the buoy itself does not have significance.

When lighted, the IALA cardinal mark is fitted with a very distinctive white light having a continuous or periodic very quick flashing

(i.e., 100 or 120 flashes per minute), quick-flashing, or long-flashing (i.e., not less than two seconds in duration) phase characteristic, as indicated on IALA plate 1. It is not planned to use many cardinal marks in conjunction with IALA System "B" in U.S. waters.

Chart Symbols for the IALA Combined System

Section L. 70 of *Chart No. 1* contains chart symbols for the buoys and lights of the IALA Combined Cardinal and Lateral Systems. As can be seen, the symbols are quite different from those used to represent the buoys and lights of the U.S. Lateral System. As is the case with lighted buoys in the U.S. system, the abbreviated light phase characteristic data is written in italics to help distinguish lighted buoys from fixed light structures. Navigators and their assistants who expect to operate in areas covered by one of the IALA Combined Systems should become very familiar with the associated chart symbology as part of their voyage-preparation process.

Use of Buoys and Beacons During Piloting

Regardless of the buoyage system in use where a navigator is operating, his first objective upon sighting a buoy or beacon in piloting waters is, as with any visual navigation aid, to identify it and correlate its position on the navigational chart of the area. By day, the shape, color scheme, top- or daymarks, and if distinguishable, any markings can all be used to identify the buoy or beacon, and determine its position relative to the channel or hazard it is marking. At night, the light on a lighted buoy, or minor light on a beacon, will usually serve as the only means of identifying it from any distance. Once he has identified the buoy or beacon, the navigator can then make full use of it as a valuable aid to navigation, subject to the restraints discussed below.

Because a buoy is moored in position by a cable, with enough slack left to allow for rising and falling tide levels, it does not always maintain an exact position directly over its anchor. For this reason, the navigator should never make exclusive use of buoys to determine the ship's position. Buoys should always be regarded as warnings or aids rather than fixed navigation marks, especially during winter months or when they are moored in exposed waters. Rough weather, ice, or even collision with a ship may carry a buoy completely away or drag it to a new location. Moreover, the light on a lighted buoy may fail, or motion-actuated sound devices may not sound in calm seas or in heavy icing conditions. A smaller buoy called a station or marker buoy is sometimes placed in close proximity to important navigation aids, such as a sea buoy, to mark its location in case the regular aid is

accidentally shifted from station. Beacons, because they are permanently mounted at a stationary location, are much preferred for LOPs if the need arises.

In practice, whenever a ship is proceeding down a channel marked with buoys or beacons, true bearings should continually be taken to the next aids ahead of the ship. If the ship were steaming in a channel marked with buoys, for example, true bearings to buoys on the right side of the channel should be growing ever larger, or in other words, the buoy should show a right-bearing drift. Left-hand buoys should show a left-bearing drift. Initially, the bearings will not drift very rapidly, if at all, while the ship is relatively distant from the buoy, but as she approaches, the drift rate should increase. If the bearings remain unchanged as the ship draws near, the ship is proceeding down a line of bearing to the buoy. Unless the course is adjusted so as to establish a bearing drift, the ship will collide with the buoy—an extremely distressing event for all concerned, especially the navigator.

When navigating in a narrow channel, buoys and especially beacons can be extremely helpful in the safe piloting of the ship. Because of the time lag inherent in the dead reckoning plot, which will be explained in Chapter 8, piloting in a narrow channel with many twists and turns would be extremely difficult without buoys or beacons to mark the channel boundaries. Although these aids cannot be exclusively relied upon to keep the vessel within the channel, use of buoys and beacons in conjunction with independently obtained fixes and an accurate dead reckoning plot can make piloting in a narrow channel no more difficult than driving an automobile down a city street.

Summary

This chapter has described the characteristics of both lighted and unlighted navigation aids in both U.S. and foreign waters that assist the navigator in the safe direction of the movements of his ship in the piloting environment. The first objective of the navigator upon sighting any visual aid to navigation is its positive identification, in order that its position relative to the channel or a hazard it may be marking can be correlated on the chart. By day, the overall appearance of an aid will usually serve to identify it, and by night, the various attributes of its light, particularly its period, will serve this function. When piloting by means of navigation lights at night, the navigator should be able to predict the time and direction in which a light should be sighted, or once sighted, lost from sight, by means of the computed visibility of the light. Piloting in a winding or narrow channel would be extremely awkward without the availability of buoys

and beacons to assist in the endeavor. The inexperienced navigator should become very familiar with the contents of this chapter, because piloting as it is currently practiced would be very difficult if not impossible without reliance upon the visual aids to navigation discussed herein.

7

Navigational Instruments

Almost all professional disciplines have a set of tools uniquely associated with them, and the practice of navigation is no different in this respect. This chapter will describe many of the more common instruments used by U.S. Navy, American Merchant Marine, and private navigators. Although navigational instruments may be classified in a number of ways, they will be considered here in groups according to the following purposes: instruments to measure direction, distance, and speed; instruments to measure depth; instruments for plotting; and instruments for miscellaneous use.

The Measurement of Direction

The horizontal direction of one terrestrial point from another, expressed as an angle from 000° clockwise to 360°, is termed a *bearing*. There are three different types of bearings with which the surface navigator is concerned, depending on which direction is used as the basic reference or 000° direction. If a bearing is measured with reference to the ship's longitudinal axis, it is termed a *relative bearing*; if measured with respect to a magnetic compass needle aligned with magnetic north, it is a *magnetic bearing*; and if measured with respect to a gyrocompass repeater having zero gyro error, or a magnetic compass corrected to true north, the angle is termed a *true bearing*. Terminology associated with direction and bearing is fully discussed in Chapter 9 of this text, which describes shipboard compasses, In this section, some of the more common instruments used in conjunction with the shipboard compass to determine one of the three types of bearings—relative, magnetic, and true—will be described.

The Azimuth Circle

The term "azimuth" is often used interchangeably with the word "bearing," although technically the former term refers to the bearing of a celestial body, while the latter pertains to the bearing of a terrestrial object. Perhaps the most common device used for obtaining either is the *azimuth* or *bearing circle*. Figure 7–1 depicts the azimuth

Figure 7–1. Azimuth circle for the standard Navy 7½-inch gyrocompass repeater.

circle found most often on Navy ships, designed to fit on the standard 7½-inch gyro repeater.

The azimuth circle consists of a nonmagnetic brass ring, formed to fit snugly over the repeater face. It can be turned to any desired direction by means of two finger lugs provided on the ring. A pair of sighting vanes, consisting of a peep vane at one end and a vertical wire at the other, are mounted on one diameter of the ring. A reflector of dark glass is also attached to the vane containing the vertical wire, for use in observing the azimuth of celestial bodies.

To observe a bearing, an observer looks through the peep vane, sometimes called the "near" vane, toward the object to be observed or "shot." He then rotates the ring until the object appears beyond the vertical wire of the opposite or "far" vane. A reflecting mirror is built into the circle to bring the portion of the compass card directly beneath the far vane into the field of vision; it will also bring the relative bearing scale inscribed around the perimeter of the compass card binnacle into view. The bearing of the observed object is then read by the position of the vertical wire on the compass card, if true bearings are required, or by its position on the relative bearing scale, if a relative bearing is desired. As an alternative method of measuring relative bearing, the ring of the azimuth circle is inscribed with bearings in counterclockwise order, from 000° at the far vane; the relative bearing to an object may be obtained by sighting it through the sight vanes and then reading the bearing that appears on the ring directly over the gyro repeater's lubber's line mark representing the ship's head.

A second set of sighting devices designed for observing the azimuth of the sun is attached to the ring at right angles to the sight vanes. One of these devices is a mirror, and the other is a housing containing a triangular reflecting prism. To observe the azimuth of the sun, the

azimuth circle is rotated until the sun is above the prism. Its rays strike the mirror and are then reflected onto the prism, which in turn reflects a narrow ray across the graduations on the compass card, where the bearing is read.

Both the far sight vane and the prism housing incorporate leveling bubbles for the purpose of horizontal alignment of the azimuth circle at the moment the bearing or azimuth is shot; if the ring is not level at this time, error may be introduced into the bearing obtained.

The Telescopic Alidade

Another piece of equipment used for observing bearings is the *telescopic alidade*. It is quite similar in construction to the azimuth circle, except that it is fitted with a telescopic observation device rather than a set of sight vanes. The optical system simultaneously projects the image of approximately 25° of the compass card and a view of a built-in level into the field of view of the telescope. The object to which a bearing is to be obtained is sighted in the telescope and its bearing is read off from the compass card. Older models of the telescopic alidade have a straight telescope attached, and newer models have the eyepiece inclined at an angle for ease in viewing. Both types are pictured in Figure 7–2.

In general, most observers seem to prefer the azimuth circle for most applications when observing a bearing, as the field of vision in the telescopic alidade is comparatively narrow. Accurate observations of a lighted navigation aid at night, however, are often facilitated by the use of the telescopic alidade.

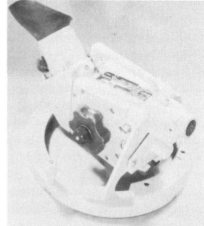

Figure 7–2. Two types of telescopic alidades found on board Navy ships.

Measurement of Distance

There are two instruments commonly used by the shipboard navigator for the measurement of distance. One of these, radar, is discussed in a separate chapter of this text; the other instrument is the hand-held stadimeter.

The Stadimeter

The stadimeter is normally used most frequently on board Navy ships by the OOD to obtain precise ranges from his ship to others in a formation. For shorter ranges up to 2,000 yards, it is generally conceded to be more accurate than most surface search radars for this purpose. In piloting, however, the stadimeter is also used as a navigational instrument by the navigator to ascertain accurate distances to navigation aids of known height above the water.

There are two kinds of stadimeters currently in use on board most larger vessels. The Fisk type, pictured in Figure 7–3, is probably the more common of the two and is used on board most Navy ships; the other, the Brandon sextant type, is shown in Figure 7–4 next page.

Both stadimeters incorporate two scales. One, located on the index arm of the frame, is the object height scale, graduated in logarithmic form for object heights between 50 and 200 feet. The other, inscribed around the index drum beneath the frame, is the distance scale; it is graduated in a spiral logarithmic scale for distances between 50 and 10,000 yards. Both types of instruments are equipped with a removable telescope fitting in the rear view finder, a reflecting mirror in the right side of the forward view finder, and an index mirror under the rear view finder. To use either instrument, the height of the object to be observed is first set on the index arm scale. Then the object is sighted in the telescope; turning the index drum causes a

Figure 7–3. The Fisk stadimeter.

Figure 7–4. The Brandon sextant stadimeter.

reflected image in the mirror in the right side of the forward view finder to move up or down relative to the direct image observed through the left side. When the top of the reflected image is superimposed alongside the bottom of the direct image, the distance to the object is read directly from the index drum scale. If the stadimeter is used to obtain the distance to an object having a height less than 50 feet, such as a YP training craft, the index arm height scale is set to some integral multiple of the object's actual height, between 50 and 200 feet. The distance obtained from the index drum scale is then divided by the amount of the height multiple to yield the actual distance. If the stadimeter were used to obtain the distance between two YPs of masthead height 35 feet, for example, the instrument height scale would be set to 70 feet (2 × 35′). If a distance of 620 yards were indicated on the index drum, the actual distance between them would be 310 yards (620 ÷ 2).

Because of the logarithmic scale, distances read on the index drum scale can be read with great precision up to 2,000 yards, but beyond that the accuracy progressively decreases. Prior to use, alignment of the instrument should always be checked by observing the sea horizon. When the reflected and direct images of the horizon appear side by side, the index drum distance scale should read "infinity." If it does not, the mirrors must be adjusted before further use. The stadimeter is a delicate instrument; if handled and stowed properly, however, it is of great value to the navigator in the piloting situation.

Measurement of Speed

There are two kinds of speed with which the navigator of a surface ship is concerned—"true" speed, or speed relative to the earth (often

called "speed over the ground" (SOG)), and ship's speed through the water. True speed is normally calculated empirically by measuring the time required for the ship to traverse a known distance. Speed through the water in which the ship is floating is measured both mechanically and empirically by methods to be discussed in this section.

One of the earliest methods developed to measure a ship's speed relative to the water in which she floats is to time the passage of a wood chip or retrievable float along the ship's length. This procedure is still in limited use today by some navigators in the piloting environment. A team is set up to periodically throw the chip or float over the bow and time its transit along the side until it reaches the stern, or until a marker knot is reached in the uncoiling retrieving line of the float. Ship's speed is then read from a precalculated table.

Fortunately, the modern navigator has several more sophisticated means of determining speed available to him. Primary among these are the impeller and pit logs, the doppler speed log, and some very recent developments in the determination of speed by the use of satellite and terrestrial-based electronic navigation systems.

The Impeller and Pit Logs

All marine instruments designed for direct measurement of speed through the water are known as *logs*. Many smaller sail and power boats are equipped with a type of log called an *impeller log* that consists of a sensing device incorporating a small propeller or paddlewheel located beneath the waterline just outside the hull. The speed of rotation of the impeller caused by the water flow past it is mechanically or electrically translated into vessel speed through the water, similar to the speedometer on an automobile.

Probably the most common of all instruments installed in larger vessels to measure speed through the water is the *pitometer* ("pit") *log*, so called because it incorporates a *pitot tube*. This is a three-foot-long tube generally located near the keel, which can be extended through the ship's hull. It contains two orifices, one of which measures dynamic pressure, and the other, static pressure. Through either a system of bellows or mercury tubes, depending on the type of equipment installed, the difference between the dynamic and static pressure is continually monitored. This pressure difference is proportional to speed. A control unit converts the pressure difference to speed units, and transmits this information to remote locations wherever required on the ship. When using the pit log, the navigator must always remember that when extended the pitot tube increases the ship's draft by about three feet. For this reason, most ships make it

a standard procedure to raise the pit log tube whenever the ship is about to transit relatively shallow water.

The Doppler Speed Log

A comparatively recent outgrowth of advances in solid state electronics and sonar technology in recent years is the *doppler speed log*. This instrument depends for its operation on one or more sonar beams projected into the water by a transducer mounted on the bottom of the hull of a vessel. By electronically analyzing the return of the sonar beam pattern reflected back either from the sea bottom or from the water itself in deeper areas, a very accurate determination of speed can be made. Depending on the model of equipment and number of beams, not only fore-and-aft speeds, but also athwartship speed can be measured to the nearest hundredth of a knot. Moreover, when bottom echos are being used, the speed determined is the speed over the ground (SOG), a very desirable quantity to have for navigational purposes. Most models of doppler speed logs also provide a readout of depth to the nearest foot, meter, or fathom whenever bottom echos are returned, and most feature a distance-run-since-last-reset odometer. The display unit for one such system is shown in Figure 7–5A.

Figure 7–5A. The Raytheon DSN–450 Doppler Speed Log. (Courtesy Raytheon Company)

Because of the current high costs of such installations, their use today is restricted for the most part to large commercial ships such as jumbo tankers and deep-draft freighters. Although a few models have peen placed aboard various Navy ships for test and evaluation purposes, none are currently installed for operational use.

Doppler sonar systems and associated equipment are discussed in more detail in *Marine Navigation 2*.

Electronic Navigation Systems

Perhaps the most esoteric of all means of determining vessel speed is that soon to be afforded by the NAVSTAR satellite *Global Positioning System* (GPS) currently being developed, and scheduled to be fully implemented by late 1987. Although the primary goal of the system will be to provide a continuous three-dimensional position-fixing capability everywhere on earth, many receivers now being designed for the system will also have the capability of continuous derivation of speed over the ground. Unfortunately, because GPS receivers must incorporate a fairly capable computer to translate the received satellite signals into meaningful positional data, the resulting high receiver costs (currently beginning at $16,000) will limit speed determination by use of the system primarily to military and commercial vessels throughout the decade of the 1980s. The GPS system is described more fully in *Marine Navigation 2*.

Two other long-range electronic navigation systems also provide continuous speed-determination capability, though not as accurately as that possible with GPS. The *Omega Navigation System*, made fully operational in 1981, is a hyperbolic radionavigation system that makes continuous position-fixing possible everywhere on earth through the generation of an extensive lattice of hyperbolic lines of position that criss-cross the entire globe. *Loran-C* is an older radionavigation system which, like Omega, generates a hyperbolic grid pattern covering most of the littoral zones of North America, Europe, and the Far East. Although both these systems are intended primarily for position-fixing, several of the more sophisticated models of Omega and Loran-C receivers incorporate a continuous readout of vessel speed over the ground as an important auxiliary feature. Both Omega and Loran-C are covered in detail in *Marine Navigation 2*.

Use of Shaft RPM to Estimate Speed

In addition to the foregoing methods of direct measurement of vessel speed, another fairly simple yet effective means of estimating approximate speed long familiar to navigators of screw-driven ships is the use of *shaft revolutions per minute* (RPM). For all larger con-

stant-draft vessels having nonvariable pitch propellers, there is a fairly consistent relationship between shaft RPM and speed through the water. When a ship is commissioned, one of her sea trials consists of the preparation of a graph showing speed versus RPM. From this graph, tables are constructed indicating the number of RPM necessary on the ship's shafts for each knot of speed desired. In fact, on many Navy ships not having any direct speed-measuring devices except a pit log, the navigator will often consider the estimation of speed by shaft RPMs of superior accuracy to this instrument, especially when the pit log has not been calibrated for some time. Since the hull form and its resistance may change over time because of planned alterations and marine fouling, most navigators will try to schedule their ships to run a measured mile at least once a year, to verify and modify the shaft RPM data as required. Measured miles are marked off by range markers on shore; most major ports have a measured mile laid off in close proximity.

Measurement of Depth

Measurement of water depth is accomplished on most modern ships not fitted with a doppler speed log primarily by means of an electronic device called the *echo sounder* or *fathometer*, a now common-use name applied to an early Raytheon model. The echo sounder most commonly found on board Navy ships is the AN/UQN–4, pictured in Figure 7–5B. Most echo sounder devices consist of a fixed transducer mounted on the underside of the vessel's hull, and an operating and display console remotely located where needed, usually in the chart-room. The output display may be either a small CRT, an LED display, a strip chart recorder, or a combination of these. In operation, the echo sounder transmits a sound pulse vertically into the water, and computes the depth by measuring the time interval from transmission of the sound signal until the return of its echo from the bottom. Most newer models of echo sounders allow the operator to display the depths thus measured in feet, fathoms, or meters, using any one of several scales.

Echo sounders and their operation, including the AN/UQN–4, and the procedures for navigation by use of the water depth measurements they provide, are elaborated upon in *Marine Navigation 2*.

When using echo sounder depths, the navigator must always remember that for most models the depths recorded are those from the position of the sonar transducer to the bottom. For actual water depths, the navigator must add the transducer depth of his ship to all readings.

The auxiliary function of the doppler speed log as a depth-finder in shallower water was mentioned in the preceding section. It should

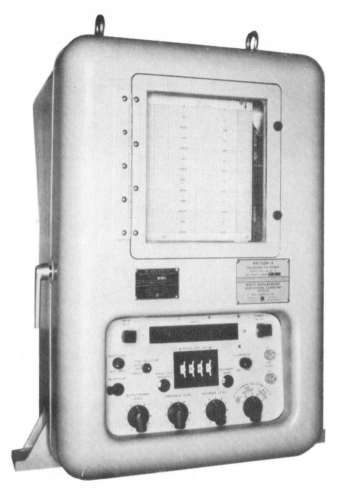

Figure 7–5B. AN/UQN–4 echo sounder.

be mentioned here that on most models this feature only operates to maximum depths of about 1,000 feet. Beyond this limit, the navigator must revert to the ship's echo sounder for depth determinations.

An alternative mechanical method of measuring depth in piloting waters is the *hand leadline* (pronounced lĕd.) Although its use has been discontinued today for the most part, the leadline was a reliable means of depth measurement even before the birth of Christ. It consists of a lead weight attached to a 25-fathom line customarily marked as follows:

2 fm	2 strips of leather
3 fm	3 strips of leather
5 fm	white rag
7 fm	red rag

10 fm	leather with hole
13 fm	same as 3 fm
15 fm	same as 5 fm
17 fm	same as 7 fm
20 fm	line with two knots
25 fm	line with one knot

An indentation in the bottom of the lead weight allows the application or "arming" of the weight with tallow for the purpose of obtaining a sample of the composition of the bottom.

On vessels making use of the hand leadline, a "leadsman" is stationed on a platform called "the chains," usually located about two-thirds of the distance aft from the bow to the bridge. In reporting depths obtained by use of the leadline, it is customary to refer to the markings simply as marks; reports such as "By the mark five," or "Mark twain," are given.

Plotting Instruments

Plotting instruments are the least sophisticated yet most fundamental of all of the navigator's tools. Of these, the ordinary *pencil* is the most basic. Most navigators prefer either a No. 2 or 3 pencil; it should be sharpened regularly so as to write all lines clearly and sharply, but lightly enough to facilitate easy erasure. A gum eraser is generally used on nautical charts, since it is less destructive to chart surfaces than the pencil-tip eraser.

The *dividers* and *drawing compass* are standard plotting instruments second only to the pencil in simplicity. The use of these instruments to measure distance on the nautical chart has already been described. Details of their use will not be repeated here, except to mention that the navigator should become practiced enough with the instrument that he can manipulate it with one hand.

The drawing compass is quite similar to the dividers, except that one of the two points is leaded to allow the drawing of an arc. The use of a drawing compass to measure the latitude and longitude of a position and to plot a given position on a chart has already been described in Chapter 4. In addition to this, the drawing compass is employed whenever it is necessary to swing a distance arc. Like the dividers, the navigator should become proficient in the manipulation of this instrument with one hand.

There are several types of instruments used for plotting direction, of which the simplest is the *parallel rulers*. The rulers consist of two parallel bars with cross pivot braces of equal length so attached that the bars are always kept parallel as they are opened and closed. In

Figure 7–6. The parallel ruler.

operation the rulers are laid on the chart's compass rose with a leading edge lying across the center of the rose in the desired direction as indicated on the periphery of the rose. By holding first one bar and then the other, the ruler can be "walked" over the surface of the chart to the location at which a line is to be drawn (Figure 7–6).

Because parallel rules are somewhat slow and are difficult to use if a compass rose is not printed nearby on the chart, a number of plotting devices have been developed which tend to eliminate the disadvantages of the parallel ruler by incorporating a protractor on the instrument. One of these is the *Weems Parallel Plotter,* shown in Figure 7–7. Printed on its surface are a semicircular protractor for measuring courses and bearings against a charted meridian, and two quarter-circle protractors for use in aligning the plotter against printed parallels of latitude. Once aligned with a meridian or latitude line, the instrument is transferred to the site of the line to be plotted by rolling it over the chart surface using two rollers mounted along one edge. Another somewhat simpler and very popular device is the *Weems Navigation Plotter No. 641,* shown in Figure 7–8. Originally designed for air navigation, this plotter incorporates a straightedge on one side

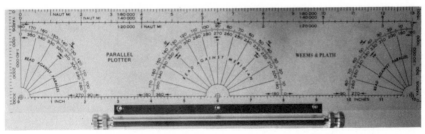

Figure 7–7. The Weems Parallel Plotter. (Courtesy Weems & Plath, Inc.)

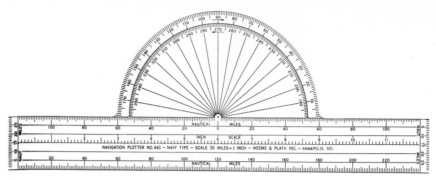

Figure 7–8. The Weems Navigation Plotter No. 641. (Courtesy Weems & Plath, Inc.)

and a protractor on the other that can be quickly aligned with any convenient meridian or parallel of latitude printed on a Mercator chart. Once aligned, the plotter is shifted along the longitude or latitude line until the straightedge is in the desired position for plotting the desired course or bearing line.

Another simple instrument for plotting direction is the *Hoey Position Plotter*, pictured in Figure 7–9. This plotter consists of a clear plastic protractor, with a drafting arm attached and pivoted at the center. The protractor is imprinted with a grid system permitting alignment with any convenient meridian or parallel of latitude on a Mercator chart. A direction line is plotted with this instrument by steadying the drafting arm in the desired position with the thumb and forefinger of one hand, while aligning the protractor grid with the other hand. Because it can be aligned with any convenient meridian or parallel of latitude, plotting direction lines with this instrument is quite fast and easily done with a few minutes of practice.

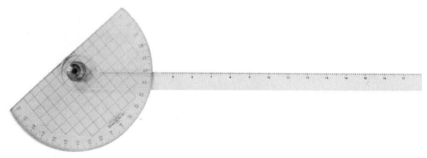

Figure 7–9. The Hoey Position Plotter.

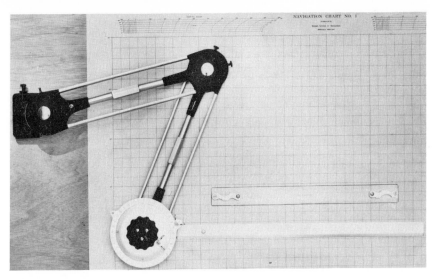

Figure 7–10. The parallel motion protractor.

A more complex yet very fast instrument for plotting direction is a type of drafting machine called a *parallel motion protractor*, or PMP; it is shown in Figure 7–10. It is made up of a rotatable protractor, graduated in degrees from 0 to 360, with a drafting arm affixed thereon, which is moved across the chart by a parallel motion linkage fastened to the chart board. The protractor can be aligned and set in position with reference to a chart meridian or latitude line, and the linkage permits movement of the protractor and its drafting arm to any part of the board without any change in the orientation of the protractor disc. The drafting arm can either be held in position on the protractor by thumb pressure or locked in place by a set screw. Drafting arms marked with various distance scales can be inserted into the protractor, for rapid distance measurement on certain scale charts and plotting sheets. The instrument is a great convenience in laying off courses and bearings and for transferring lines from one location to another on the chart.

It is possible to obtain a very accurate position by measuring the two angles between three adjacent objects on shore. The device used to represent the two angles on a chart is called a *three-arm protractor*. It consists of a central circular disc graduated in 360 degrees, with a central fixed arm and two movable arms attached at the center. After an observation of the two horizontal angles is made, usually by a sextant, the three-arm protractor is set to represent the two angles, one on the right and the other on the left of the fixed arm. The protractor is then set onto the chart, and the central fixed arm is

aligned with the central of the three shore objects, as they are represented on the chart. The protractor is moved around, always with the central arm in place on the central object, until the other two arms fall on their respective objects. The center of the protractor then represents the position of the observer. The resultant fix is of very high accuracy, and for this reason, the instrument is extensively employed in chart survey work. Under normal conditions of piloting, however, the three-arm protractor is used primarily as a back-up instrument for determining a fix if the ship's gyro fails.

With the exception of the parallel motion protractor, all of these instruments are relatively inexpensive, with their prices for the most part being under $10. They are available commercially from almost every nautical supply store.

Miscellaneous Instruments

In addition to the various devices described heretofore, there are several miscellaneous types of instruments usually found on board ship that are of great value to the surface navigator. These can be broadly classified for purposes of this text as weather instruments, speed/time/distance calculators, and timing devices.

Weather Instruments

On Navy ships not having a separate meteorological division on board, the duties of observing and recording the weather fall to the navigator and his staff of quartermasters. A complete round of weather observations is made hourly on board ship when it is under way, and a synoptic report of these conditions is made to a Naval Oceanography Command Center four times daily. The basic instruments installed on Navy ships for the purpose of weather observation are the *barometer*, for measuring atmospheric pressure, the *thermometer*, for measuring air temperature, the *psychrometer*, for measuring dry- and wet-bulb temperatures from which the relative humidity and, if desired, the dew point can be calculated, and the *anemometer*, for measuring relative wind speed and direction from which the true wind can be derived.

Complete descriptions and directions for use of all these instruments can be found in the U.S. Navy *NAVOCEANCOMINST* 3144.1 series, *Manual for Ship's Surface Weather Observations*, and in the naval training manual *Quartermaster 3 & 2*. Another excellent reference book for both military and civilian navigators on the subject of weather instruments and their use is *Weather for the Mariner* by W. J. Kotsch (Naval Institute Press, 1983), available at most nautical supply stores.

Speed-Time-Distance Calculators

In the course of his plotting, the navigator has occasion for numerous calculations of speed, time, and distance, based on the formula $D = S \times T$, where D is distance in nautical miles, S is speed in knots, and T is time expressed in hours. If he had to work each calculation out mathematically, the navigator would have time for little else. Fortunately, several devices have been developed that greatly facilitate the calculation of one of these quantities if the other two are known. These are the three- and six-minute rules, the logarithmic speed-time-distance scale, the nautical slide rule, and the electronic calculator.

The first of the "devices" mentioned, the three- and six-minute rules, are not really devices at all, but rather are simply special case mathematical formulas for calculating the amount of distance a vessel will traverse at a given speed in knots in three or six minutes.

The *three-minute rule* can be simply stated as follows:

Distance traveled in yards in three minutes
$$= \text{Ship speed in knots} \times 100.$$

The rule can be used to compute quickly the distance a ship proceeding at a given speed will travel in three minutes, or, if distance traveled in three minutes is given, it will easily yield the ship's speed in knots. For example, if a ship is proceeding at 15 knots, it will travel $15 \times 100 = 1,500$ yards in three minutes. If a ship traverses 600 yards in three minutes, its speed is $600 \div 100 = 6$ knots.

The *six-minute rule* can be stated as follows:

Distance traveled in miles in six minutes
$$= \text{Ship's speed in knots} \times \frac{1}{10}$$

Like the previous rule, the six-minute rule is of great value because it instantaneously yields the number of miles a vessel traveling at a given speed will traverse in six minutes. For instance, if a ship were traveling at an ordered speed of 17 knots, it would traverse a distance of $17 \times \frac{1}{10} = 1.7$ miles in six minutes. The navigator makes extensive use of both rules, especially during the practice of dead reckoning in the piloting environment, which is discussed in the next chapter.

The *logarithmic speed-time-distance scale* is an extremely useful tool for finding any one of these quantities if the other two are given. The scale is usually incorporated on all large-scale charts and plotting sheets, and is also found in nomogram form on the maneuvering board. Figure 7–11 on the next page illustrates a typical logarithmic speed-time-distance scale.

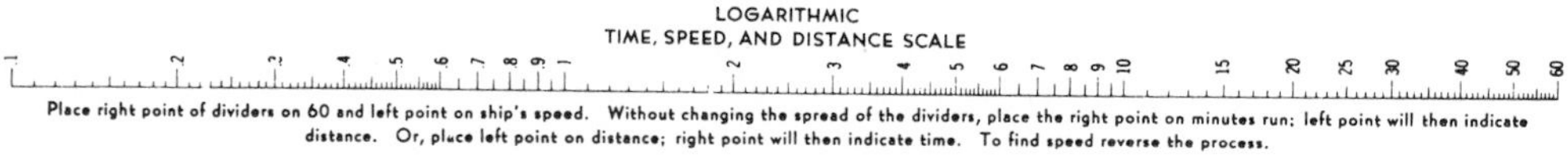

Place right point of dividers on 60 and left point on ship's speed. Without changing the spread of the dividers, place the right point on minutes run; left point will then indicate distance. Or, place left point on distance; right point will then indicate time. To find speed reverse the process.

Figure 7–11. Logarithmic speed-time-distance scale.

This type of scale is always used in conjunction with a pair of dividers. In using it, speed is always represented as the distance a ship would travel in 60 minutes; the distance can be expressed either in yards or nautical miles. If the time in minutes to traverse a certain distance at a given speed is required, for instance, one point of the dividers—the "time" point—is placed at 60 on the scale, and the other—the "distance" point—is set at the distance the ship would proceed in 60 minutes. Without changing the divider spread, the distance point is then moved to the desired distance, and the time required in minutes is read off under the other divider point. Distance is expressed either in miles or yards; once the units of time and distance are chosen for any given problem, they must be consistently used throughout.

As another example, suppose that the distance in miles that a ship will travel in 10 minutes at a speed of 21 knots is required. First, the divider spread is established by placing the "time" point at 60, and the "distance" point on 21 miles. Then the time point is moved to 10, and the corresponding distance, 3.5 miles, appears under the distance point. If the distance in yards were required, the distance point could have been initially placed at 42, representing 42,000 yards traveled in 60 minutes at 21 knots. After moving the time point to 10, the number "7" representing 7,000 yards would have appeared under the distance point.

For a final example, imagine that the speed of a ship which has traveled 4,000 yards in 10 minutes is required. The divider spread may be established in either of two ways. One point may be set at 2, representing two miles traveled in 10 minutes, and the other at 10. Or one point may be set at 4, representing 4,000 yards, and the other on 10. After the spread has been thus established, the dividers are then moved so that the time point is placed on 60. If the distance point were originally placed on 2 miles, the speed, expressed as 12 miles per 60 minutes, or 12 knots, can be read directly under the distance point. If the distance point had been originally set on 4,000 yards, however, the speed will be expressed as 24,000 yards traveled in 60 minutes. Speed in knots may be obtained simply by dividing this figure by 2,000 yards. As an alternative method in this last case,

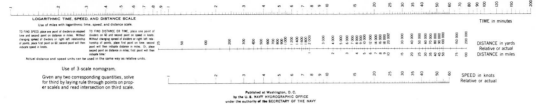

Figure 7–12. The logarithmic speed-time-distance nomogram on a maneuvering board.

the properties of the log scale may be taken advantage of to eliminate the division by 2,000 by setting the time point on 30, vice 60. The digits appearing under the distance point then represent the ship's speed in knots.

By using the logarithmic speed-time-distance scale in conjunction with a pair of dividers in the fashion described above, any one of the three quantities of the speed-time-distance equation can be found if the other two quantities are given.

The maneuvering board contains a set of three logarithmic scales used in conjunction with one another to solve the equation; the arrangement is referred to as a *nomogram*. The topmost of the three scales is for time, the second is for distance, and the bottom scale is for speed, as shown in Figure 7–12.

To use the nomogram, pencil marks are placed over the two given quantities on the appropriate scales, and the desired third quantity is read by placing a straightedge over the two marks and observing the point of intersection on the third scale. As an alternative method, the topmost time scale can be used as a single logarithmic speed-time-distance scale, as previously explained. Use of the nomogram in this alternative way is considered to produce more accurate results than those obtained by using all three scales in conjunction with one another.

The *nautical slide rule* is an inexpensive device very widely used by surface navigators for the purpose of obtaining a rapid solution of the speed-time-distance equation. In principle it is much like the maneuvering board nomogram described above, except that the three scales have been bent into circular form on a plastic base and covered by a faceplate. Figure 7–13 depicts a nautical slide rule. Distance is expressed both in yards and miles, time is expressed in minutes and hours, and speed is expressed in knots. To use the instrument, the two known factors are set by rotating the slide rule to the appropriate positions, and the third factor appears by the appropriate arrow.

The speed-time-distance equation can be solved using the nautical

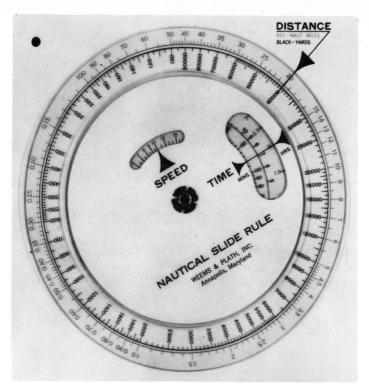

Figure 7–13. The nautical slide rule. (Courtesy Weems & Plath, Inc.)

slide rule in the time it takes to set the two known quantities on it. Its speed, and the fact that it requires no other instrument manipulations, makes it a favorite with all navigators.

A newer device that is beginning to be used by more and more marine navigators for performing speed-time-distance conversions, as well as for solving many other types of quantitative navigational problems, is the hand-held *electronic calculator*. With a few minutes' practice, the navigator can quickly and accurately solve for any one quantity of interest in the basic speed-time-distance formula, given the other two.

More information on the electronic calculator and its steadily increasing marine navigation applications appears in Appendix A of *Marine Navigation 2*.

Shipboard Timepieces

The accurate determination of time is of great importance to the navigator both for piloting and celestial navigation. The ship's speed, distance made good, and position are all functions of time, as are most aspects of the daily shipboard routine at sea. It follows, therefore,

that the navigator must have reliable timepieces at his disposal on board ship.

The *chronometer*, considered one of the most accurate mechanical timepieces devised by man, is the principal navigational timepiece normally found on board most ocean-going vessels. Older chronometers such as the Size 85 Hamilton chronometer found on board most Navy surface ships (Figure 7–14) contain spring-driven mechanical movements of extremely high precision. Newer chronometers contain movements built around a quartz crystal oscillator powered by a flashlight battery. Regardless of their mechanisms, marine chronometers are built to withstand shock, vibration, and variations of temperature. Commerical models range in price from about $150 to $600 or more, depending on their size, casing materials, and optional features such as bells or chimes.

Figure 7–14. The Hamilton chronometer.

Not even a chronometer can keep absolutely correct time, but the feature that distinguishes a chronometer from ordinary watches and clocks is its rate of gain or loss of time, which is constant over long periods of time. This characteristic allows the chronometer error to be determined with precision, by methods presented in *Marine Navigation 2*, so that the navigator can arrive at the correct time when desired. The correct time precise to the nearest second is an important requirement for optimal accuracy in celestial navigation.

Most navigators who perform a great deal of celestial navigation at sea will keep at least two chronometers on board for comparison purposes. Most Navy ships have an allowance for three; they are usually kept together in a special box in the chartroom.

Further information on chronometers and their use is given in *Marine Navigation 2*.

Other timepieces normally found on board most vessels at sea are the *stopwatch* and the *wall clock*. A stopwatch is a necessity for several purposes, among which are the timing of the periods of navigational lights, as discussed in the preceding chapter of this text, and the recording of the time of observation of a celestial body. A special type of stopwatch called a *comparing watch* is used for this latter purpose; its use is explained in *Marine Navigation 2*.

Since it would be quite inconvenient to have to go to the storage location of the ship's chronometers every time a crew member desired the correct time, wall clocks are a common feature on board almost every type of ocean-going vessel. Although not as accurate as a chronometer, many newer battery-powered quartz-crystal models are nearly so. There is a wide variety of marine wall clocks available commercially. Two fairly standard models found on board most Navy ships are a spring-driven 24-hour wall clock that must be periodically hand wound with a key, and a newer electric model with a 12-hour dial that will operate for several months on a single flashlight battery.

Summary

This chapter has described many of the more common navigational instruments found on board most seagoing Navy, commercial, and private vessels. The navigator should become very familiar with all of them, as well as with the other more complex navigational aids described elsewhere in this text, such as radar, the gyro, the magnetic compass, the sextant, and the various pieces of electronic gear associated with electronic navigation. Just as a surgeon would be lost without his scalpel, the navigator will be quite literally lost unless he understands and skillfully uses the instruments of his profession.

8

Dead Reckoning

The major concern of the navigator operating at sea is the accurate determination of his position. Not only must he be continually concerned with the present position of his ship, but equally as important, the navigator must also be able to calculate the ship's probable position at future times. To put it in simple terms, the navigator continually seeks answers to the following questions: "Where is the ship now?" and, "Where is the ship going to be in the next minute; the next few minutes; the next hour; or the next day?"

In the eighteenth and nineteenth centuries, and for the first twenty-five years of this century, the navigator calculated his ship's current and future positions by a mathematical process known as deduced—often abbreviated to "ded"—reckoning. This process involved the use of laborious trigonometric computations based on a known point of departure; the technique was necessitated by the inaccuracy and the scarcity of contemporary ocean charts. Although graphic methods made possible by inexpensive yet highly accurate mass-produced modern charts have for the most part replaced the earlier mathematical solutions for the ship's position, the slightly altered term *dead reckoning* is still applied to the process.

Dead reckoning as it is practiced today may be defined as the process of determining a ship's approximate position geometrically by applying to her last established charted position a vector or series of vectors representing all true courses and speeds subsequently ordered. The dead reckoning technique is essential because it is often impossible to obtain an accurate determination or "fix" of the ship's position when required, despite the growing availability of various radionavigation and satellite navigation systems. Moreover, even after a fix has been obtained and plotted, it does not represent the location of a moving ship either at the present time or a future time, but rather it only reflects where the ship was at a certain time in the recent past. To obtain the ship's approximate present or future position, the navigator must rely on the dead reckoning plot.

In this chapter the procedures used by the navigator in maintaining

the dead reckoning or "DR" plot will be examined in detail. Inasmuch as the initial element in every plot is the fix, the first few sections will describe the fix and the techniques for obtaining it. Next, the principles of keeping the DR plot will be fully discussed and illustrated, and finally a variation of the DR plot called the track will be explained.

Determining the Fix

The initial element of the ship's DR plot is the *fix*, which may be simply defined as the ship's position on the earth's surface at some given point in time. A fix is determined by the intersection of at least two simultaneous *lines of position* (LOPs), each of which may be thought of as a locus of points along which a ship's position must lie. Although the intersection of two such LOPs would be sufficient to obtain a discrete position, it is usually the practice to obtain at least three simultaneous LOPs in determining a fix to guard against the possibility of one or more of them being in error. These LOPs may be obtained in any one of several ways, including but not limited to visual observation of land or seamarks, use of electronic equipment, and observation of celestial bodies.

The Line of Position

The most accurate LOP possible is obtained by visually observing two or more objects in line, as in Figure 8–1. LOPs of this type are called *visual ranges;* two or more objects are described as being "in range" if they are sighted in line with one another. The visual range LOP is plotted on the chart by placing a straightedge along an imaginary line drawn through the objects sighted in line, and drawing a short segment of the line near the approximate position of the ship

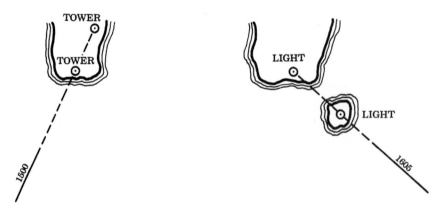

Figure 8–1. Two illustrations of a visual range LOP.

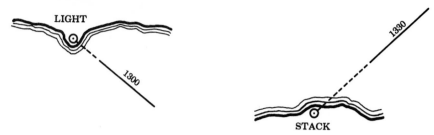

Figure 8–2. Two illustrations of a visual gyro bearing LOP.

on the chart at the time of the observation. The dashed construction lines appearing in Figure 8–1 and in other figures in this chapter are included only for clarity, and should not appear on the chart in practice. LOPs are never extended to the object or navigation aid observed in order to avoid the possibility of the land or seamark being erased with repeated use of the chart. These, as well as all other single lines of position, are always labeled with the time of observation above the line segment, as shown in Figure 8–1.

Unfortunately, it is seldom possible to observe a range at the precise time at which a fix is desired or required. Consequently, the visual LOP is usually plotted by observing a bearing to a single object by means of the gyro repeater, as in the examples in Figure 8–2.

The procedure for plotting the visual bearing LOP is basically the same as that used to plot the visual range, except that the straightedge is laid down on the chart along the true line of bearing to the object sighted or "shot"; this bearing is plotted with reference to true north on the chart by means of the chart compass rose. If it is not possible to obtain the true bearing to an object from the gyro repeater because the gyro is inoperative, the bearing taker may shoot a relative bearing to the object. In such cases, the navigator must first determine the

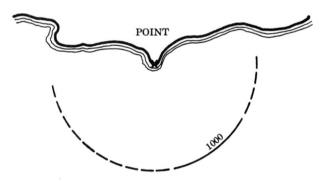

Figure 8–3. A distance or range LOP.

Figure 8–4. A single celestial LOP.

ship's true heading, and then convert the relative bearing to a true bearing before it can be plotted. As before, only the segment of the line near the ship's approximate charted position is drawn; it is labeled above the line with its time of observation.

If the distance to an object can be found either visually by use of a stadimeter, or electronically by the use of radar, the ship must lie somewhere on a circle centered on the object, with radius equal to the distance measured. An LOP obtained in this manner is technically termed a *distance line of position*, but in normal usage it is generally referred to simply as a "range." A sample range LOP is plotted in Figure 8–3. The range LOP is plotted using a drawing compass set to the appropriate distance on the chart scale. The pivot point of the instrument is placed over the object or landmark shot, and an arc is drawn with the other leaded point. Again, only a small portion of the arc in the vicinity of the approximate position of the ship on the chart is drawn; it is labeled with the time of the observation above the line.

If a single line of position is obtained by observation of a celestial body, by methods explained in *Marine Navigation 2: Celestial and Electronic*, it should be labeled with the time of the observation, and the name of the body observed, as in Figure 8–4.

The Fix

As was mentioned earlier, if two or more simultaneous lines of position are plotted, the point at which they intersect is the ship's position for that time. Furthermore, it makes no difference how the LOPs were obtained—radar ranges may be crossed with visual bearings, several radar ranges may be crossed, or a radar range and radar bearing to the same object may be used, though this latter technique is not recommended because of the possible radar bearing inaccuracies discussed in Chapter 10 of this text.

Even though only two simultaneous LOPs are necessary and sufficient to obtain a fix, it is the usual practice to obtain at least three LOPs whenever a fix is to be determined. The third LOP can not only resolve possible ambiguities in selecting the proper fix location when a range LOP is plotted, but it also will immediately point out any possible errors in the selection, observation, and plotting of the

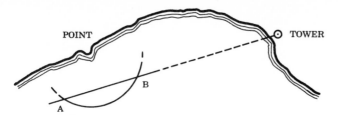

Figure 8–5. *Ambiguity resulting from crossing only two LOPs.*

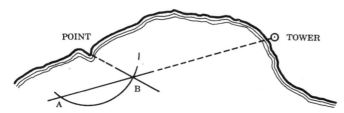

Figure 8–6. *Resolving ambiguity with a third LOP.*

objects shot. Consider the example in Figure 8–5, in which a radar range is simultaneously taken and plotted with a visual bearing.

Is the ship at position A or position B? In the absence of any other information, the navigator should assume his position to be the one in closest proximity to any navigational hazards present, and make a recommendation for a safe course accordingly. If a third LOP had been obtained at the same time, however, there would be no doubt as to the ship's position (Figure 8–6).

Occasionally, due to uncompensated bearing errors in the gyro repeater, ambiguity will arise even when three simultaneous lines of position are plotted. In such situations, the navigator should again assume the ship to be in the worst possible position and make recommendations accordingly. He should then attempt to resolve the ambiguity by immediately shooting another round of bearings. In the situation illustrated in Figure 8–7, the navigator should assume that

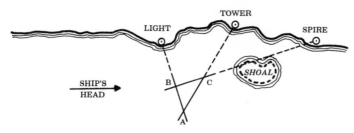

Figure 8–7. *Ambiguity caused by an unknown gryo error.*

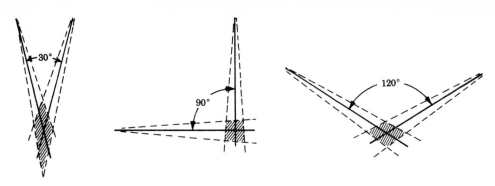

Figure 8–8. Effect of a ±5° gyro error on two LOPs.

the ship is located at position C, and recommend a bold alteration of course to starboard to avoid the shoal. If he was in fact at position A or B, he would thereby increase the margin of safety.

In order to minimize the effect of possible errors in observed bearings when plotting a fix, the navigator should attempt to optimize the angular spread of the objects shot. If two objects are used, they should be as close as possible to 90° apart; if three objects are shot, they should optimally be 120° apart. The illustrations in Figure 8–8 depict the reasoning behind this rule of thumb by showing the effects of a possible ±5° error in the bearings of two objects 30°, 90°, and 120° apart. The solid lines represent the true bearings to the objects, and the dashed lines, the bearings with a ±5° error applied. As the angle between the objects increases or decreases from the 90° optimum, the shaded areas of uncertainty become larger; at an angle of 90°, the effect of the 5° error is minimized. In similar fashion, it could be shown that for three bearings the optimum angular spread is 120°; if the three bearings were shot to objects all located within a 180° arc, as might be the case when transiting a coast, the optimum angle of intersection between adjacent bearings would be 60°.

The symbol used to represent a fix on the navigator's dead reckoning plot is a circle, about one-eighth inch in diameter, placed over the intersection of the LOPs used to determine the fix. Since it can be assumed that these lines of position were shot simultaneously at the time of the fix, it is customary to label only the fix circle with the time at which the LOPs were shot, written horizontally as shown in Figure 8–9. The individual LOPs on which the fix is based are not labeled with times. If the fix was shot on the half minute, a prime sign (') is used to denote this fact. If the LOPs were obtained by means of observations of celestial bodies, the names of the bodies should be printed above or below the individual LOP segments.

 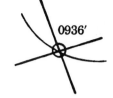 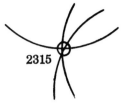

Figure 8–9. Examples of fix labels.

Principles of the Dead Reckoning Plot

After he has obtained and plotted a fix, the navigator is ready to proceed with the DR plot. Through the years, the principles of keeping the dead reckoning plot have been formalized into a set of rules known as the *Six Rules of DR*. If faithfully followed, these rules, used in conjunction with the standard labeling procedures described in this chapter, will result in the navigator's plot being understood by anyone familiar with navigation:

1. A DR position will be plotted every hour on the hour.
2. A DR position will be plotted at the time of every course change.
3. A DR position will be plotted at the time of every speed change.
4. A DR position will be plotted for the time at which a fix or running fix (described later in this chapter) is obtained.
5. A DR position will be plotted for the time at which a single line of position is obtained.
6. A new course line will be plotted from each fix or running fix as soon as it is plotted on the chart.

For the purposes of the DR plot, each time a fix or running fix is plotted a vector representing the ordered course and speed is originated from it; in practice, this vector is usually referred to simply as a "course line." The direction in which this vector is drawn represents the ordered true course, referenced to the chart compass rose. Its length represents the distance that the ordered speed would have carried the ship in the time interval under consideration, measured against the chart distance scale. A DR position for any time is always labeled with a semicircle, one-eighth inch in diameter, and its time is printed nearby at an oblique angle. As a simple example, consider the plot in Figure 8–10 on the following page.

Here, the navigator has obtained and plotted a fix at 0830, and he desires to plot a 0900 DR position based on it. A DR course line or vector was drawn in the direction 085°T from the fix, representing the ship's true course to be steered between 0830 and 0900. The

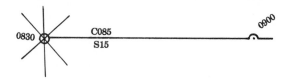

Figure 8–10.   Laying down a DR plot.

ship's ordered speed during this period is to be 15 knots, so the length of the vector measured on the chart distance scale is 7½ miles. The resulting DR position is thus located on the course line and labeled with a semicircle. Note that the course line is labeled with the ordered true course above the line and the ordered speed below. Each time a new course line is laid down in accordance with one of the Six Rules of DR, it should be so labeled.

As a more complicated example of the utilization of the rules of DR in maintaining the DR plot, consider the following excerpt from a ship's *Deck Log:*

0800. With Pollock Rip Light bearing 270°T at 6 miles, departed anchorage L–21 on course 090°T, speed 15 knots.

0930. Changed speed to 10 knots to avoid a sailboat.

1000. Changed course to 145°T, increased speed to 20 knots.

1030. Changed course to 075°T.

1106. Obtained radar fix on lighthouse, bearing 010°T, range 7 miles.

1115. Changed course to 090°T, changed speed to 18 knots to arrive at operating area S–2 on time at 1215.

The corresponding DR track appears in Figure 8–11.

As previously mentioned, fix positions are shown by one-eighth-inch diameter circles, and DR positions are indicated by semicircles. Fixes and DR positions are further differentiated by printing the times

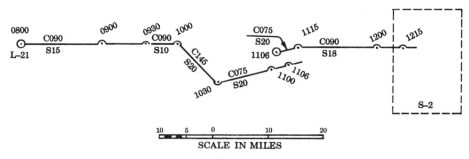

Figure 8–11.   An extended DR plot.

for fixes horizontally, while times for DR positions are printed at an oblique angle.

Beginning at the 0800 position, the navigator plotted the ship's true course line in the direction 090°T, the ordered course. This line was labeled with the ordered course above the line, and the ordered 15-knot speed below. Since the ship traveled 15 miles in one hour at 15 knots, the prescribed 0900 DR position was plotted on the 090° course line 15 miles as measured on the chart distance scale from the 0800 position. Due to the change in speed at 0930, a DR position for this time was plotted 7.5 miles further on. At 1000 the navigator plotted the 1000 DR position, changed the direction of the course line to 145°T, and labeled it with the new ordered course and speed. At 1030 the direction of the course line was changed to 075°T; the 1100 DR position was then plotted to scale on this line. At 1106, a new fix was obtained; hence the new DR course line commenced at that point. Note that the old course line ends at the 1106 DR position. To enhance the neatness of the plot, any portion of the old course line extending beyond the 1106 DR position should be erased. To complete the plot, the DR position for the 1115 course and speed change was plotted, and the plot was then extended to 1215 to obtain an estimate of the ship's position for that time.

Notice in the example above that it took the navigator 9 minutes from his fix at 1106 to decide on his new course and speed to make the rendezvous position on time. It was not until 1115 that the ship was placed on its new heading and speed. Because of the time lag from the time of obtaining and plotting a fix until the information it depicts can be acted upon, the navigator should lay out a DR course line from his new fix or running fix as soon as he plots it. In confined waters, the navigator usually updates his position by taking a fix at least every 3 minutes; he then runs a DR course line out from each fix the distance the ship would travel in the three minutes until his next fix. This particular interval is chosen in order to allow the navigator to use the 3-minute rule described in the preceding chapter to instantly calculate the distance the ship will proceed along the DR course line from his new fix in 3 minutes. Plotting 3-minute DR positions in this manner not only allows the navigator to see where the ship's approximate position is at frequent intervals, but it also allows a comparison to be made between the calculated and true ship's position every three minutes. If the ship is being set to the right or left of its DR position by the effects of a current, this fact becomes immediately apparent to the navigator, and he can recommend a small change of course or speed to compensate for it. In less restricted waters, the navigator may plot a fix only once every 15

minutes; each time he plots a new fix, he should lay out a 15-minute DR plot from it. If for some reason the navigator cannot obtain a new fix at the time he expected, he simply extends his dead reckoning plot until such time as he is able to obtain a new fix. Proper labeling and neatness are always of paramount importance in keeping the DR plot. A sloppy plot almost always leads to errors in the plot.

It must be emphasized that the DR position is only an approximation, because the DR plot intentionally ignores the effect of any possible current acting on the ship as it proceeds through the water. If a considerable amount of time has elapsed since the last fix, the ship may actually be far from the DR position. In piloting waters, reliance on a DR plot under these conditions could lead to disaster. When operating far at sea, however, the frequency of fixes is not nearly so critical. Often in the case of vessels lacking any long-range electronic navigation equipment, many hours or even days may elapse between successive fixes. Under these circumstances, the DR plot assumes especially great importance, for it may be the only means of estimating the ship's position with any degree of accuracy.

The Running Fix

It sometimes happens, especially when piloting in low visibility, that the navigator can obtain a line of position from only one object at a time as the vessel proceeds. In such circumstances the navigator may desire to advance a line of position obtained earlier to the time at which he shot a later LOP. The position thus produced is termed a *running fix*, because the ship has proceeded or "run" a certain distance during the time interval between the two LOPs. The earlier LOP is advanced to the time of the later LOP by using the ship's DR positions plotted for the two times. This is the reason for Rule 5 of the Six Rules of DR, which requires a DR position to be plotted at the time of obtaining a single LOP.

Two examples of the procedure for plotting a running fix are described below. The first involves advancing a bearing LOP, with no intervening course change, and the second concerns the advancement of a range LOP, with a course and speed change occurring in the interval between the times of the first and second LOPs.

For the first example, consider the DR plot in Figure 8–12A. In this figure, a single line of bearing was obtained and plotted on the chart at 1805, together with a DR position for this time, as shown. At 1830, a second line of bearing was obtained. To produce a running fix, the inexperienced navigator will normally draw a construction line connecting the two DR positions involved. In this example the DR

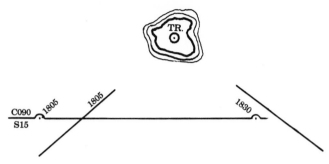

Figure 8–12A. A 25-minute DR plot, with two LOPs. The 1830 LOP, plotted to form the 1830 running fix, is not labeled.

course line may be used, since there was no course change between 1805 and 1830. Next, a point on the 1805 LOP is selected and advanced parallel to the line connecting the two DR positions, an amount of distance equal to the distance between them. At this stage, the plot appears as in Figure 8–12B. Finally, the remainder of the advanced 1805 line of bearing is drawn through this advanced point, parallel to the original LOP. To denote the fact that it is an advanced LOP, it is labeled as in Figure 8–12C with the original time and the time at which it was advanced. The running fix is located at the point at'which the advanced LOP and the second LOP intersect; it is labeled with a one-eighth-inch diameter circle and the time, as shown. To ensure differentiation between a regular fix and a running fix, the abbreviation "R Fix" is always placed near the running fix symbol in addition to the time, which is written horizontally.

Since the time of the running fix and the time of the second LOP are the same, it is unnecessary to label the second LOP. After plotting

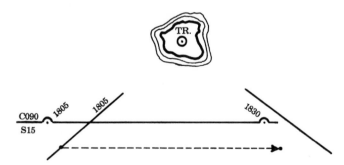

Figure 8–12B. Advancing a point on the original line of bearing, in preparation for advancing the line itself.

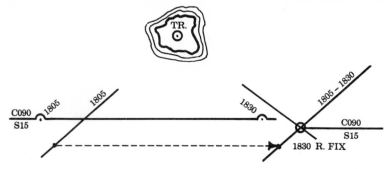

Figure 8–12C. Completed and labeled running fix using an advanced line of bearing.

the running fix, a new DR course line should be originated from it. The old course line ends at the DR position plotted for the time of the second LOP.

As an example of the technique used to obtain a running fix by advancing a range LOP, consider the DR plot shown in Figure 8–13A. In Figure 8–13A, a radar range was obtained and plotted at 1805, along with the corresponding DR position, as shown. At 1820 the ship changed course and speed; at 1830 a second LOP was obtained, this time a line of bearing. The same basic procedure previously described is used to obtain a running fix. A construction line may first be drawn connecting the two DR positions for 1805 and 1830. Since at least three points on the range arc would have to be advanced in order to reconstruct it, the simpler technique of advancing its center is used instead. Since the arc is centered on the tower symbol, this symbol is advanced parallel to the construction line connecting the two DR positions, a distance equal to the distance between them. The plot now appears as in Figure 8–13B. To complete the running fix, the arc is reproduced using the advanced center and

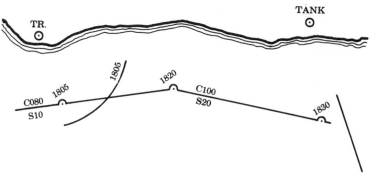

Figure 8–13A. A 25-minute plot, with a course and speed change. Again, the 1830 LOP is not labeled, as it was plotted to form the 1830 running fix.

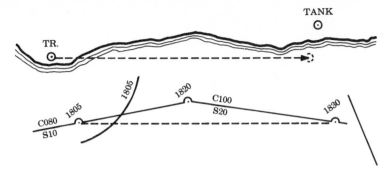

Figure 8–13B. Advancing the center of a distance arc.

the same radius that was used in constructing the original range arc. As was the case in the first example, only the advanced LOP and the running fix symbol are labeled with the time. The completed running fix for this example appears in Figure 8–13C.

Notice in the second example that the basic procedure for obtaining the running fix did not vary, even though there was a course and speed change between the times of the two LOPs. In plotting the running fix, only the DR positions for the times of the two LOPs are considered, regardless of the number of intervening course or speed changes. The earlier LOP is always advanced to the time of the later LOP through a distance parallel and equal to the distance between the two DR positions.

The fact that the DR plot plays a significant part in the determination of each running fix leads to two important considerations. First, the plot must be kept as accurately as possible during the interval between the times at which the two LOPs are obtained, since the DR positions for these times determine the distance and direction through which the earlier LOP will be advanced. Second, in the

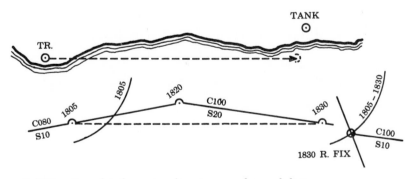

Figure 8–13C. Completed running fix using an advanced distance arc.

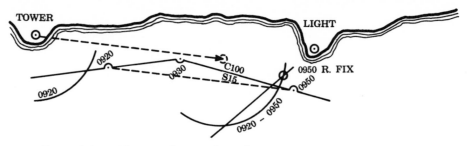

Figure 8–14. Choosing the most hazardous position.

piloting environment if more than 30 minutes have elapsed after a single LOP is obtained, it should not normally be advanced to form a running fix. Even with a comparatively weak one-knot current flowing, a DR position could be as much as a half-mile from the actual ship's position after a half-hour of travel, since the DR plot does not take current into account. Because of the precision required, in piloting any DR position recorded more than 30 minutes after the time of obtaining a single LOP is considered too inaccurate to use as a basis for advancing a line of position. Even when a line of position is advanced for lesser amounts of time, the resulting running fix is merely a better approximation of position than the corresponding DR position—better, because the ship must lie somewhere along the second LOP. In the confined piloting environment, if more than 30 minutes elapse without the navigator being able to obtain a second LOP or fix, he should not hesitate to recommend stopping the ship until such time as he can obtain a good fix or estimated position (described in the following section).

If an ambiguity develops when working with a range arc as either the advanced or subsequent LOP, the running fix position chosen should be the one that places the ship closest to any existing navigation hazards. The navigator should then recommend the best course of action for this position, just as was the case in similar circumstances when simultaneous LOPs resulted in a large triangle. An illustration of a situation of this type appears in Figure 8–14.

The Estimated Position

As alluded to in the preceding section, sometimes in the routine practice of piloting, it proves impossible to obtain more than one LOP at a time within the recommended half-hour time limit required to form a running fix. Situations of this nature can often arise when piloting at night, in foul weather, or in fog. In such circumstances, it may be possible for the navigator to determine an *estimated position*

(EP) based on whatever incomplete positioning information might be available, in conjunction with the DR plot. For example, a single LOP of good reliability may be obtainable in the form of a bearing to an aid to navigation, or distance to the nearest land. Or a series of echo sounder depth readings might be recorded that would tend to corroborate (or in some cases conclusively invalidate) the position indicated by the DR plot.

One fairly well-established method of obtaining an estimated position when a single LOP is available is to draw a construction line from the DR position corresponding to the time of the LOP to the closest point on it. In the case of a straight-line LOP such as a line of bearing, this construction line would by definition be a perpendicular drawn from the DR position to the LOP, as shown in Figure 8–15. The intersection of the construction line with the LOP is the estimated position, labeled with a ⅛-inch square and the time.

A questionable running fix, or a fix based on LOPs of low confidence, can also be treated as an estimated position, depending on the circumstances of the case. In Chapter 13, procedures are presented for determining an estimated position that compensates for the effects of any expected current upon the position indicated by the DR plot or a running fix.

On the reliability scale, an EP is normally considered to be about midway between a good running fix and an unsubstantiated DR position. The exact degree of confidence that can be placed in an EP is a matter of judgment based on experience and the circumstances of each case. A new DR plot is normally not originated from an EP during piloting, although in some cases a line representing the estimated course and speed may be extended from it, to determine whether any hazards might be encountered if the vessel were in fact at the position estimated.

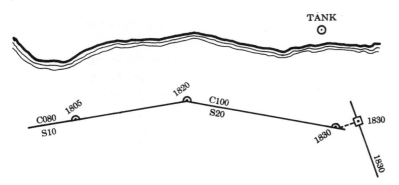

Figure 8–15. Forming an estimated position based on a single LOP.

The Track

Whenever his ship is to get under way, the navigator always lays down on his chart a kind of preplanned DR plot called a *track*, which is the intended path that the ship should follow over the ground. This track is in fact a form of DR plot, with its course and speed vectors representing the intended course and planned speed, rather than the ordered true course and speed. Figure 8–16 shows a portion of a track that a navigator has laid down to exit Norfolk, Virginia. Note that the direction of each of the vectors or "legs" of the track are labeled with the abbreviation TR above the intended track line, and the intended speed, called the *speed of advance*, is labeled by the letters SOA below the line. If the navigator knows the time at which the ship will get under way, he can compute in advance the times at which the ship should reach each junction point on the track, and he can label the track as shown in Figure 8–16.

In addition to plotting an exit track from a port or anchorage, the navigator also plots a track to show the planned course across open

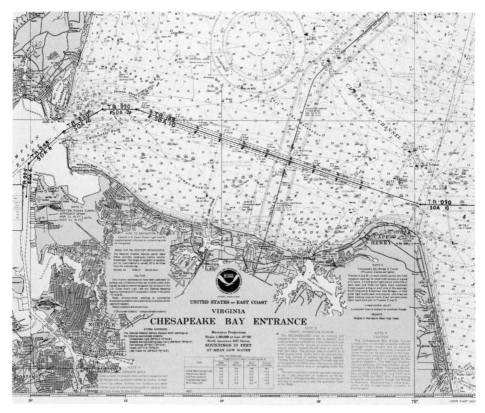

Figure 8–16. An exit track from Norfolk, Virginia.

water, or through the entrance to a port or approach to an anchorage. In these circumstances, the navigator may use as the initial point of the track some preplanned point of departure or arrival positioned outside the confines of the harbor or anchorage area. The technique of planning a track across open water will be discussed in detail in Chapter 15 of this text dealing with voyage planning; the procedure for laying down an approach track to an anchorage is described in Chapter 14.

In practice, a DR plot is always plotted in conjunction with and relative to some preplanned track; rarely, if ever, does a navigator lay out a DR plot with no preconceived notion of where the ship is supposed to proceed. If a fix places the ship's position 50 yards to the right of her intended track at some time, for instance, the navigator will extend the DR course line from the fix for a two- or three-minute distance in the direction of the ordered true course, and plot a DR position. From this future DR position, he will lay down a new course line in a direction suitable to bring the ship back on track, usually by aiming directly for the next junction point. He will then recommend that this course be ordered at the time of the future DR position. If the ship arrives at a junction point early, he may recommend that speed be decreased so as to arrive at the next point on time. Conversely, if the ship arrives late, he may recommend an increase in speed.

Danger Bearings

In conjunction with plotting the intended track, the navigator should clearly mark the safe limits of navigable water on either side of a channel by means of a precomputed visual bearing to a prominent landmark or navigation aid known as a *danger bearing*. Hatching is always applied on the hazardous side of the bearing, and the side on which the hazard exists is indicated by labeling the bearing NLT for not less than, or NMT for not more than, the indicated bearing. Consider the track shown in Figure 8–17. In this example, the navigator has laid a track down the center of a narrow channel, and he has drawn danger bearings marking the shallow water to the right of the 075° leg and the shoal to the left of the 100° leg. If the actual bearing to the first light were anything less than 075°T as the ship approached on the 075° track leg—say 070°T—she would be in danger of running through the shallow water. On the other hand, if she were so far to the left of track as she approached the junction point of the two legs that the actual bearing to the second light were greater then 082°T, she would be in danger of running onto the shoal. As long as bearings to the first light remain greater than 075°T and bearings to

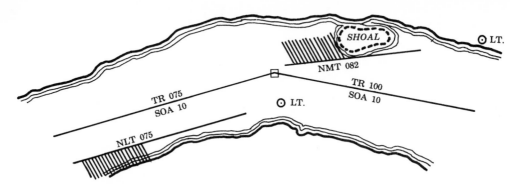

Figure 8–17. Two examples of danger bearings.

the second light remain less than 082°T as the ship approached, she would be proceeding in safe water.

When danger bearings are drawn in relation to an intended track, those marking dangers ahead and to the left of track are always labeled NMT, and those to the right, NLT. If danger bearings are required relative to a ship that is under way, they may be either NMT or NLT, depending on whether the objects upon which they are based lie ahead or astern of the ship. If the chart on which the danger bearings are drawn is intended for daytime use only, the danger bearings may be drawn on the chart with a red pencil. If the possibility exists that the chart will be used at night, some dark color other than red should be used, such as blue or black, so that the danger bearings will be visible when viewed under a red light.

In addition to danger bearings, some navigators take the additional precaution of shading all shoal water areas with a blue pencil, so that navigable areas of water will stand out.

Summary

This chapter has examined in some detail the procedures used by the navigator to maintain the dead reckoning plot and to plot the intended track. As an aid in labeling the various components of both, Figure 8–18 presents in visual form a summary of all of the different labeling procedures described herein.

The secret of success in dead reckoning can be summed up in two words: practice and neatness. Although the basic principles of plotting can be learned and practiced in the classroom environment, it is quite another thing to actually do it on a moving platform in the middle of the night. While working with sample problems in the classroom cannot ensure success, the standards of labeling and neatness which

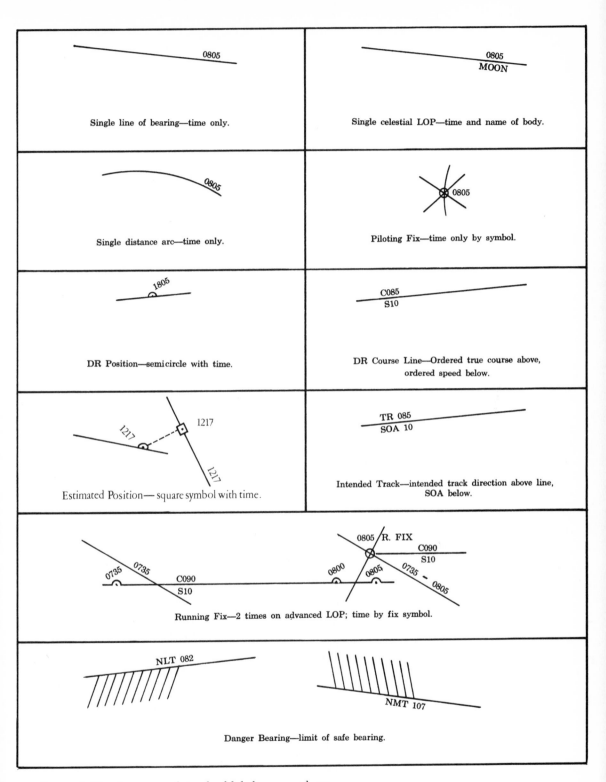

Figure 8–18. *Summary of standard labeling procedures.*

can be acquired and practiced therein can prepare the inexperienced navigator in large measure for the practical application awaiting him at sea.

9

Shipboard Compasses

On board a vessel at sea, there are three principal references for direction: the ship's longitudinal axis, the magnetic meridian, and the true or geographic meridian. The horizontal direction of one terrestrial point from another, expressed as an angle from 000° clockwise to 360°, is termed a *bearing*. Bearings measured using the ship's longitudinal axis as the reference direction are called *relative bearings*, indicated by the abbreviation R following the bearing. Those based upon the magnetic meridian, determined by use of the magnetic compass, are referred to as *magnetic bearings*, abbreviated M. And bearings given with reference to the geographic meridian, determined by the use of the ship's gyrocompass, are *true bearings*, abbreviated by the letter T. The ship's head, or heading, can be thought of as a special bearing denoting the direction in which the ship is pointing; it can be expressed either with reference to magnetic or true north, or with respect to the north axes of the magnetic or gyrocompasses. No matter what reference direction the navigator uses for the ship's head and other bearings, however, they must first be converted to true bearings before they can be used in the navigation plot.

In practice, relative bearings are not normally used for navigation purposes; they find their most extensive use in relating an object's position relative to the ship's bow, for purposes of visualizing the physical relationships involved. Ordinarily they are estimated visually, but if the ship's gyro system becomes inoperative, exact relative bearings to land or seamarks can be shot with a bearing circle or alidade. In order to use relative bearings in a navigation plot, the navigator must have a method of determining the ship's true head when the gyro is inoperative, so as to be able to convert the relative bearings to true.

The Magnetic Compass

Virtually all vessels, from the smallest of recreational craft to the jumbo tanker and aircraft carrier, are fitted with at least one magnetic compass. On most small boats and vessels that operate mainly in

inland and coastal waters, the magnetic compass is the primary reference for direction and course headings. Most ocean-going vessels, including almost all Navy warships, have one or more gyrocompass systems installed to serve this purpose, but even on these ships, the magnetic compass serves as an important back-up in case of gyro failure, and as a primary means of checking the accuracy of the gyrocompass at regular intervals while under way. Even though the modern gyrocompass is extremely accurate, highly reliable, and easy to use, it is nevertheless a highly complex instrument requiring periodic expert maintenance, dependent on an electrical power supply, and subject to electronic and mechanical failures of its component parts. The magnetic compass, on the other hand, is a comparatively simple, self-contained mechanism that operates independent of any electrical power supplies, requires little or no maintenance, and is not easily damaged.

Most older commercial ships and Navy warships having a secondary conning station carry two magnetic compasses. The main one, located in close proximity to the helmsman's station in the pilot house or bridge is called the *steering* (abbreviated "stg") *compass*. The other, usually located in or near the secondary conning station, is often called the *standard* ("std") *compass*. They are usually the same configuration of magnetic compass, differing only in name for reference purposes. Many newly constructed merchant and Navy ships, however, carry only one steering compass, because they are fitted with two redundant gyrocompass systems for which the probability of simultaneous failure is so low that a second magnetic compass is considered to be unnecessary. Figure 9–1A is a photograph of a U.S. Navy standard No. 1

Figure 9–1A. The Navy standard No. 1 7-inch magnetic compass.

Figure 9–1B. Binnacle for a standard Navy magnetic compass.

7-inch magnetic compass. The compass is mounted in a stand called a binnacle, pictured in Figure 9–1B.

The compass itself consists of a circular card, graduated with 360 degrees around the face; this card floats within a bowl containing a compass fluid. A pair of magnets is attached to the underside of the card, beneath its north-south axis. The card is suspended within the fluid, thereby reducing friction on the pivot about which the card rotates and damping vibrations and oscillations of the card as the ship moves. The compass bowl assembly is supported externally by a set of gimbals, hinged on both the longitudinal and athwart-ships axes, which permits the compass to remain nearly horizontal at all times.

In operation, the compass magnets tend to align themselves with the earth's magnetic lines of force existing at that location. If these lines coincided exactly with the earth's meridians, the north position on the compass card would always point toward true north, and any compass directions read with reference to the card would be true directions. Unfortunately, this ideal situation never occurs, because of two effects that come into play—variation and deviation.

Variation

Variation can be defined as the angle between the magnetic line of force or magnetic meridian, and the geographic meridian, at any

location on the earth's surface; it is expressed in either degrees east or west to indicate on which side of the geographic meridian the magnetic meridian lies. Variation is caused primarily by the fact that the earth's magnetic and geographic poles do not coincide, and to a lesser extent, by certain magnetic anomalies in the earth's crust. Thus, if we are at position A on the earth's surface in Figure 9–2, our compass would tend to align itself with the magnetic lines of force flowing from the south to the north magnetic pole, as shown. But true north, as represented by the geographic meridian passing through A, would lie off to the right. The angle formed at position A between the magnetic and geographic meridians, angle V, is the variation; in this case the variation is west, since the north axis of the compass card is pointing to the west of geographic or true north.

If we were to change our position on the diagram below in a random manner, we would find that the angle V would in most cases change each time we moved—at some locations, the angle would become smaller, while at others it would become larger. If we moved to the back side of the globe, the magnetic lines of force would for the most part tend off to the right of the geographic meridian. The variation would now be east. Finally, if we moved near to the geographic north pole, we would observe that the lines of force flowed in a southerly direction; for this reason, the magnetic compass provides only a general estimate of direction in high latitudes.

As we shifted our position about on the globe, we would find that at some locations the values of the variation would be the same as at other locations. If we moved in such a way that the angle V between

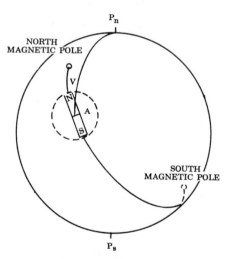

Figure 9–2.   Illustration of the variation angle.

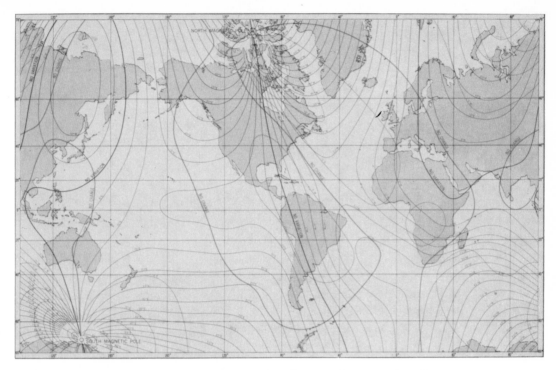

Figure 9–3. Simplified chart of the world with selected isogonic lines, Epoch 1975.

the meridians remained constant, we would trace a path known as an *isogonic line*—a line along which the measured variation is the same. Figure 9–3 depicts a series of isogonic lines drawn on a simplified world chart.

There is another characteristic associated with variation that concerns the navigator. The magnetic field of the earth does not remain constant, but rather it is continually slowly changing both in direction and in intensity; the magnetic poles actually wander slightly over the earth's surface from year to year. Because of this phenomenon, the variation at any given location on the earth changes a small amount from year to year. The navigator can determine the value for the variation at his position by referring to a chart of the area in which he is located. In the centers of the compass roses inscribed on every chart, the value of the variation as measured at that position is given for the date on which the chart was printed. In addition, the annual rate of change of the variation is also supplied; this allows the navigator to compute the variation for the date with which he is concerned, if different from the year in which the chart was printed. The navigator always uses the information contained in the compass rose closest to his DR position on the chart for his computations. As an example, consider the compass rose in Figure 9–4 on the following page.

Figure 9–4. A sample chart compass rose.

The variation at that location in 1980, when the chart was printed, was 14° 45′ West, with a 2′ annual increase. If the navigator wished to compute the variation for 1983, he would multiply the 2′ annual increase by 3 (1983 minus 1980) to get 6′ of increase from 1980 to 1983. He would then add this 6′ to the 1980 value, 14° 45′ West, to get the 1983 variation, 14° 51′ West. In practice, the navigator always rounds off the value he computes to the nearest half degree; in the example above, therefore, he would round off the 1983 value to 15.0° West. The reason for this practice will be discussed later. Note also that the inner bearing circle of the compass rose has been rotated 14° 45′ to the west, corresponding to the variation. This is done for the convenience of the small boatman, who navigates for the most part by the use of his magnetic compass; he can instantly convert magnetic bearings to true bearings or a true heading to a magnetic heading by laying a straightedge across the compass rose.

Deviation

In addition to variation, the navigator must take one other effect into account when working with his magnetic compass—*deviation*. Deviation is defined as the angle between the magnetic meridian and the north axis of the magnetic compass card; like variation, it is expressed in either degrees east or west to indicate on which side of the magnetic meridian the compass card north lies. The effect is caused by the interaction of the ship's metal structure and electrical currents with the earth's magnetic lines of force and the compass magnets. Essentially, a ship is like a metal bar moving through a magnetic field; the magnetic effect on the bar varies depending on the angle it makes with respect to the lines of force comprising the

field. Likewise, the deviation on a ship varies with the ship's heading. The deviation will also change periodically as metal is moved from one place to another on the ship, e.g., when making equipment alterations; it can even be affected by semipermanent magnetism induced in the ship as it remains alongside a pier for an extended time, such as during overhaul. A Navy ship's degaussing system also has an effect. The degaussing system consists of a number of horizontal wire coils wrapped around a ship on the underside of its hull. The system is an attempt to lessen the strength of the ship's magnetic "signature" as a countermeasure against magnetic mines. On most headings, the deviation of a ship will normally change if its degaussing system is energized.

The effect of deviation on a shipboard compass can be regulated somewhat by adjusting or "compensating" it for the influence of the ship's metal structure, but it can never be completely eliminated. It is necessary, therefore, to have some means of determining what value of deviation to use for all possible ship headings, so as to be able to convert from a compass to a true heading or bearing. In practice, a table is constructed by the navigator in which he enters the values of deviation for every fifteen degrees of the ship's head magnetic, starting with 000°M. The table is based on the ship's head magnetic so that it can be used regardless of the variation at any particular location. An example of a Navy ship's magnetic compass deviation table appears in Figure 9–5 next page. Note that in the table there is one column for degaussing off (DG OFF) and a second for degaussing on (DG ON), a necessity because of the effect of the system on the compass. If the ship's head magnetic lies between two tabulated entries, the value of the deviation must be interpolated by methods discussed below. The result, like variation, is normally rounded off to the nearest half degree.

It usually takes about three hours to make up the table, as the ship must be "swung" around 15 degrees at a time while under way in order to record the data. *Swinging ship*, as the procedure is called, is required by Navy instructions whenever the deviation on any heading exceeds three degrees. Since the deviation on a ship is always changing because of the causes previously mentioned, it is usually necessary to swing ship about twice each year.

Having described variation and deviation, it is now possible to discuss the technique for converting a steering compass heading or bearing to a true heading or bearing, and vice versa. Conversion of a ship's heading will be stressed herein, because as has been explained, the deviation depends on the ship's magnetic head. To convert any other bearing obtained by use of a shipboard magnetic com-

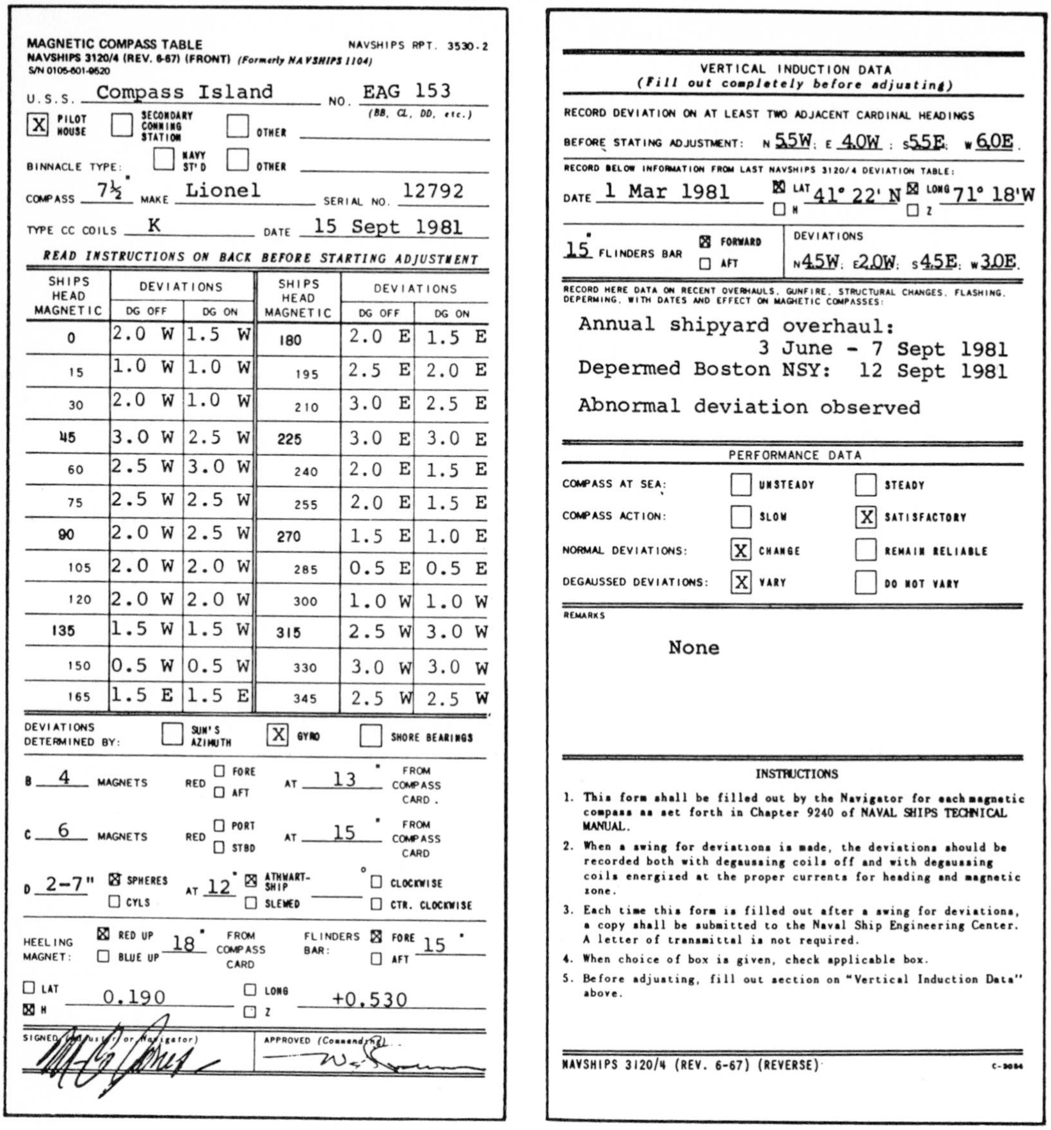

MAGNETIC COMPASS TABLE NAVSHIPS RPT. 3530-2
NAVSHIPS 3120/4 (REV. 6-67) (FRONT) *(Formerly NAVSHIPS 1104)*
S/N 0105-601-9620

U.S.S. Compass Island NO. EAG 153 *(BB, CL, DD, etc.)*

[X] PILOT HOUSE [ ] SECONDARY CONNING STATION [ ] OTHER

BINNACLE TYPE: [ ] NAVY STD [ ] OTHER

COMPASS 7½" MAKE Lionel SERIAL NO. 12792

TYPE CC COILS K DATE 15 Sept 1981

READ INSTRUCTIONS ON BACK BEFORE STARTING ADJUSTMENT

SHIPS HEAD MAGNETIC	DEVIATIONS		SHIPS HEAD MAGNETIC	DEVIATIONS	
	DG OFF	DG ON		DG OFF	DG ON
0	2.0 W	1.5 W	180	2.0 E	1.5 E
15	1.0 W	1.0 W	195	2.5 E	2.0 E
30	2.0 W	1.0 W	210	3.0 E	2.5 E
45	3.0 W	2.5 W	225	3.0 E	3.0 E
60	2.5 W	3.0 W	240	2.0 E	1.5 E
75	2.5 W	2.5 W	255	2.0 E	1.5 E
90	2.0 W	2.5 W	270	1.5 E	1.0 E
105	2.0 W	2.0 W	285	0.5 E	0.5 E
120	2.0 W	2.0 W	300	1.0 W	1.0 W
135	1.5 W	1.5 W	315	2.5 W	3.0 W
150	0.5 W	0.5 W	330	3.0 W	3.0 W
165	1.5 E	1.5 E	345	2.5 W	2.5 W

DEVIATIONS DETERMINED BY: [ ] SUN'S AZIMUTH [X] GYRO [ ] SHORE BEARINGS

B 4 MAGNETS RED [ ] FORE [ ] AFT AT 13" FROM COMPASS CARD.

C 6 MAGNETS RED [ ] PORT [ ] STBD AT 15" FROM COMPASS CARD

D 2-7" [X] SPHERES [ ] CYLS AT 12" [X] ATHWARTSHIP [ ] SLEWED ° [ ] CLOCKWISE [ ] CTR. CLOCKWISE

HEELING MAGNET: [X] RED UP [ ] BLUE UP 18" FROM COMPASS CARD FLINDERS BAR: [X] FORE [ ] AFT 15"

[ ] LAT [X] H 0.190 [ ] LONG [ ] Z +0.530

SIGNED (Adjuster or Navigator) APPROVED (Commanding)

VERTICAL INDUCTION DATA
(Fill out completely before adjusting)

RECORD DEVIATION ON AT LEAST TWO ADJACENT CARDINAL HEADINGS

BEFORE STATING ADJUSTMENT: N 5.5W; E 4.0W; S 5.5E; W 6.0E.

RECORD BELOW INFORMATION FROM LAST NAVSHIPS 3120/4 DEVIATION TABLE:

DATE 1 Mar 1981 [X] LAT [ ] H 41° 22' N [X] LONG [ ] Z 71° 18'W

15" FLINDERS BAR [X] FORWARD [ ] AFT DEVIATIONS N 4.5W; E 2.0W; S 4.5E; W 3.0E.

RECORD HERE DATA ON RECENT OVERHAULS, GUNFIRE, STRUCTURAL CHANGES, FLASHING, DEPERMING, WITH DATES AND EFFECT ON MAGNETIC COMPASSES:

Annual shipyard overhaul:
3 June - 7 Sept 1981
Depermed Boston NSY: 12 Sept 1981

Abnormal deviation observed

PERFORMANCE DATA

COMPASS AT SEA: [ ] UNSTEADY [ ] STEADY
COMPASS ACTION: [ ] SLOW [X] SATISFACTORY
NORMAL DEVIATIONS: [X] CHANGE [ ] REMAIN RELIABLE
DEGAUSSED DEVIATIONS: [X] VARY [ ] DO NOT VARY

REMARKS

None

INSTRUCTIONS

1. This form shall be filled out by the Navigator for each magnetic compass as set forth in Chapter 9240 of NAVAL SHIPS TECHNICAL MANUAL.
2. When a swing for deviations is made, the deviations should be recorded both with degaussing coils off and with degaussing coils energized at the proper currents for heading and magnetic zone.
3. Each time this form is filled out after a swing for deviations, a copy shall be submitted to the Naval Ship Engineering Center. A letter of transmittal is not required.
4. When choice of box is given, check applicable box.
5. Before adjusting, fill out section on "Vertical Induction Data" above.

NAVSHIPS 3120/4 (REV. 6-67) (REVERSE) C-9084

Figure 9–5. A sample ship's deviation table.

pass to a true bearing, the ship's head magnetic must first be determined as an intermediate step. In practice, it is necessary to convert a steering or standard compass heading to a true heading primarily when it is desired to convert relative bearings to true bearings, for purposes of the navigation plot, with the gyrocompass system inoperative. Occasionally, it is desirable to compute a true ship's head from a steering or standard compass heading in order to obtain the error of an operable gyro. Situations which necessitate going the other

way, from true to magnetic compass headings, include the recommendation of a suitable steering compass course to steer in order to achieve a desired true course when the gyro is inoperative, and the calculation of the proper magnetic headings on which to set the ship during the process of swinging ship.

Converting from Compass to True

To convert a steering compass heading to a true heading, the navigator must take both variation and deviation into account. The amount by which a steering compass direction differs from a true direction is often referred to as the *compass error*, sometimes abbreviated to CE, which is nothing more than the algebraic sum of the variation and deviation. The sequence of operations in converting a steering or standard compass heading to true is as follows: first, apply the deviation to the steering compass heading to obtain the magnetic heading; second, apply the variation to the magnetic heading to produce the desired true heading. In the process, westerly errors must be subtracted and easterly errors added. Through the years, mariners have used the following memory aid to keep the proper sequence in mind when making such conversions:

Can	**Dead**	**Men**	**Vote**	**Twice**	**At**	**Elections**
Compass head	Deviation	Magnetic head	Variation	True head	Add	Easterly error

The most difficult part of the conversion of a magnetic compass heading to a true heading is to obtain the proper value of deviation for the given steering compass heading. Recall that the standard deviation table of the kind shown in Figure 9–5 is based on the ship's head magnetic, or M in the formula above. Since this is the case, when converting from compass to true headings it is necessary to interpolate *twice* to be certain of obtaining the correct value of the deviation D. For the first interpolation, the steering compass heading can be considered an approximation of the magnetic head. The deviation so obtained allows a better approximation to the magnetic head to be made, by applying this deviation to the original steering compass heading. In so doing, easterly deviation is added, westerly subtracted.

The magnetic head so computed is now a better approximation than the steering compass heading. It remains only to interpolate a second time to extract the correct deviation; this result is then rounded to the nearest half degree. As an example of this process, consider the following question:

A ship's head is 305° per steering compass (p stg c). What is the ship's magnetic heading? Degaussing is OFF.

Referring to the deviation table in Figure 9–5, the steering compass heading is bracketed by 300°M and 315°M, for which the corresponding deviations are 1.0°W and 2.5°W. Writing them in tabular form, we have:

$$305° \xrightarrow{\hspace{2cm}} \begin{array}{ll} 300° & 1.0°W \\ 315° & 2.5°W \end{array}$$

The desired deviation, therefore, must be $\frac{5}{15} = \frac{1}{3}$ of the difference between 1.0°W and 2.5°W:

$$\frac{5}{15} \times (2.5 - 1.0) = \frac{1}{3} \times 1.5° = .5°; D = 1.5°W.$$

The first estimate of deviation is, therefore, 1.5°W.

Applying 1.5°W to 305° p stg c, the result 303.5° M is obtained. Now the second interpolation must be performed:

$$\frac{3.5}{15} \times (2.5 - 1.0) = \frac{3.5}{15} \times 1.5 = .4; D = 1.4°W.$$

The required deviation, rounded to the nearest .5°, then, is 1.5°W.

This results in a ship's head magnetic of 305° − 1.5° = 303.5°M.

After having determined M, the ship's head magnetic, the next step in the conversion procedure is to apply the variation. Its value, rounded to the nearest half degree, is either added to or subtracted from the magnetic heading to obtain the true heading.

The values for variation and deviation are rounded to the nearest half degree because the magnetic compass graduations are so small that it is impossible to read a smaller subdivision. Thus, it would be meaningless to calculate a value for either the variation or deviation to the nearest tenth or beyond.

Because of the construction and placement of the magnetic compass on most ships, it is virtually impossible to shoot a steering compass bearing to an external object. Some ships, however, are fitted with a periscopic device that does make it possible to observe such a bearing. In converting such a magnetic compass bearing, other than

the ship's head, to a true bearing, care must be exercised to use only the ship's magnetic head in determining the proper value for the deviation. Deviation depends only on the ship's head, and not on any other bearing shot by use of the compass. Once the compass error is determined for a given ship's head, it can be applied to any bearing obtained by the use of that compass, as long as the ship remains on the given course. When the ship changes course, a new compass error incorporating the new value of deviation must be figured. An example of this type of problem follows:

While steaming on a heading of 305° p stg c, the following bearings were observed using the ship's steering compass:

Lighthouse   102° p stg c
Reef Light   329° p stg c

What are the true bearings of the lighthouse and Reef Light? Degaussing is OFF, and variation is 9°E.

From the previous example, the deviation on course 305° p stg c is 1.5°W, and ship's head magnetic is 303.5°M. Knowing the variation, the conversion formula can be written thus:

C	D	M	V	T	A	E
305°	1.5°W	303.5°	9°E			

The true heading, T, is 303.5° + 9° = 312.5°T. The compass error on this heading is the algebraic sum of the deviation and the variation:

$$(-)\ 1.5°W\ (+)\ 9°E\ =\ 7.5°E$$

This compass error (CE) can now be applied to the observed bearings to convert them to true:

Object	°p stg c	CE	°T
Lighthouse	102°	(+) 7.5E	109.5°
Reef Light	329°	(+) 7.5E	336.5°

If the ship changed course, the compass error would change, because of the resulting change in the deviation. If further magnetic compass bearings were observed and converted to true, a new value of the compass error would have to be computed.

Converting from True to Compass

To convert a true heading to a steering compass heading, for the purpose of steering some true course through the water with the gyro inoperative, for example, corrections are applied in reverse order according to the sequence

T   V   M   D   C   A   W

which many navigators are fond of remembering by means of the old ditty "*True Virgins Make Dull Companions At Weddings.*" The last two letters indicate the rule to add westerly errors, and subtract easterly errors, when converting from true to compass headings. Since in this type of problem the ship's head magnetic is an intermediate step found by applying the variation to the true heading, only one interpolation for deviation is required. As a typical example of a problem requiring this methodology, consider the following:

While steaming on a heading of 149°T, the ship's gyro suddenly tumbled. What steering compass course should be steered to keep the ship on the same true course? Assume a variation of 9°E, with degaussing OFF.

Writing down the known quantities in the format given above, we obtain the following:

$$T \quad V \quad M \quad D \quad C$$
$$149° \quad 9°E$$

Since easterly variation is subtracted when proceeding from true to compass heading, the ship's head magnetic, M, is determined to be:

$$149° \ (-) \ 9° \ = \ 140°M.$$

Referring to the deviation table of Figure 9–5, this magnetic heading is bracketed by 135°M and 150°M, for which the respective deviations with degaussing OFF are 1.5°W and 0.5°W. Writing them in tabular form, we have

$$\begin{array}{ccc} & 135° & 1.5°W \\ 140° \longrightarrow & & \\ & 150° & 0.5°W \end{array}$$

A one-step interpolation, rounded to the nearest half-degree,

yields a value of 1.0°W for the deviation. The conversion can now be completed as follows:

T	V	M	D	C
149°	(−)9°E	140°	(+)1.0°W	141°

Thus, the ship must steer 141° p stg c in order to remain on a heading of 149°T.

The Gyrocompass

A gyrocompass is essentially a north-seeking gyroscope; it is encased in a housing fitted with various electronic components that keep the spin axis of the gyro aligned with terrestrial meridians, and sense the angle between the ship's head and the gyro spin axis. A detailed explanation of the gyroscope and the theory of its operation are beyond the scope of this text, but the navigator should be aware of the basic principles by which the gyro operates.

Basically, a classical gyro consists of a comparatively massive, wheel-like rotor balanced in gimbals that permit rotation in any direction about three mutually perpendicular axes through the center of gravity of the rotor. The three axes are pictured in Figure 9–6A; they are called the spin axis, the torque axis, and the precession axis.

Once a gyroscope rotor is made to rotate, its spin axis would remain forever oriented toward the same point in space unless it were acted upon by an outside force. In the marine gyrocompass, the spin axis

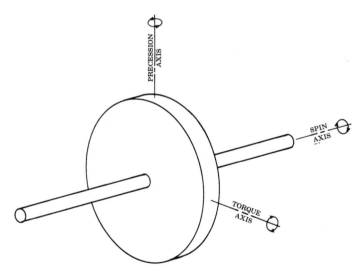

Figure 9–6A. *The three axes of a gyroscope.*

is kept aligned with a terrestrial meridian in a plane tangent to the earth's surface by a directive force derived from the tangential velocity component of the earth's rotational motion. Because this tangential velocity component is a maximum at the equator and diminishes to zero at the poles, the directive force is great in lower and mid-latitudes, but diminishes in strength as the earth's poles are approached. In latitudes beyond 70° north or south, the ship's velocity may become so great in relation to the earth's tangential velocity that large errors can be introduced into the directive force of the gyrocompass. For this reason, the gyrocompass must be continually checked for error beyond 70° north and south latitudes. Beyond 75° to 80° latitude, most standard shipboard gyrocompass systems become so slow to respond to correcting forces that extreme errors on the order of ten to twenty degrees or more are routinely experienced. Most shipboard gyrocompasses become virtually useless beyond about 85° latitude.

Fortunately, however, ice conditions in the polar regions are such that even in summer it is extremely rare that either commercial or Navy surface ships would venture much beyond 70° north or south latitudes in the normal course of operations, so most marine navigators will never be confronted with this problem. Should an occasion to operate in these polar regions arise, navigation is accomplished by employing a number of rather specialized piloting techniques and instruments developed over the years for use there. If required, further information on polar navigation can be found in *Dutton's Navigation and Piloting* and in the *American Practical Navigator* (Bowditch).

The gyrocompass currently installed on many Navy ships is the Mark 14 Mod 2 produced by the Sperry Rand Corporation; it is shown with a side panel removed in Figure 9–6B.

As was stated earlier, many newer Navy and merchant ships have two redundant gyrocompass systems installed. Even on older ships, the reliability of the gyrocompass is great enough that the magnetic compass is infrequently used for the most part. The gyrocompass itself on all but the very smallest ships is usually placed well down in the interior of the ship's hull, where it receives minimum exposure to roll, pitch, and yaw, and in the case of Navy ships, maximum protection from battle damage. It is connected by cables to *gyrocompass repeaters*, positioned where required throughout the ship. These repeaters use electronic servo-mechanisms that reproduce the master gyrocompass readings at their remote locations.

A gyrocompass repeater in design is similar to a magnetic compass. It consists basically of a movable card, graduated in 360°, which is mounted within a case called a binnacle. Gyrocompass repeaters lo-

Figure 9–6B. The Mark 14 Mod 2 shipboard gyrocompass. (Courtesy Sperry Marine Systems)

cated on bridge wings and other locations at which they will be used for obtaining visual bearings for navigational and other purposes are mounted in gimbals that keep the repeater horizontal, atop a stand of sufficient height to allow comfortable observation of bearings. A picture of a gyro repeater and its stand appears in Figure 9–7. The ship's head is indicated by a mark called the lubber's line on the rim of the binnacle; relative bearings are inscribed around the interior of the binnacle alongside the compass card for use if the gyro fails. To aid in shooting bearings with the repeater, it is fitted with an azimuth circle (see Figure 7–1 on page 111), which can be rotated about the face of the binnacle; its use was previously described in Chapter 7.

In operation, the gyro is made to spin in such a way that its axis is nearly always aligned with the earth's geographic meridians, with small errors caused by the combined effects of the ship's motion, friction, and malfunctions of the gyro's electronic circuitry. Electronic systems within the gyro translate the angular difference between the spin axis and the ship's head into an electronic signal that drives the compass cards in all the gyro repeaters, keeping the north axis on the cards aligned with the spin axis of the gyro. As long as there is no error in either the alignment of the gyro with true north or in the transmission system, true bearings can be read directly off the gyro

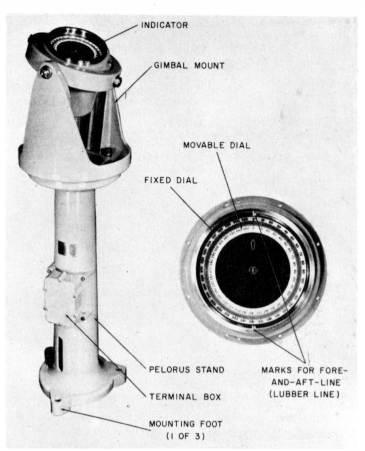

Figure 9–7. A gyrocompass repeater.

repeaters and used in the navigation plot. The gyrocompass, then, has several advantages over the magnetic compass:

It seeks the true or geographic meridian instead of the magnetic meridian.

It can be used near the earth's magnetic poles, where the magnetic compass is useless.

It is not affected by surrounding material.

Its signal can be fed into inertial navigation systems, automatic steering systems, and in warships, into fire control equipment.

Being an intricate electronic instrument, however, it is also subject to certain disadvantages:

It requires a constant source of electrical power and is sensitive to power fluctuations.

It requires periodic maintenance by qualified technicians.

As has been mentioned, the gyro is inherently a very accurate instrument, but there are several sources of error within the system. Consequently, it is rare that a gyro repeater is free from all error, especially that caused by the transmission network, but the error in most modern gyrocompass systems is so small (less than .1°–.2°) as to be considered insignificant for most applications. Even if the error is relatively large, it causes no particular problem for the navigator, as long as it remains within manageable limits and is constant. The navigator must always take any significant error (greater than ~.3°) into account while keeping his DR plot and when making a course recommendation to the conning officer. Navy instructions require the gyro error to be determined at least once a day. In practice, the navigator will probably make this determination much more frequently, especially in piloting waters. As explained in other chapters of this text, the Navy navigator must record the error on the 0800, 1200, and 2000 position reports that he submits to the commanding officer each day when under way; he must enter the error on each page of the *Bearing Book*, and, as is discussed in this section, he must apply the error each time he plots a gyro bearing or makes a course recommendation.

Methods of Determining Gyrocompass Error

There are several methods of determining gyro error available to the navigator. In the following paragraphs, only those procedures applicable to pilot waters will be described; the determination of gyro error on the open sea will be covered in *Marine Navigation 2* as part of celestial navigation. Gyro error, like compass error, is expressed in degrees east or west. If the gyro is rotated to the west of true north, the error is west, and conversely, if rotated to the east, the error is east. The diagrams in Figures 9–8A and 9–8B may help to visualize this. In the situation depicted below, a gyro error caused the gyro repeater card to be rotated away from true north toward the west by an amount of 5°. Any bearings shot on this repeater, then,

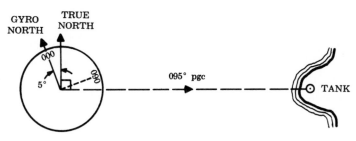

Figure 9–8A. A 5° westerly gyro error illustrated.

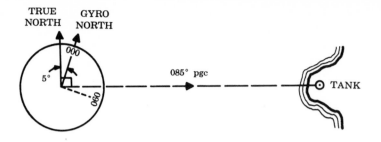

Figure 9–8B. A 5° easterly gyro error illustrated.

would be too high. Here, the tower at a right angle to true north should bear 090°T; with reference to this repeater, however, it would bear 095° per gyro compass (pgc).

Now let the card be rotated by an easterly error 5° to the east of true north, as in Figure 9–8B. Now the same object would bear 085° pgc, vice 090°T. Any other bearings shot using this repeater would, in similar fashion, be 5° too low.

One of the best methods of determining gyro error is to observe an artificial or natural visual range. The range is first shot visually by use of the azimuth circle on the gyro repeater, and its gyro bearing is noted. Then the observed bearing is compared to the true bearing of the visual range, as determined from a chart showing the objects, or in the case of an established lighted range, from the *Light List* or *List of Lights*. The difference in the two bearings is the amount of the gyro error. If the gyro repeater bearing is higher than the actual true bearing, the error is west; if lower, the error is east.

If the ship is at a known location, such as a pier or an anchorage, a gyro error can be obtained by comparing a known bearing to an object ashore, as measured on a chart, with the bearing as observed from the gyro repeater. Once again, the difference in bearings is the amount of the error; if the gyro bearing is too high, the error is west; if too low, the error is east.

In the same manner, the ship's heading while tied up alongside and parallel to a pier can be compared with a known pier heading. The difference is the gyro error.

If the ship is not under way, a favorite method of obtaining gyro error is trial and error adjustment of three or more simultaneous lines of position until a point fix results. If the three LOPs meet at a point when initially plotted, there is no gyro error. If they form a triangle, the lines are adjusted by successive additions or subtractions of 1°, then if necessary, .5° to the bearings, until they meet at a point fix. The total correction thus applied to any one LOP is the gyro error.

If the correction had to be subtracted, the error is west; if added, the error is east.

Finally, if no other method is available, the gyro can be compared to another compass of known error. If, for example, the navigator had a reliable deviation table available, he could use the ship's magnetic compass for this purpose by converting the compass heading to true and then comparing the true ship's head to the gyro heading. If a second gyrocompass system of known error were installed, he could compare the headings of the two gyros. If the gyro heading in question is too high, the error is west; if too low, the error is east.

There are two memory aids to help the navigator decide whether the gyro error is east or west:

> If the compass is best (higher), the error is west;
> if the compass is least (lower), the error is east,

and,

G.E.T.—Gyro + East = True.

When plotting a gyro bearing or series of bearings, the navigator must always remember to subtract the amount of any westerly gyro errors, and add the amount of any easterly errors prior to plotting them, since only true bearings should be plotted on a chart. In practice, Navy navigators usually take gyro error into account by rotating the compass rose attached to the parallel motion protractor (PMP), described in Chapter 7, to compensate for the error. Since the conning officer is conning and the helmsman is steering by reference to the gyro, he must also remember to incorporate the gyro error into any course recommendations he makes.

Summary

There are three references for horizontal direction on board a vessel at sea: the ship's head (relative bearings), the ship's magnetic compass (magnetic bearings), and the ship's gyrocompass (gyrocompass bearings, equivalent to true bearings if the gyro error is negligible). Because most modern ocean-going ships are fitted with one or more gyrocompass systems, the gyroscope is normally the main reference for direction for the surface navigator. When properly used, serviced, and maintained, the modern gyrocompass is an extremely reliable and accurate instrument, but as is the case with all electronic instruments, it is subject to error and damage. In the event of complete gyro failure, the navigator has an excellent back-up system available

in the form of the magnetic compass. Even when the gyro is fully operational, however, the navigator must be constantly aware of the possible existence of error. It is the unknown, unobserved error that contributes to marine disasters. There are on file numerous reports of vessels having been put aground and lost because of a navigator's adherence to a course laid in safe waters, while the actual track was an unknown path leading to danger. With the ever-increasing volume of super-tanker and liquid natural gas carrier traffic at sea, the modern consequences of such a mishap can range, and all too often in the recent past have ranged, far beyond the loss of the ship herself. Now, more than ever, the watchword of the navigator must be constant vigilance at sea.

10

Radar, a word derived from the terms *radio detection and ranging*, is of great practical value to the navigator in the piloting environment. Since its orginal development in World War II as a detection device for enemy ships and aircraft, improved technology in electronics and electronic circuitry has made possible quantum jumps in the state of the art of this device, so that today there are many different types of radars, each with particular qualities that adapt it to specific applications. Almost all modern ocean-going vessels, and many coastal craft, are equipped with some variety of navigational radar. It is, in fact, required by Coast Guard regulations for all commercial vessels over 1,600 tons operating in U.S. waters; vessels over 10,000 tons must have two independent radar systems. Naval surface warships will often have as many as four different varieties of radar equipment, depending on the size and type of ship. They are air-search, fire-control, surface-search, and navigational radars. Often, a single surface-search radar will also perform the function of a separate navigational radar, especially on ships the size of a destroyer and smaller. As an important auxiliary function, navigational radars can not only be used to locate navigational aids and perform radar navigation, but they are also useful for tracking other vessels in the vicinity so as to avoid risk of collision.

This chapter will briefly examine the more important characteristics of a navigational or surface-search radar and the fundamental techniques of its operation with which the navigator must be familiar in order to use this device effectively. An excellent reference for a more detailed discussion of this subject is *Publication No. 1310*, the *Radar Navigation Manual*, published by the Defense Mapping Agency Hydrographic/Topographic Center. This publication contains a very thorough yet uncomplicated discussion of the theory and application of radar as it applies to all aspects of surface navigation.

Characteristics of a Surface Search/Navigational Radar

The basic principle of radar is the determination of range to an object or "target" by the measurement of the time required for an

extremely short burst or "pulse" of radio-frequency (RF) energy, transmitted in the form of a wave, to travel from a reference source to a target and return as a reflected echo. Most surface-search and navigational radars use very short high-frequency electromagnetic waves formed by an antenna into a beam very much like that of a searchlight. The antenna is usually parabolic in shape, and rotates in a clockwise direction to scan the entire surrounding area. Bearings to the target are determined by the orientation of the antenna at the moment the reflected echo returns.

In marine surface radars, the beam is fairly narrow in horizontal width, but may be quite broad in vertical height. Figure 10–1 illustrates a typical radar beam, and two side lobes, which are additional beams of low energy unavoidably radiated out in many types of surface radars due to limitations in the antenna size and shape.

The standard radar set is made up of five components—the transmitter, modulator, antenna, receiver, and indicator. The transmitter consists of an oscillator that produces the RF waves. The modulator is essentially a timing device that regulates the transmitter so that it sends out relatively short pulses of energy, separated by relatively long periods of rest. The antenna performs two functions—it forms the outgoing pulse train into a beam during the time the transmitter is on, and it collects returning target echoes during the interval the transmitter is at rest. In the receiver, the reflected radio energy collected by the antenna is converted into a form that may be presented visually on an output display device or indicator, usually a form of a cathode ray tube.

A radar system that operates in this way is termed "pulse-modulated"; if the system radiated energy continuously, the strong output signal would completely mask the weaker incoming reflected echo, making range determination impossible. This gives rise to the military electronic countermeasure (ECM) technique known as "jamming," whereby an enemy radar set is neutralized by flooding it with continuous RF energy having the same frequency as the set's radar pulse.

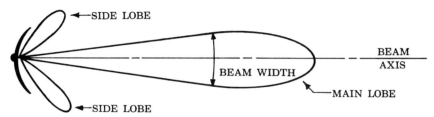

Figure 10–1. A navigation or surface-search radar beam and its side lobes, from above.

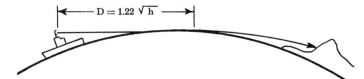

Figure 10–2.   Refraction of a radar beam.

Because it is made up of high-frequency RF energy of very short wave length, usually between three and ten centimeters long for most surface-search and navigational radars, the radar beam acts much like a beam of light. Energy comprising the beam travels out and is reflected back at nearly the speed of light; for this reason, pulse lengths and rest periods between them are measured in microseconds. Radars are commonly described by listing their frequency, number of pulses per second (pulse repetition rate), length of their pulse in microseconds (pulse length), and antenna rotation rate.

Radar beams are also like light waves in that they travel in straight lines for the most part, although refraction does cause them to bend downward and follow the curvature of the earth to some extent, especially at lower frequencies. For this reason, radars are limited in range to the distance of their *radar horizon*, which for most standard surface radars can be closely approximated by the formula

$$D = 1.22 \sqrt{h}$$

where D is the horizon distance in miles, and h is the height of the radar antenna in feet. Objects beyond the horizon can be detected only if they are of sufficient height above the earth's surface, as is the case with the mountain range in Figure 10–2. In this figure, the low shore line located under the radar horizon would not be detected, but the high mountain range located inland would be.

Radar antennas are usually placed high on the superstructure of the ship to extend the radar horizon as far as possible.

The Radar Output Display

Although there are several different types of output display devices, almost all incorporate a cathode ray tube in some form, located within a console called a *radar repeater*. The most common of these displays is called the plan position indicator scope, or PPI; this is the type used in most marine surface radar sets. In this presentation, the observer's ship is located at the center of a circular scope, and external objects within the range of the radar presentation are depicted at scaled distances outward from the center. Bearing on the PPI scope

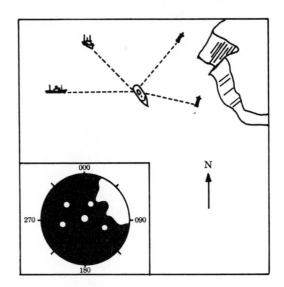

Figure 10–3A. A PPI presentation oriented to true north.

is indicated around the periphery of the screen, from 000° at the top clockwise to 360°. On ships having a gyrocompass, the scope has a gyro input, and the presentation is oriented so that the true north direction lies under the 000° mark. If gyro failure occurs, the radar presentation automatically reorients to a relative picture, with the ship's head at the 000° position at the top of the scope. Figure 10–3A shows a typical PPI presentation with the scope oriented by gyro input to true north, and Figure 10–3B depicts the same presentation oriented relative to the ship's head or longitudinal axis.

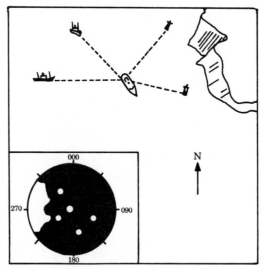

Figure 10–3B. A PPI presentation oriented relative to ship's head.

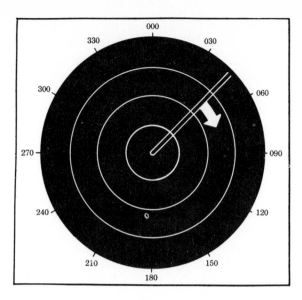

Figure 10–4. A PPI radarscope, with range rings illuminated.

As the antenna rotates, its beam is represented on the scope by a thin line that "sweeps" around the center, much like a spoke of a turning wheel. This line, called the *sweep*, illuminates or "paints" any objects within range of the radarscope onto its face. This presentation of an object on the radarscope is called a "pip" or "blip." On many radarscopes, the sweep can also be made to illuminate range rings at selected intervals outward from the center of the scope, thus allowing range estimates to be made. Bearing estimates can be made by referring to the bearings surrounding the scope face. A drawing of a PPI scope with its sweep and range rings appears in Figure 10–4.

To assist in the determination of range and bearing, most radar repeaters are fitted with devices called a *bearing cursor* and a *range strobe*. The bearing cursor is a narrow radial beam of light, usually centered on the ship's position on the scope; it can be rotated around by the operator through 360°. The range strobe is a pinpoint of light, which is moved in and out along the bearing cursor. On some older radar repeaters, the range strobe moves in and out along the sweep, and creates a variable range ring as the sweep rotates around the scope. Each indicator is controlled by a separate hand crank incorporating an appropriate distance or bearing readout.

To obtain a "mark"—a range and bearing to a target—the operator first rotates the bearing cursor so that it bisects the pip representing the object on the scope, and then he positions the range strobe so that it touches the inner edge of the pip. The range and bearing are read off from dials near the two hand cranks.

The size of the physical area to be depicted on the scope is variable on most radar repeaters, and is determined by the operator by the selection of a suitable range scale. Some repeaters only allow selection of specific scales such as 2, 4, 8, or 10 miles, whereas others allow selection of any scale between an arbitrary upper and lower limit, such as 1 to 50 miles. The scale number refers to the radius of the presentation; if a scale of 8 miles were selected, for example, the most distant object that could be shown on the scope would be eight miles away, and it would appear near the perimeter of the presentation.

Interpretation of a Radarscope Presentation

Interpretation of information presented on a radarscope is not always easy; much experience is often required on the part of the operator to obtain correct readings, especially during unfavorable meteorological conditions or when the radar is operating in a degraded state owing to partial failure of its electronic components. Even in the best of conditions with a finely tuned radar set, many factors tend to produce errors in interpretation of radar information. Among these are bearing resolution, range resolution, radar shadows, multiple echoes, and false echoes.

Bearing resolution is the minimum difference in bearing between two targets at the same range that can be discerned on the radarscope. The radar beam width causes a target to appear wider than it is in actuality. If two or more objects are close together at about the same range, their pips may merge together, giving the impression to the

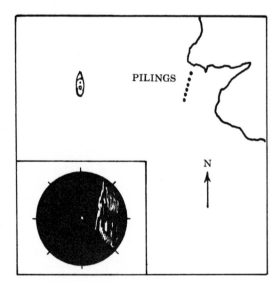

Figure 10–5A. A false shoreline caused by lack of bearing resolution.

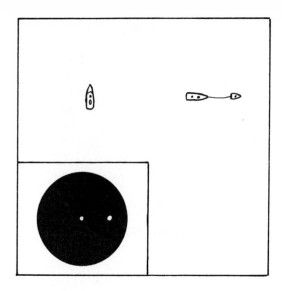

Figure 10–5B.   Tug and tow merged because of lack of range resolution.

operator that only one target is present. Such erroneous presentations often appear in coastal areas, where numbers of rocks, piles, and boats located offshore may present a false impression of the location of the shoreline, called a "false shoreline" (Figure 10–5A).

Range resolution is the minimum difference in range between two objects at the same bearing that can be discerned by radar. Pulse length and frequency both affect the range resolution of a particular radar; they can be adjusted on some sets to improve resolution at either long or short ranges. False interpretation of a radar presentation may occur as a result of this cause if two or more targets appear as one, or if objects such as rocks or small boats near the shore are merged with the shoreline (Figure 10–5B).

Radar shadows occur when a relatively large radar target masks another smaller object positioned behind it, or when an object beyond the radar horizon is obscured by the curvature of the earth. The shoreline in Figure 10–2 on page 173 lies in such a radar shadow zone.

The *multiple echo* is caused by pulses within a radar beam bouncing back and forth between the originating ship and a relatively close-in target, especially another ship. A multiple echo is a false pip which appears on the scope at the same bearing as the real target but at some multiple of the actual target range (Figure 10–6). If only one false pip appears at twice the actual range, it is often termed a *double echo;* if a second false pip appears at three times the range, it is a *triple echo.* Ordinarily, not more than one or two false pips appear in this fashion.

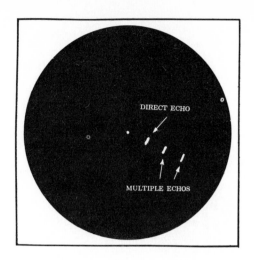

Figure 10–6. A multiple echo.

Another type of false pip sometimes encountered is the *false echo;* like the multiple echo, this is a pip that appears on the scope where there is no target in actuality. One type of false echo results when a portion of the reflected energy from a target returns to the antenna by bouncing off part of the ship's structure, as illustrated in Figure 10–7. The resulting false pip, sometimes called an *indirect echo,* always appears at the same range as the actual target, but at the bearing of the intermediate reflective surface.

A second type of false echo is the *side-lobe effect.* As explained in the first part of this chapter, the parabolic antenna usually radiates

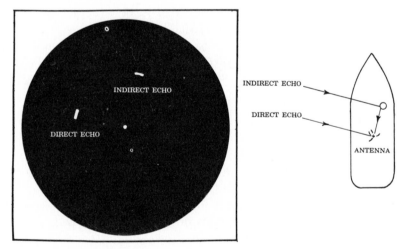

Figure 10–7. An indirect echo.

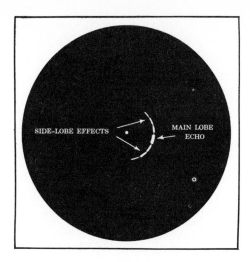

Figure 10–8. Side-lobe effects.

several side lobes, in addition to the main lobe of the radar beam. If energy from these side lobes is reflected back by a target, an echo will appear on the radarscope on each side of the main lobe echo. If the target is close enough, a semicircle or even a complete circle may be produced, with a radius equal to the target range. Normally, a target must be fairly close to produce a side-lobe effect, because of the low energy of the side lobes (Figure 10–8).

Use of Radar During Piloting

Because of the way in which a radar set is constructed, most marine surface radar bearings are accurate only to within 3° to 5°. Accuracy in range is usually much better, however; a well-tuned radar should give ranges precise to within ± 100 yards out to the radar horizon, with slowly increasing inaccuracy beyond that point to the limits of the extreme range.

It follows that for piloting applications, radar range lines of position (LOPs) are much preferred over radar bearings. In fact, some navigators make it a rule never to use radar bearings unless absolutely no other information is available, and then only with a high degree of suspicion.

Almost any object that is fixed in position, appears in symbolic form on the chart, and is visible on the radarscope presentation can be used to obtain a radar range LOP. Care must be taken, however, especially when using landmarks, to be certain of which features are being painted on the radar set. The operator must be constantly aware of the pitfalls associated with radar range resolution and radar shadow zones, described in the preceding section.

As was explained in Chapter 3, which describes the operation of the piloting team on board a Navy ship, the CIC radar navigation team is usually relied upon to guide the ship when poor visibility conditions make piloting by visual LOPs impossible. Even then, however, the navigator is still ultimately responsible to the commanding officer for the safe navigation of the ship. He will normally keep a parallel plot on the bridge, using radar information from his pilot house repeater and any visual LOPs of opportunity.

The technique of plotting and labeling a range LOP has already been discussed in detail in Chapter 8, and will not be repeated here. A radar range LOP may be crossed with a simultaneous LOP from any other source to obtain a fix, or two or more simultaneous radar ranges may be used for this purpose. If radar information alone is being used for plotting fixes, a minimum of three LOPs should always be obtained at any given time to guard against the ever-present possibility that one of them was misinterpreted by the radar operator.

If, as a last resort, radar bearings must be used in the navigation plot, normally only relatively small, discrete objects should be used for LOPs. When none are available, bearings should then be taken on tangents. Since these bearings will be exaggerated because of the beam width, it is necessary to decrease right tangent bearings by half the beam width, and to increase left tangents by the same amount. Consider the example in Figure 10–9. In the figure, and actual tangent bearings to the island pictured should be 340° T and 002° T. Because

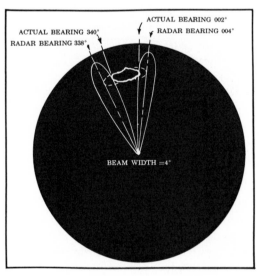

Figure 10–9. Effect of beam width on radar tangent bearings.

of the 4° radar beam width, however, energy begins to be reflected from the left tangent of the island when the centerline of the beam is directed toward 338° T. Thus, the left edge of the blip representing the island on the radarscope presentation would be located at 338°T—two degrees too low. Likewise, energy would still be reflected back from the right tangent when the radar beam is 2° beyond it, leading to a bearing for the right tangent on the radarscope of 004°, which is two degrees too high.

If the radarscope presentation is oriented to true north by gyro input, all bearings obtained should also be corrected for any known gyro error.

Summary

Radar is an extremely important tool for the navigator operating in any environment, and it is especially valuable during piloting. It is the only instrument widely available that is capable of providing precise LOPs even in conditions of poor visibility, such as fog or during hours of darkness. The primary advantages associated with radar from the navigator's point of view may be summarized as follows:

Radar can be used at night and during periods of reduced visibility when visual means of navigation are limited or impossible to use.

It is possible (though not recommended) to obtain a fix from a single object, because both range and bearing are obtainable.

Radar is available for use at greater distances from land than are most other navigational methods.

Radar fixes are obtained quickly and accurately.

Radar can be used to locate and track other shipping and storms.

Like any other highly sophisticated equipment, however, radar also has its limitations. Among the more serious of these are the following:

It is a complex electronic instrument, dependent upon a power source, and is subject to mechanical and electrical failure.

There is a minimum range limitation, resulting from returning echoes from nearby wave crests (sea return), and a maximum range limitation.

Interpretation of the radarscope display is difficult at times, even for a trained and experienced operator.

LOPs from radar bearings are inaccurate.

Radar is susceptible to both natural and deliberate interference.

Radar shadows and sea return may render objects undetectable by radar.

Although radar is not a panacea for the navigator, intelligent use of its capabilities certainly will aid him in safely directing the movements of his vessel.

11

The vertical rise and fall of the surface of a body of water on earth, caused primarily by the differences in gravitational attraction of the moon, and to a lesser extent of the sun, upon different parts of the earth is called *tide*. This phenomenon is of great interest and importance to the navigator, especially when he is piloting in relatively shallow coastal waters, because the height of tide determines the total depth of water at any given location and time. In some places, passages that could be safely negotiated by a deep-draft seagoing vessel during periods of high tide levels would be difficult or impossible to transit during periods of low water. This chapter will discuss the causes and effects of tide and will describe the procedures used by the navigator in computing and allowing for them during piloting.

Causes of Tide

As mentioned above, tide is caused primarily by the interaction of the gravitational forces of the moon, the sun, and the rotating and revolving earth upon the waters comprising the earth's oceans. Because the moon is many times closer to the earth than the sun, the effect of its gravitational pull is some two-and-a-quarter times more pronounced, even though the sun has a mass thousands of times greater. For purposes of illustration, the earth can be visualized as a spherical core covered by water. Although a detailed theoretical explanation of tide is beyond the scope of this text, in essence the strong gravitational pull of the moon on the side of the earth nearest the moon, together with the strong outward centrifugal force generated by the earth-moon system on the opposite side of the earth, cause the surrounding water to bulge out in the form of high tides on both sides, as depicted in an exaggerated fashion in Figure 11–1. The same effect on a much smaller scale also takes place with respect to the earth-sun system, but because the sun is so much further away, the effect is insignificant in comparison to that of the earth-moon system.

The moon revolves completely around the earth once each month.

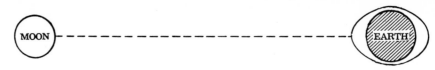

Figure 11–1. Effect of the moon on the earth's oceans.

Since the earth rotates beneath it in the same direction as the moon revolves, it takes 24 hours and 50 minutes for the earth to complete one revolution with respect to the moon. It follows that every location on earth should experience four tides every 24 hours and 50 minutes: two high tides, and two intervening low tides. This is in fact the usual tidal pattern. The 24 hour and 50 minute period is called a *tidal day,* and each successive high and low tide is said to constitute one *tide cycle.* At some locations, however, the tide patterns that actually occur are distorted from this norm because of the effects of land masses, constrained waterways, friction, the Coriolis effect (an apparent force acting on a body in motion on the earth's surface, caused by the rotation of the earth), and other factors. Altogether there are three major tidal patterns experienced at various locations on earth; each will be described in the following section.

As the moon rotates about the earth, it happens that at certain times its effects on the earth's oceans are reinforced by those of the sun, but at other times the effects are opposed. The tidal effects of the sun and moon act in concert twice each month, once near the time of the new moon, when this body is on the same side of the earth as the sun, and again near the time of the full moon, when it is at the opposite side of the earth from the sun. Tides produced at these times are abnormally high and unusually low, and are called *spring tides.* Figure 11–2A illustrates the two possible relationships

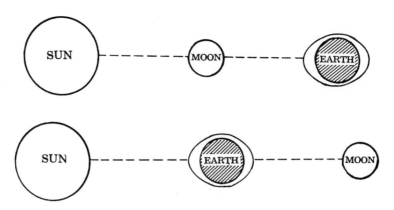

Figure 11–2A. The earth, sun, moon system at spring tide.

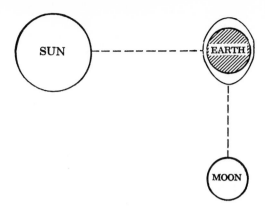

Figure 11–2B. The earth, sun, moon system at neap tide.

of the earth, sun, and moon during spring tide. The tidal effects of the sun and the moon are in opposition to one another when the moon is at quadrature—the first and last quarter—at which times the moon is located at right angles to the earth-sun line. At these times, high tides are lower and low tides are higher than usual; these are referred to as *neap tides*. Figure 11–2B depicts the relationship of the earth, sun, and moon at neap tide.

Types of Tides

Before progressing further with the discussion of tide, it is necessary to define some of the terms associated with its description:

Sounding datum   An arbitrary reference plane to which both heights of tides and water depths expressed as soundings are referenced on a chart. The sounding datum planes in most common use will be discussed later in this chapter.

High tide or *high water*   The highest level normally attained by an ascending tide during a given tide cycle. Its height is expressed in feet or meters relative to the sounding datum.

Low tide or *low water*   The lowest level normally attained by a descending tide during a tide cycle. Like high tide, its height is expressed in feet or meters relative to the sounding datum.

Range of tide   The vertical difference between the high- and low-tide water levels during any given tide cycle. It is normally expressed in feet or meters.

Stand   The brief period during high and low tides when no change in the water level can be detected.

As was mentioned in the preceding section, the physical laws governing movement of the large water masses located in the ocean basins

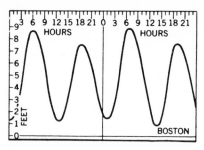

Figure 11–3A. Semidiurnal tide pattern.

and constricted by the geography of the world's land masses result in three major types of tides experienced at various locations throughout the world. They are classified according to their characteristics as semidiurnal, diurnal, and mixed.

The *semidiurnal* tide is the basic type of tide pattern observed over most of the world. There are two high and two low tides each tidal day, and these occur at fairly regular intervals a little more than six hours in length. Usually there are only relatively small variations in the heights of any two successive high or low waters. Tides at most locations on the U.S. Atlantic coast are representative of this variety.

The *diurnal* tide is a pattern in which only a single high and a single low water occur each tidal day; high and low tide levels on succeeding days usually do not vary a great deal. Tides of this type appear along the northern shore of the Gulf of Mexico, in the Java Sea, and in the Tonkin Gulf.

The *mixed* tide pattern is characterized by wide variation in heights of successive high and low waters, and by longer tide cycles than those of the semidiurnal tide. This blend of the semidiurnal and diurnal tides is prevalent on the U.S. Pacific coast, and on many of the Pacific islands.

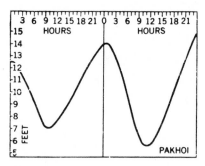

Figure 11–3B. Diurnal tide pattern.

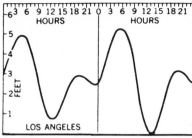

Figure 11–3C. Mixed tide pattern.

Tidal Reference Planes

In order for water depths, heights, and elevations of topographical features, nav aids, and bridge clearances to be meaningful when printed on a chart, standard reference planes for these measurements have been established. In general, heights and elevations are given on a chart with reference to a standard *high*-water reference plane, and heights of tide and charted depths of water are given with respect to a standard *low*-water reference plane. The reference planes used on charts produced by the United States are listed and described below, in the order of their relative heights from the sea bottom. All are derived from various computed mean or average water levels, measured with respect to the bottom, as observed and recorded over a number of years.

First, the high-water reference planes:

Mean high-water springs (MHWS)   The highest of all high-water reference planes, is the average height of all spring tide high-water levels.

Mean higher high water (MHHW)   The average of the higher of the high-water levels occurring during each tidal day at a location, measured over a 19-year period.

Mean high water (MHW)   The average of all high-tide water levels, measured over a 19-year period. It is the high-water reference plane used on most charts produced by the United States for the basis of measurement of heights, elevations, and bridge clearances.

Mean high-water neaps (MHWN)   The lowest of all high-water reference planes in common use; the average recorded height of all neap tide high-water levels.

Next, the low-water reference planes:

Mean low water neaps (MLWN)   The highest of all common low-water reference planes; the average height of all neap tide low-water levels.

Mean low water (MLW)   The average height of all low-tide water levels observed over a period of 19 years. Until 1980 it was the low-water reference plane used on charts of the U.S. Atlantic and Gulf coasts.

Mean lower low water (MLLW)   The average of the lower of the low water levels experienced at a location over a 19-year period. It has been the low-water reference plane used for many years for charts of the U.S. Pacific coast, and beginning in 1980, for charts of the U.S. Atlantic and Gulf coasts, as a basis of measurement of charted depth and height of tide.

Mean low water springs (MLWS)   The lowest of all low-water reference planes; the average of all spring tide low-water levels. It is the sounding datum on which most water depths of foreign charts are based.

As an example to help visualize the relationship between these reference planes, the charted depth, height of tide, and clearance under a bridge, suppose the navigator had a chart that used mean high water for the high-water reference plane from which charted heights and bridge clearances were reckoned, and mean low water, for water depths and heights of tide. The physical relationships existing would then be as illustrated in Figure 11–4.

By definition, the mean high-water and mean low-water reference planes represent the average limits within which the water level

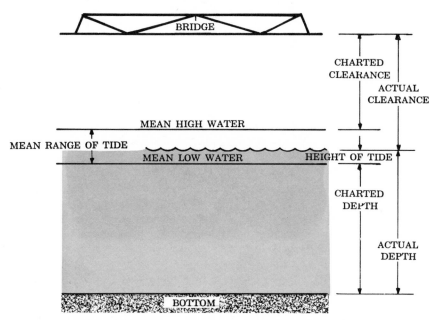

Figure 11–4.   *Relationship of terms applying to water depth and vertical clearance.*

would normally be located. The *mean range of tide* is the vertical distance between these two planes; it would represent the average range of tide at that location. The term *mean sea level* is sometimes used to denote the average depth of a body of water; it lies about midway between mean high and mean low water.

The navigator must always remember that the actual water level will occasionally fall below the mean low-water plane, particularly around the time of spring tides. Height of tide in this situation is *negative*, usually denoted by placing a minus sign in front of the height of tide figure. Thus, if he is using a chart having mean low water as the sounding datum, the actual water depth will sometimes be *less* than that indicated on the chart. Likewise, actual vertical clearance will sometimes be smaller than that indicated on a chart using mean high water as its reference plane for heights and clearances. On the other hand, if his chart is based on water levels at mean high and mean low water springs, the navigator will tend to regard charted clearances and depths with more confidence, since the actual water level will seldom exceed these limits.

Predicting Height of Tide

There are many situations that arise during the practice of piloting that require the navigator to predict the height of tide at a given time in the future. When a ship is entering port, the navigator must know the minimum depth of water through which the ship will pass; some ports must be entered over sand bars or shoals that can be safely transited by a deep draft vessel only at high water. If a ship must pass under a bridge, the navigator will want to know exactly how much vertical clearance exists at the expected time of passage. The ship's deck officers will want to know the depth of water at an intended anchorage so they can recommend the proper scope of chain to be let out, or the height and range of tide expected at a pier mooring so as to be able to allow the proper amount of slack in the mooring lines. For these and other purposes, the navigator refers to a set of four publications known as the *Tide Tables*, previously introduced in Chapter 5 on navigational publications.

The *Tide Tables* are arranged geographically, with one volume covering each of the following four areas: East Coast, North and South America; West Coast, North and South America; Europe and West Coast of Africa; and Central and Western Pacific and Indian oceans. A new set of tables is published each year by the National Ocean Service, and distributed to Navy users by the Defense Mapping Agency Office of Distribution Services. Collectively, the *Tide Tables* contain daily predictions of the times of high and low tides at some 190 major

reference ports throughout the world, and in addition, they list differences in times of tides from specific reference ports for an additional 5,000 locations referred to as *subordinate stations*. The navigator can construct a daily tide table, compute the height of tide at a given time, or compute the time frame within which the tide will be above or below a desired level, for each one of the 5,000 subordinate stations, by using the tabulated tide-difference data in conjunction with the daily predictions at the proper reference port.

Layout of the *Tide Tables*

Each volume of the *Tide Tables* is made up of seven tables. The first three are of primary interest for making tide predictions; the latter four consist of astronomical tables for predicting times of sunrise and sunset and moonrise and moonset at selected latitudes and reference ports, and a feet-to-meters conversion table useful when tides have been computed in feet for use with charts expressing soundings in meters.

Table 1 lists the times and heights of tide in both feet and meters at each high water and low water, in chronological order, for each day of the year for each of the reference locations, called *reference stations*, used in that volume. The datum or reference plane from which the predicted heights are reckoned is the same as that used for the major large-scale charts of the area. Figure 11–5 opposite is a typical page from Table 1 of the volume of the *Tide Tables* covering the East Coast of North and South America. High and low tides are identified by a comparison of consecutive heights of tide listed. Usually, two high and two low tides will appear on any given day, and there will be a high or low tide about every six hours. Occasionally, because the tidal day is 24 hours and 50 minutes long, there will be only three listings for a given day, as on 10 October in Figure 11–5. Here, a low tide appears just before midnight on the preceding day, and just after midnight on the succeeding day, with the result that only one low tide occurs on the 10th. The negative sign appearing before the height of tide figure indicates that this low tide falls below the tidal reference plane, which for this volume of the *Tide Tables* is mean low water. Times of tides given in Table 1 are those of the standard time zone in which the port is located. If the port is keeping daylight savings time, each tide time prediction must be adjusted by adding one hour to the time listed.

Table 2 of each volume of the *Tide Tables* contains a listing of tide time and height-difference data, as well as other useful information, for each of the *subordinate stations* located within the area of coverage of that particular volume. Data in Table 2 is arranged in geographical

NEW YORK (The Battery), N.Y.,

Times and Heights of High and Low Waters

OCTOBER

Day	Time h.m.	ft.	m.	Day	Time h.m.	ft.	m.
1 Th	0400	0.1	0.0	16 F	0353	-0.6	-0.2
	1019	4.9	1.5		1014	5.8	1.8
	1627	0.3	0.1		1635	-0.6	-0.2
	2235	4.3	1.3		2245	4.8	1.5
2 F	0429	0.4	0.1	17 Sa	0441	-0.3	-0.1
	1056	4.7	1.4		1112	5.6	1.7
	1703	0.6	0.2		1728	-0.3	-0.1
	2314	4.0	1.2		2346	4.5	1.4
3 Sa	0456	0.7	0.2	18 Su	0533	0.0	0.0
	1136	4.5	1.4		1213	5.4	1.6
	1741	0.8	0.2		1833	0.0	0.0
	2357	3.8	1.2				
4 Su	0520	1.0	0.3	19 M	0049	4.3	1.3
	1215	4.4	1.3		0642	0.4	0.1
	1829	1.1	0.3		1313	5.1	1.6
					1945	0.2	0.1
5 M	0044	3.6	1.1	20 Tu	0151	4.2	1.3
	0549	1.2	0.4		0802	0.6	0.2
	1258	4.3	1.3		1414	4.9	1.5
	1950	1.2	0.4		2054	0.2	0.1
6 Tu	0135	3.5	1.1	21 W	0254	4.2	1.3
	0705	1.4	0.4		0912	0.6	0.2
	1347	4.2	1.3		1516	4.8	1.5
	2100	1.1	0.3		2154	0.1	0.0
7 W	0234	3.5	1.1	22 Th	0359	4.3	1.3
	0904	1.3	0.4		1013	0.4	0.1
	1452	4.2	1.3		1620	4.7	1.4
	2156	0.9	0.3		2246	-0.1	0.0
8 Th	0343	3.7	1.1	23 F	0458	4.5	1.4
	1007	1.1	0.3		1108	0.2	0.1
	1558	4.4	1.3		1718	4.8	1.5
	2246	0.5	0.2		2334	-0.2	-0.1
9 F	0444	4.0	1.2	24 Sa	0551	4.7	1.4
	1057	0.7	0.2		1158	0.0	0.0
	1657	4.6	1.4		1808	4.8	1.5
	2331	0.2	0.1				
10 Sa	0534	4.4	1.3	25 Su	0019	-0.3	-0.1
	1148	0.3	0.1		0635	5.0	1.5
	1750	4.9	1.5		1245	-0.1	0.0
					1853	4.8	1.5
11 Su	0016	-0.1	0.0	26 M	0102	-0.3	-0.1
	0621	4.9	1.5		0717	5.1	1.6
	1235	-0.1	0.0		1328	-0.2	-0.1
	1835	5.2	1.6		1931	4.8	1.5
12 M	0059	-0.4	-0.1	27 Tu	0143	-0.3	-0.1
	0704	5.3	1.6		0754	5.1	1.6
	1323	-0.4	-0.1		1410	-0.2	-0.1
	1921	5.3	1.6		2012	4.6	1.4
13 Tu	0144	-0.6	-0.2	28 W	0221	-0.2	-0.1
	0746	5.6	1.7		0831	5.1	1.6
	1412	-0.6	-0.2		1449	-0.2	-0.1
	2007	5.4	1.6		2048	4.5	1.4
14 W	0226	-0.8	-0.2	29 Th	0257	-0.1	0.0
	0831	5.8	1.8		0906	5.0	1.5
	1458	-0.8	-0.2		1527	-0.1	0.0
	2054	5.3	1.6		2126	4.2	1.3
15 Th	0310	-0.8	-0.2	30 F	0329	0.1	0.0
	0921	5.9	1.8		0940	4.8	1.5
	1544	-0.8	-0.2		1602	0.1	0.0
	2146	5.1	1.6		2203	4.0	1.2
				31 Sa	0358	0.3	0.1
					1014	4.6	1.4
					1637	0.3	0.1
					2243	3.8	1.2

NOVEMBER

Day	Time h.m.	ft.	m.	Day	Time h.m.	ft.	m.
1 Su	0422	0.6	0.2	16 M	0520	-0.2	-0.1
	1050	4.5	1.4		1154	5.2	1.6
	1712	0.5	0.2		1813	-0.4	-0.1
	2327	3.6	1.1				
2 M	0445	0.8	0.2	17 Tu	0034	4.2	1.3
	1127	4.3	1.3		0625	0.2	0.1
	1751	0.7	0.2		1254	4.9	1.5
					1919	-0.1	0.0
3 Tu	0013	3.5	1.1	18 W	0133	4.2	1.3
	0517	1.0	0.3		0740	0.4	0.1
	1213	4.2	1.3		1351	4.7	1.4
	1852	0.8	0.2		2026	0.0	0.0
4 W	0105	3.4	1.0	19 Th	0233	4.1	1.2
	0605	1.2	0.4		0851	0.5	0.2
	1305	4.1	1.2		1451	4.4	1.3
	2009	0.8	0.2		2125	-0.1	0.0
5 Th	0159	3.5	1.1	20 F	0333	4.2	1.3
	0809	1.2	0.4		0951	0.4	0.1
	1401	4.1	1.2		1551	4.3	1.3
	2114	0.6	0.2		2217	-0.1	0.0
6 F	0300	3.7	1.1	21 Sa	0430	4.3	1.3
	0929	0.9	0.3		1045	0.2	0.1
	1509	4.2	1.3		1649	4.2	1.3
	2205	0.3	0.1		2305	-0.2	-0.1
7 Sa	0401	4.1	1.2	22 Su	0523	4.5	1.4
	1026	0.6	0.2		1133	0.0	0.0
	1615	4.4	1.3		1740	4.2	1.3
	2252	0.0	0.0		2350	-0.3	-0.1
8 Su	0457	4.5	1.4	23 M	0609	4.7	1.4
	1119	0.1	0.0		1220	-0.1	0.0
	1715	4.6	1.4		1824	4.2	1.3
	2339	-0.3	-0.1				
9 M	0547	5.0	1.5	24 Tu	0032	-0.3	-0.1
	1210	-0.3	-0.1		0650	4.9	1.5
	1806	4.8	1.5		1304	-0.2	-0.1
					1906	4.2	1.3
10 Tu	0025	-0.6	-0.2	25 W	0112	-0.3	-0.1
	0635	5.4	1.6		0728	4.9	1.5
	1301	-0.7	-0.2		1347	-0.3	-0.1
	1856	5.0	1.5		1946	4.2	1.3
11 W	0112	-0.8	-0.2	26 Th	0152	-0.2	-0.1
	0723	5.8	1.8		0803	4.9	1.5
	1351	-0.9	-0.3		1426	-0.3	-0.1
	1944	5.0	1.5		2023	4.1	1.2
12 Th	0200	-0.9	-0.3	27 F	0231	-0.1	0.0
	0810	5.9	1.8		0838	4.8	1.5
	1441	-1.1	-0.3		1505	-0.3	-0.1
	2036	4.9	1.5		2102	3.9	1.2
13 F	0248	-0.9	-0.3	28 Sa	0305	0.0	0.0
	0900	5.9	1.8		0913	4.7	1.4
	1530	-1.1	-0.3		1543	-0.2	-0.1
	2131	4.8	1.5		2139	3.7	1.1
14 Sa	0336	-0.8	-0.2	29 Su	0337	0.2	0.1
	0955	5.8	1.8		0944	4.6	1.4
	1620	-0.9	-0.3		1618	-0.1	0.0
	2232	4.6	1.4		2221	3.6	1.1
15 Su	0426	-0.5	-0.2	30 M	0403	0.3	0.1
	1055	5.5	1.7		1016	4.4	1.3
	1714	-0.7	-0.2		1653	0.1	0.0
	2336	4.4	1.3		2303	3.5	1.1

DECEMBER

Day	Time h.m.	ft.	m.	Day	Time h.m.	ft.	m.
1 Tu	0429	0.5	0.2	16 W	0014	4.2	1.3
	1057	4.3	1.3		0603	0.0	0.0
	1728	0.2	0.1		1231	4.7	1.4
	2348	3.4	1.0		1847	-0.4	-0.1
2 W	0501	0.6	0.2	17 Th	0108	4.1	1.2
	1143	4.2	1.3		0711	0.2	0.1
	1807	0.4	0.1		1324	4.4	1.3
					1949	-0.2	-0.1
3 Th	0034	3.5	1.1	18 F	0203	4.1	1.2
	0546	0.8	0.2		0820	0.4	0.1
	1229	4.1	1.2		1418	4.1	1.2
	1908	0.4	0.1		2049	-0.1	0.0
4 F	0122	3.6	1.1	19 Sa	0259	4.0	1.2
	0706	0.8	0.2		0921	0.4	0.1
	1324	4.1	1.2		1513	3.8	1.2
	2021	0.3	0.1		2143	-0.1	0.0
5 Sa	0217	3.8	1.2	20 Su	0354	4.1	1.2
	0848	0.7	0.2		1016	0.3	0.1
	1426	4.0	1.2		1612	3.7	1.1
	2122	0.1	0.0		2231	-0.1	0.0
6 Su	0317	4.1	1.2	21 M	0449	4.2	1.3
	0955	0.3	0.1		1105	0.1	0.0
	1532	4.1	1.2		1706	3.7	1.1
	2215	-0.2	-0.1		2317	-0.1	0.0
7 M	0419	4.5	1.4	22 Tu	0535	4.4	1.3
	1052	-0.1	0.0		1153	0.0	0.0
	1639	4.2	1.3		1755	3.7	1.1
	2305	-0.5	-0.2				
8 Tu	0518	4.9	1.5	23 W	0000	-0.2	-0.1
	1147	-0.4	-0.1		0620	4.5	1.4
	1740	4.4	1.3		1238	-0.2	-0.1
	2355	-0.7	-0.2		1840	3.8	1.2
9 W	0612	5.4	1.6	24 Th	0045	-0.2	-0.1
	1240	-0.8	-0.2		0701	4.6	1.4
	1835	4.5	1.4		1322	-0.3	-0.1
					1922	3.8	1.2
10 Th	0048	-0.9	-0.3	25 F	0127	-0.2	-0.1
	0702	5.7	1.7		0739	4.7	1.4
	1333	-1.1	-0.3		1404	-0.4	-0.1
	1928	4.6	1.4		2002	3.8	1.2
11 F	0140	-1.0	-0.3	26 Sa	0206	-0.2	-0.1
	0753	5.8	1.8		0815	4.7	1.4
	1424	-1.2	-0.4		1445	-0.5	-0.2
	2022	4.6	1.4		2042	3.7	1.1
12 Sa	0230	-1.1	-0.3	27 Su	0245	-0.1	0.0
	0844	5.8	1.8		0851	4.7	1.4
	1515	-1.3	-0.4		1524	-0.5	-0.2
	2118	4.5	1.4		2120	3.7	1.1
13 Su	0321	-1.0	-0.3	28 M	0319	-0.1	0.0
	0939	5.6	1.7		0926	4.6	1.4
	1605	-1.2	-0.4		1600	-0.4	-0.1
	2216	4.4	1.3		2200	3.6	1.1
14 M	0411	-0.7	-0.2	29 Tu	0351	0.0	0.0
	1037	5.4	1.6		1000	4.5	1.4
	1656	-1.0	-0.3		1632	-0.3	-0.1
	2316	4.3	1.3		2243	3.6	1.1
15 Tu	0504	-0.4	-0.1	30 W	0419	0.1	0.0
	1135	5.1	1.6		1037	4.3	1.3
	1749	-0.7	-0.2		1704	-0.2	-0.1
					2322	3.6	1.1
				31 Th	0453	0.2	0.1
					1120	4.2	1.3
					1741	-0.1	0.0

Time meridian 75° W. 0000 is midnight. 1200 is noon.
Heights are referred to mean low water which is the chart datum of soundings.

Figure 11–5. A daily prediction page for New York, Tide Tables, *Table 1.*

TABLE 2. - TIDAL DIFFERENCES AND OTHER CONSTANTS

NO.	PLACE	POSITION Lat.	POSITION Long.	DIFFERENCES Time High Water	DIFFERENCES Time Low Water	DIFFERENCES Height High Water	DIFFERENCES Height Low Water	RANGES Mean	RANGES Spring	Mean Tide Level
		° ' N	° ' W	h. m.	h. m.	ft	ft	ft	ft	ft
	New York, East River Time meridian, 75°W			on WILLETS POINT, p.52						
1283	Lawrence Point--------------------------	40 47	73 55	-0 03	+0 13	-0.7	0.0	6.4	7.6	3.2
1285	Wolcott Avenue--------------------------	40 47	73 55	-0 03	+0 13	-1.0	0.0	6.1	7.2	3.0
				on NEW YORK, p.56						
1287	Pot Cove, Astoria----------------------	40 47	73 56	+2 20	+2 29	+0.8	0.0	5.3	6.3	2.6
1289	Hell Gate, Hallets Point---------------	40 47	73 56	+2 00	+2 04	+0.6	0.0	5.1	6.1	2.5
1291	Horns Hook, East 90th Street-----------	40 47	73 57	+1 50	+1 30	+0.3	0.0	4.8	5.8	2.4
1293	Welfare Island, north end--------------	40 46	73 56	+1 45	+1 25	+0.3	0.0	4.8	5.8	2.4
1295	37th Avenue, Long Island City----------	40 46	73 57	+1 30	+1 10	0.0	0.0	4.5	5.5	2.2
1297	East 41st Street, New York City--------	40 45	73 58	+1 20	+0 56	-0.2	0.0	4.3	5.2	2.1
1299	Hunters Point, Newtown Creek-----------	40 44	73 57	+1 18	+0 53	-0.4	0.0	4.1	4.9	2.0
1301	English Kills entrance, Newtown Creek---	40 43	73 55	+1 30	+1 04	-0.3	0.0	4.2	5.0	2.1
1303	East 27th Street, Bellevue Hospital-----	40 44	73 58	+1 08	+1 03	-0.3	0.0	4.2	5.0	2.1
1305	East 19th Street, New York City--------	40 44	73 58	+1 02	+0 58	-0.4	0.0	4.1	4.9	2.0
1307	North 3d Street, Brooklyn--------------	40 43	73 58	+0 55	+0 42	-0.4	0.0	4.1	4.9	2.0
1309	Williamsburg Bridge--------------------	40 43	73 58	+0 52	+0 38	-0.4	0.0	4.1	4.9	2.0
1311	Wallabout Bay-------------------------	40 42	73 59	+0 50	+0 35	-0.4	0.0	4.1	4.9	2.0
1313	Brooklyn Bridge-----------------------	40 42	74 00	+0 13	+0 07	-0.2	0.0	4.3	5.2	2.1
	Harlem River									
1315	East 110th Street, New York City----	40 47	73 56	+1 52	+1 35	+0.6	0.0	5.1	6.1	2.6
1317	Willis Avenue Bridge--------------	40 48	73 56	+1 47	+1 30	+0.5	0.0	5.0	6.0	2.5
1319	Madison Avenue Bridge-------------	40 49	73 56	+1 52	+1 35	+0.4	0.0	4.9	5.9	2.4
1321	Central Bridge--------------------	40 50	73 56	+1 52	+1 35	+0.2	0.0	4.7	5.7	2.3
1323	Washington Bridge-----------------	40 51	73 56	+1 52	+1 35	-0.1	0.0	4.4	5.2	2.2
1325	207th Street Bridge---------------	40 52	73 55	+1 40	+1 30	-0.5	0.0	4.0	4.8	2.0
1327	Broadway Bridge-------------------	40 52	73 55	+1 20	+1 20	-0.7	0.0	3.8	4.6	1.9
1329	Spuyten Duyvil Bridge-------------	40 53	73 56	+1 01	+1 03	-0.9	0.0	3.6	4.3	1.8
	Long Island Sound, South Side			on WILLETS POINT, p.52						
1331	WILLETS POINT--------------------------	40 48	73 47	Daily predictions			0.0	7.1	8.3	3.5
1333	Hewlett Point--------------------------	40 50	73 45	-0 03	-0 03	0.0	0.0	7.1	8.3	3.5
1335	Port Washington, Manhasset Bay---------	40 50	73 42	-0 01	+0 11	+0.2	0.0	7.3	8.6	3.6
1337	Execution Rocks------------------------	40 53	73 44	-0 06	-0 08	+0.2	0.0	7.3	8.6	3.6
1339	Glen Cove, Hempstead Harbor------------	40 52	73 39	-0 11	-0 06	+0.2	0.0	7.3	8.6	3.6
	Oyster Bay			on BRIDGEPORT, p.48						
1341	Oyster Bay Harbor-------------------	40 53	73 32	+0 08	+0 11	+0.6	0.0	7.3	8.4	3.6
1343	Bayville Bridge--------------------	40 54	73 33	+0 13	+0 18	+0.7	0.0	7.4	8.5	3.7
1345	Cold Spring Harbor-----------------	40 52	73 28	+0 08	+0 06	+0.7	0.0	7.4	8.5	3.7
1347	Eatons Neck Point----------------------	40 57	73 24	+0 03	+0 06	+0.4	0.0	7.1	8.2	3.6
1349	Lloyd Harbor entrance, Huntington Bay---	40 55	73 26	+0 03	+0 01	+0.7	0.0	7.4	8.5	3.7
1351	Northport, Northport Bay---------------	40 54	73 21	+0 03	+0 06	+0.6	0.0	7.3	8.4	3.6
1353	Nissequogue River entrance-------------	40 54	73 14	-0 03	-0 06	+0.3	0.0	7.0	8.0	3.5
1355	Stony Brook, Smithtown Bay-------------	40 55	73 09	+0 08	+0 08	-0.6	0.0	6.1	7.0	3.0
1357	Stratford Shoal------------------------	41 04	73 06	-0 05	-0 09	-0.1	0.0	6.6	7.6	3.3
1359	Port Jefferson Harbor entrance---------	40 58	73 05	+0 03	-0 01	-0.1	0.0	6.6	7.6	3.3
1361	Port Jefferson-------------------------	40 57	73 05	+0 06	+0 03	-0.1	0.0	6.6	7.6	3.3
1363	Setauket Harbor------------------------	40 57	73 06	+0 04	+0 09	0.0	0.0	6.7	7.7	3.3
1365	Conscience Bay entrance (Narrows)------	40 58	73 07	+0 02	+0 02	0.0	0.0	6.7	7.7	3.3
1367	Mount Sinai Harbor--------------------	40 58	73 02	+0 05	+0 16	-0.7	0.0	6.0	6.9	3.0
1369	Herod Point---------------------------	40 58	72 50	-0 07	-0 16	-0.8	0.0	5.9	6.8	2.9
1370	Northville----------------------------	40 59	72 45	-0 02	-0 05	-1.3	0.0	5.4	6.2	2.7
1371	Mattituck Inlet-----------------------	41 01	72 34	+0 05	-0 06	-1.5	0.0	5.2	6.0	2.6
1373	Horton Point--------------------------	41 05	72 27	-0 20	-0 35	*0.60	*0.60	4.0	4.6	2.0
1374	Hashamomuck Beach----------------------	41 06	72 24	+0 04	-0 15	*0.63	*0.63	4.2	4.8	2.1
1375	Truman Beach--------------------------	41 08	72 19	-0 42	-0 52	*0.51	*0.51	3.4	3.9	1.7
				on NEW LONDON, p.44						
1377	Plum Gut Harbor, Plum Island----------	41 10	72 12	+0 27	+0 16	0.0	0.0	2.6	3.1	1.3
1379	Little Gull Island--------------------	41 12	72 06	+0 12	-0 22	-0.4	0.0	2.2	2.6	1.1
	Shelter Island Sound									
1381	Orient-----------------------------	41 08	72 18	+0 36	+0 36	-0.1	0.0	2.5	3.0	1.2
1383	Greenport--------------------------	41 06	72 22	+1 04	+0 49	-0.2	0.0	2.4	2.9	1.2
1385	Southhold--------------------------	41 04	72 25	+1 43	+1 33	-0.3	0.0	2.3	2.7	1.1
1387	Noyack Bay--------------------------	41 00	72 20	+2 05	+1 44	-0.3	0.0	2.3	2.7	1.1
1389	Sag Harbor--------------------------	41 00	72 18	+0 59	+0 48	-0.1	0.0	2.5	3.0	1.2
1391	Cedar Point-------------------------	41 02	72 16	+0 44	+0 27	-0.1	0.0	2.5	3.0	1.2
	Peconic Bays									
1393	New Suffolk-------------------------	41 00	72 28	+2 26	+2 11	0.0	0.0	2.6	3.1	1.3
1395	South Jamesport---------------------	40 56	72 35	+2 32	+2 40	+0.1	0.0	2.7	3.2	1.3
1397	Shinnecock Canal--------------------	40 54	72 30	+2 33	+2 31	-0.2	0.0	2.4	2.9	1.2
1399	Threemile Harbor ent., Gardiners Bay----	41 02	72 11	+0 21	+0 02	-0.2	0.0	2.4	2.9	1.2
1401	Promised Land, Napeague Bay----------	41 00	72 05	-0 14	-0 08	-0.3	0.0	2.3	2.7	1.1
1403	Montauk Harbor entrance--------------	41 04	71 56	-0 25	-0 16	-0.7	0.0	1.9	2.3	0.9
1405	Montauk, Fort Pond Bay---------------	41 03	71 58	-0 29	-0 24	-0.5	0.0	2.1	2.5	1.1
1407	Montauk Point, north side-----------	41 04	71 52	-1 13	-1 31	-0.6	0.0	2.0	2.4	1.0

Endnotes can be found at the end of table 2.

Figure 11–6. A subordinate data page, Tide Tables, Table 2.

	Time from the nearest high water or low water														
h. m.	h. m.	h. m.	h. m.	h. m.	h. m.	h. m.	h. m.	h. m.	h. m.	h. m.	h. m.	h. m.	h. m.	h. m.	h. m.
4 00	0 08	0 16	0 24	0 32	0 40	0 48	0 56	1 04	1 12	1 20	1 28	1 36	1 44	1 52	2 00
4 20	0 09	0 17	0 26	0 35	0 43	0 52	1 01	1 09	1 18	1 27	1 35	1 44	1 53	2 01	2 10
4 40	0 09	0 19	0 28	0 37	0 47	0 56	1 05	1 15	1 24	1 33	1 43	1 52	2 01	2 11	2 20
5 00	0 10	0 20	0 30	0 40	0 50	1 00	1 10	1 20	1 30	1 40	1 50	2 00	2 10	2 20	2 30
5 20	0 11	0 21	0 32	0 43	0 53	1 04	1 15	1 25	1 36	1 47	1 57	2 08	2 19	2 29	2 40
5 40	0 11	0 23	0 34	0 45	0 57	1 08	1 19	1 31	1 42	1 53	2 05	2 16	2 27	2 39	2 50
6 00	0 12	0 24	0 36	0 48	1 00	1 12	1 24	1 36	1 48	2 00	2 12	2 24	2 36	2 48	3 00
6 20	0 13	0 25	0 38	0 51	1 03	1 16	1 29	1 41	1 54	2 07	2 19	2 32	2 45	2 57	3 10
6 40	0 13	0 27	0 40	0 53	1 07	1 20	1 33	1 47	2 00	2 13	2 27	2 40	2 53	3 07	3 20
7 00	0 14	0 28	0 42	0 56	1 10	1 24	1 38	1 52	2 06	2 20	2 34	2 48	3 02	3 16	3 30•
7 20	0 15	0 29	0 44	0 59	1 13	1 28	1 43	1 57	2 12	2 27	2 41	2 56	3 11	3 25	3 40
7 40	0 15	0 31	0 46	1 01	1 17	1 32	1 47	2 03	2 18	2 33	2 49	3 04	3 19	3 35	3 50
8 00	0 16	0 32	0 48	1 04	1 20	1 36	1 52	2 08	2 24	2 40	2 56	3 12	3 28	3 44	4 00
8 20	0 17	0 33	0 50	1 07	1 23	1 40	1 57	2 13	2 30	2 47	3 03	3 20	3 37	3 53	4 10
8 40	0 17	0 35	0 52	1 09	1 27	1 44	2 01	2 19	2 36	2 53	3 11	3 28	3 45	4 03	4 20
9 00	0 18	0 36	0 54	1 12	1 30	1 48	2 06	2 24	2 42	3 00	3 18	3 36	3 54	4 12	4 30
9 20	0 19	0 37	0 56	1 15	1 33	1 52	2 11	2 29	2 48	3 07	3 25	3 44	4 03	4 21	4 40
9 40	0 19	0 39	0 58	1 17	1 37	1 56	2 15	2 35	2 54	3 13	3 33	3 52	4 11	4 31	4 50
10 00	0 20	0 40	1 00	1 20	1 40	2 00	2 20	2 40	3 00	3 20	3 40	4 00	4 20	4 40	5 00
10 20	0 21	0 41	1 02	1 23	1 43	2 04	2 25	2 45	3 06	3 27	3 47	4 08	4 29	4 49	5 10
10 40	0 21	0 43	1 04	1 25	1 47	2 08	2 29	2 51	3 12	3 33	3 55	4 16	4 37	4 59	5 20

Duration of rise or fall, see footnote

	Correction to height														
Ft.	Ft.	Ft.	Ft.	Ft.	Ft.	Ft.	Ft.	Ft.	Ft.	Ft.	Ft.	Ft.	Ft.	Ft.	Ft.
0.5	0.0	0.0	0.0	0.0	0.0	0.0	0.1	0.1	0.1	0.1	0.1	0.2	0.2	0.2	0.2
1.0	0.0	0.0	0.0	0.0	0.1	0.1	0.1	0.2	0.2	0.2	0.3	0.3	0.4	0.4	0.5
1.5	0.0	0.0	0.0	0.1	0.1	0.1	0.2	0.2	0.3	0.4	0.4	0.5	0.6	0.7	0.8
2.0	0.0	0.0	0.0	0.1	0.1	0.2	0.3	0.3	0.4	0.5	0.6	0.7	0.8	0.9	1.0
2.5	0.0	0.0	0.1	0.1	0.2	0.2	0.3	0.4	0.5	0.6	0.7	0.9	1.0	1.1	1.2
3.0	0.0	0.0	0.1	0.1	0.2	0.3	0.4	0.5	0.6	0.8	0.9	1.0	1.2	1.3	1.5
3.5	0.0	0.0	0.1	0.2	0.2	0.3	0.4	0.6	0.7	0.9	1.0	1.2	1.4	1.6	1.8•
4.0	0.0	0.0	0.1	0.2	0.3	0.4	0.5	0.7	0.8	1.0	1.2	1.4	1.6	1.8	2.0
4.5	0.0	0.0	0.1	0.2	0.3	0.4	0.6	0.7	0.9	1.1	1.3	1.6	1.8	2.0	2.2
5.0	0.0	0.1	0.1	0.2	0.3	0.5	0.6	0.8	1.0	1.2	1.5	1.7	2.0	2.2	2.5
5.5	0.0	0.1	0.1	0.2	0.4	0.5	0.7	0.9	1.1	1.4	1.6	1.9	2.2	2.5	2.8
6.0	0.0	0.1	0.1	0.3	0.4	0.6	0.8	1.0	1.2	1.5	1.8	2.1	2.4	2.7	3.0
6.5	0.0	0.1	0.2	0.3	0.4	0.6	0.8	1.1	1.3	1.6	1.9	2.2	2.6	2.9	3.2
7.0	0.0	0.1	0.2	0.3	0.5	0.7	0.9	1.2	1.4	1.8	2.1	2.4	2.8	3.1	3.5
7.5	0.0	0.1	0.2	0.3	0.5	0.7	1.0	1.2	1.5	1.9	2.2	2.6	3.0	3.4	3.8
8.0	0.0	0.1	0.2	0.3	0.5	0.8	1.0	1.3	1.6	2.0	2.4	2.8	3.2	3.6	4.0
8.5	0.0	0.1	0.2	0.4	0.6	0.8	1.1	1.4	1.8	2.1	2.5	2.9	3.4	3.8	4.2
9.0	0.0	0.1	0.2	0.4	0.6	0.9	1.2	1.5	1.9	2.2	2.7	3.1	3.6	4.0	4.5
9.5	0.0	0.1	0.2	0.4	0.6	0.9	1.2	1.6	2.0	2.4	2.8	3.3	3.8	4.3	4.8
10.0	0.0	0.1	0.2	0.4	0.7	1.0	1.3	1.7	2.1	2.5	3.0	3.5	4.0	4.5	5.0
10.5	0.0	0.1	0.3	0.5	0.7	1.0	1.3	1.7	2.2	2.6	3.1	3.6	4.2	4.7	5.2
11.0	0.0	0.1	0.3	0.5	0.7	1.1	1.4	1.8	2.3	2.8	3.3	3.8	4.4	4.9	5.5
11.5	0.0	0.1	0.3	0.5	0.8	1.1	1.5	1.9	2.4	2.9	3.4	4.0	4.6	5.1	5.8
12.0	0.0	0.1	0.3	0.5	0.8	1.1	1.5	2.0	2.5	3.0	3.6	4.1	4.8	5.4	6.0
12.5	0.0	0.1	0.3	0.5	0.8	1.2	1.6	2.1	2.6	3.1	3.7	4.3	5.0	5.6	6.2
13.0	0.0	0.1	0.3	0.6	0.9	1.2	1.7	2.2	2.7	3.2	3.9	4.5	5.1	5.8	6.5
13.5	0.0	0.1	0.3	0.6	0.9	1.3	1.7	2.2	2.8	3.4	4.0	4.7	5.3	6.0	6.8
14.0	0.0	0.2	0.3	0.6	0.9	1.3	1.8	2.3	2.9	3.5	4.2	4.8	5.5	6.3	7.0
14.5	0.0	0.2	0.4	0.6	1.0	1.4	1.9	2.4	3.0	3.6	4.3	5.0	5.7	6.5	7.2
15.0	0.0	0.2	0.4	0.6	1.0	1.4	1.9	2.5	3.1	3.8	4.4	5.2	5.9	6.7	7.5
15.5	0.0	0.2	0.4	0.7	1.0	1.5	2.0	2.6	3.2	3.9	4.6	5.4	6.1	6.9	7.8
16.0	0.0	0.2	0.4	0.7	1.1	1.5	2.1	2.6	3.3	4.0	4.7	5.5	6.3	7.2	8.0
16.5	0.0	0.2	0.4	0.7	1.1	1.6	2.1	2.7	3.4	4.1	4.9	5.7	6.5	7.4	8.2
17.0	0.0	0.2	0.4	0.7	1.1	1.6	2.2	2.8	3.5	4.2	5.0	5.9	6.7	7.6	8.5
17.5	0.0	0.2	0.4	0.8	1.2	1.7	2.2	2.9	3.6	4.4	5.2	6.0	6.9	7.8	8.8
18.0	0.0	0.2	0.4	0.8	1.2	1.7	2.3	3.0	3.7	4.5	5.3	6.2	7.1	8.1	9.0
18.5	0.1	0.2	0.5	0.8	1.2	1.8	2.4	3.1	3.8	4.6	5.5	6.4	7.3	8.3	9.2
19.0	0.1	0.2	0.5	0.8	1.3	1.8	2.4	3.1	3.9	4.8	5.6	6.6	7.5	8.5	9.5
19.5	0.1	0.2	0.5	0.8	1.3	1.9	2.5	3.2	4.0	4.9	5.8	6.7	7.7	8.7	9.8
20.0	0.1	0.2	0.5	0.9	1.3	1.9	2.6	3.3	4.1	5.0	5.9	6.9	7.9	9.0	10.0

Range of tide, see footnote

Obtain from the predictions the high water and low water, one of which is before and the other after the time for which the height is required. The difference between the times of occurrence of these tides is the duration of rise or fall, and the difference between their heights is the range of tide for the above table. Find the difference between the nearest high or low water and the time for which the height is required.

Enter the table with the duration of rise or fall, printed in heavy-faced type, which most nearly agrees with the actual value, and on that horizontal line find the time from the nearest high or low water which agrees most nearly with the corresponding actual difference. The correction sought is in the column directly below, on the line with the range of tide.

When the nearest tide is high water, subtract the correction.

When the nearest tide is low water, add the correction.

Figure 11–7.   Table 3 of a Tide Table.

sequence, and an alphabetical index of subordinate station names is located in the back of each volume. A sample page from Table 2 of an *East Coast Tide Tables* appears in Figure 11–6 on page 192. As can be seen by an inspection of this table, information given for each subordinate station includes its latitude and longitude, differences in times and heights of tide between the subordinate and reference stations, and its mean range of tide in feet.

Table 3 of the *Tide Tables* is used primarily to find the height of tide at a given time, after daily tide predictions for a given location have been obtained; it can also be used to find the time frame within which the tide will be either above or below a desired height. A typical page from Table 3 is depicted in Figure 11–7 on page 193.

Use of the *Tide Tables*

To solve problems involving the predicted height of tide, the navigator makes use of the three tables of the applicable volume of the *Tide Tables*, in conjunction with a standard tide form such as that which appears in Figure 11–8. The top part of the tide form is designed for construction of a daily tide table for a given reference or subordinate station, making use of Tables 1 and 2. The bottom part is for computing the height of the tide at a given time at the designated location, using information from Table 3.

Suppose that the navigator wished to compute the height of tide at Brooklyn Bridge, New York City, at 0900 on 9 November. To solve this problem, the navigator must first construct a daily tide table for Brooklyn Bridge for 9 November, and then use this table in conjunction with Table 3 to compute the height of tide at 0900.

The first step in constructing the complete tide table for Brooklyn Bridge for 9 November is to locate the subordinate station difference data by referring to the index for Table 2 in the back of the proper *Tide Table* volume, in this case the *East Coast of North and South America*. A portion of the index appears in Figure 11–9 on page 196. In the index, the subordinate station number of Brooklyn Bridge is found to be 1313. Turning to this number in Table 2, illustrated in Figure 11–6, the Brooklyn Bridge time and height-difference data is noted and recorded onto the top of the blank tide form. The reference station for Brooklyn Bridge and other nearby subordinate stations appears in bold-faced-type in the "Differences" column above the subordinate station data; it is New York (at the Battery), located on page 56 of Table 1.

Opening Table 1 to the proper page, the daily predictions for the month of November for New York at the Battery are found, as re-

NAVIGATION DEPARTMENT _____

COMPLETE TIDE TABLE

Date: _____

Subordinate Station   _____

Reference Station   _____

HW Time Difference   _____

LW Time Difference   _____

Difference in height of HW   _____

Difference in height of LW   _____

Reference Station   Subordinate Station

HW _____ _____   _____ _____

LW _____ _____   _____ _____

HW _____ _____   _____ _____

LW _____ _____   _____ _____

HW _____ _____   _____ _____

LW _____ _____   _____ _____

HEIGHT OF TIDE AT ANY TIME

Locality: _____ Time: _____ Date: _____

Duration of Rise or Fall: _____

Time from Nearest Tide: _____

Range of Tide: _____

Height of Nearest Tide: _____

Corr. from Table 3:   _____

Height of Tide at: _____

Figure 11–8. A standard form for tide calculations.

produced in Figure 11–5, and the daily predictions for 9 November are recorded on the tide form. In addition, the last tide event on the preceding day is also recorded, for reasons that will become obvious later. At this point, the partially completed tide form now appears as illustrated in Figure 11–10A on page 197.

Figure 11–9. A portion of the Index to Table 2, Tide Tables.

As the final step in preparing a complete tide table for Brooklyn Bridge for 9 November, the tide time and height-difference data are added algebraically to the daily prediction data for New York, with care being taken to use high-water difference figures for high water time and height conversions, and low-water difference figures for low water conversions. At this time, it is noted that when the low water time difference, + 7 minutes, is added to the last tide on the 8th—a low tide at 2339—it does not result in this tide appearing at the bridge on the 9th; this entry is now dropped from further consideration. The reason for recording the last tide event on the preceding day at the reference station should now be apparent. If the subordinate station tide time differences are positive, they may result in a prior day's tide event at the reference station occurring on the day in question at the subordinate station, when added to the reference station time predictions. Conversely, if subordinate station time differences are negative, then the first tide event on the succeeding day at the reference station should be recorded, because the algebraic addition of a negative time correction could well result in a tide occurring on the day in question at the subordinate station.

COMPLETE TIDE TABLE

Date: __9 NOVEMBER__

Subordinate Station ___BROOKLYN BRIDGE___

Reference Station ___NEW YORK___

HW Time Difference ___+ 0 13___

LW Time Difference ___+ 0 07___

Difference in height of HW ___- 0. 2'___

Difference in height of LW ___0.0'___

Reference Station			Subordinate Station	
HW ____ ____			____ ____	
LW __2339__ __-0.3'__ (8 NOV)			____ ____	
HW __0547__ __5.0'__			____ ____	
LW __1210__ __-0.3'__			____ ____	
HW __1806__ __4.8'__			____ ____	
LW ____ ____			____ ____	

Figure 11–10A. Reference station tide table, 9 November.

After all tide time and height conversions have been made, the complete tide table for 9 November at Brooklyn Bridge appears as shown in Figure 11–10B on the next page. Note that only three tides would occur on 9 November at the bridge.

The height of tide at Brooklyn Bridge at 0900 is now ready to be calculated. To do this, the quantities indicated on the bottom of the tide form are computed for the tide cycle containing the time in question, and recorded on the form for use with Table 3. First, the duration of the falling tide from the 0600 high water to the 1217 low water is computed and found to be 6 hours 17 minutes. Second, the time from the nearest tide comprising the tide cycle—in this case, the 0600 high—is figured; it is 3 hours 0 minutes. Third, the range of tide between the 0600 high and the 1217 low water is found—5.1 feet. Fourth, Table 3 in Figure 11–7 is entered to find the correction to height of the nearest tide, as described in the following paragraph.

Table 3 is entered at the top left with the tabulated duration of rise or fall that is closest to the computed value; in this case, the entering argument is 6 hours 20 minutes. Next, the tabulated time from the

COMPLETE TIDE TABLE

Date: 9 NOVEMBER

Subordinate Station   BROOKLYN BRIDGE

Reference Station   NEW YORK

HW Time Difference   + 0   13

LW Time Difference   + 0   07

Difference in height of HW   - 0.2'

Difference in height of LW   0.0'

Reference Station		Subordinate Station	
HW ____ ____		____ ____	
LW 2339 -0.3' (8 NOV)		— —	
HW 0547 5.0'		0600 4.8'	
LW 1210 -0.3'		1217 -0.3'	
HW 1806 4.8'		1819 4.6'	
LW ____ ____		— —	

Figure 11–10B. Brooklyn Bridge tide table, 9 November.

nearest high or low water closest to the computed value is found directly across to the right from the entering argument; for this problem, it is 2 hours 57 minutes. The correction sought is in the bottom half of the table, directly beneath this second argument, and opposite the tabulated range of tide that most nearly agrees with the actual range; the correction found is − 2.2 feet. Finally, following the instructions on the bottom of Table 3, the correction is subtracted from the 0600 high water to find the height of tide at 0900 of 2.6 feet. The completed table appears in Figure 11–10C.

The preceding problem was worked under the assumption that the location of interest, Brooklyn Bridge, was observing standard zone time, the time on which the predictions in the *Tide Tables* are based. If daylight savings time were in effect, however, it would first be necessary to convert all reference station daily predictions to daylight savings times by adding one hour. The complete tide table for the subordinate station would then be computed in the same manner as described above.

NAVIGATION DEPARTMENT

COMPLETE TIDE TABLE

Date: 9 NOVEMBER

Subordinate Station	BROOKLYN BRIDGE
Reference Station	NEW YORK
HW Time Difference	+ 0 13
LW Time Difference	+ 0 07
Difference in height of HW	- 0.2′
Difference in height of LW	0.0′

Reference Station			Subordinate Station	
HW				
LW	2339	-0.3′ (8 NOV)	—	—
HW	0547	5.0′	0600	4.8′
LW	1210	-0.3′	1217	-0.3′
HW	1806	4.8′	1819	4.6′
LW			—	—

HEIGHT OF TIDE AT ANY TIME

Locality: BROOKLYN BRIDGE Time: 0900 R Date: 9 NOV

Duration of Rise or Fall:	6 h 17 m
Time from Nearest Tide:	3 h 00 m
Range of Tide:	5.1′
Height of Nearest Tide:	4.8′
Corr. from Table 3:	-2.2′
Height of Tide at: 0900	2.6′

Figure 11–10C. Height of tide at a given time.

For some subordinate stations listed in Table 2, height of tide differences are expressed as multiplicative ratios sometimes with an algebraic correction factor, rather than algebraic differences. This is indicated by the placement of an asterisk (*) before the ratio number.

If such a ratio is given, the height of tide at the subordinate station is obtained by multiplying the heights of high and low tide given at the reference station by the respective subordinate station ratios, and then applying any correction factors listed.

The Bridge Problem

During the course of piloting in coastal waters or rivers, it occasionally becomes necessary for a vessel to pass under an overhanging obstruction such as a bridge or cable. In this situation, the navigator can be faced with two problems. First, he must ascertain whether or not the ship can pass safely under the obstruction at the time she is nominally scheduled to reach it. If she cannot, he must then determine the time frame within which the vessel can pass.

Suppose that in the example worked in the previous section, the charted clearance of the Brooklyn Bridge was 97 feet, and the masthead height of the navigator's ship, measured from the waterline to the highest point on the mast, was 100 feet. In order for the ship to pass beneath the bridge, then, the actual vertical clearance must be equal to or greater than the masthead height of 100 feet. In practice an additional safety margin would also be required—several feet in this case—to ensure safe passage under the bridge, but for the sake of simplicity in working this sample problem, such a safety margin will be disregarded.

As the first step in determining whether or not the ship can pass under the bridge at 0900, a diagram such as that appearing in Figure 11–11A is drawn to depict the physical relationships existing at 0900 at Brooklyn Bridge. It should be apparent from an inspection of the diagram that the actual clearance at 0900 could be computed if the height of tide at that time and the mean range of tide were known. In the preceding section, the height of tide at 0900 at the bridge was calculated to be + 2.6 feet. The mean range of tide can quickly be found by referring to the Brooklyn Bridge data in Table 2 of the *Tide Tables*, which appears in Figure 11–6; it is 4.3 feet. The actual bridge clearance at 0900, then, is given by the algebraic sum

$$97 + (4.3 - 2.6) = 98.7 \text{ feet.}$$

Since the masthead height of the ship is greater than 98.7 feet, the ship cannot pass under the Brooklyn Bridge at 0900.

Having reached this conclusion, the navigator is then faced with the more involved problem of determining when the ship can pass beneath the bridge. Once again, as a first step, a diagram is drawn to aid in visualizing the situation. For clarity, a representation of the ship is also included; the diagram appears in Figure 11–11B.

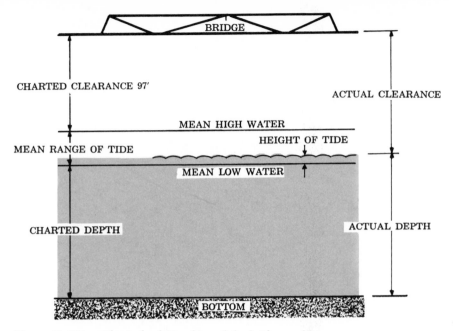

Figure 11–11A. Physical relationships of the bridge problem.

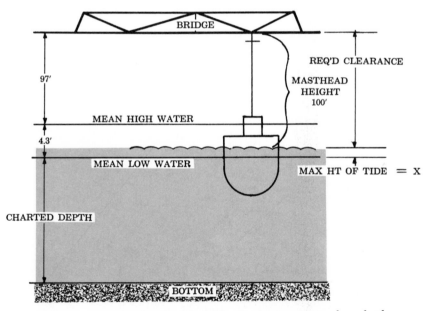

Figure 11–11B. Determining height of tide allowing passage under a bridge.

From an inspection of this diagram it can be seen that the actual vertical clearance under the bridge must be equal to or greater than 100 feet for the ship to pass, disregarding any safety factor. If *x*

represents the maximum acceptable height of tide, then the algebraic inequality

$$100 \leqslant 97 + (4.3 - x)$$

must be satisfied in order for the ship to pass under the bridge. Solving the inequality for the value of the height of tide providing clearance of exactly 100 feet yields a value of 1.3 feet. As this figure is rather low compared to the mean range of tide, 4.3 feet, it is obvious that passage under the bridge must be made at some time fairly close to the time of a low tide (see Figure 11–12). Referring to the tide table computed previously for 9 November at Brooklyn Bridge, the low tide nearest the desired 0900 time of passage is the low tide occurring at 1217. The height of this tide is −.3 foot, 1.6 feet below the tide height at which the mast would strike the bridge.

To find the time frame around the 1217 low tide within which the height of tide will be at or below 1.3 feet, it is necessary to work through Table 3 of the *Tide Tables* "in reverse." The entering arguments will be the maximum acceptable "correction" to height of the 1217 tide, and the ranges of tide and durations of rise or fall for the tide cycles on each side of the 1217 low tide. A plot of the height of tide versus time might help to clarify the problem at this point. In order to find the two times on either side of the 1217 low tide within which the tide will be below the maximum acceptable level

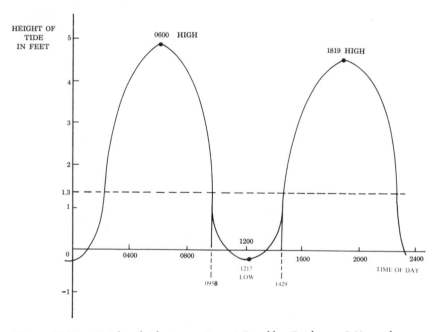

Figure 11–12. Height of tide versus time at Brooklyn Bridge on 9 November.

of 1.3 feet, it is necessary to work through the table twice: one to find the earlier time, and again to find the later time.

The correction to height will be the same in both cases, 1.3 (the maximum acceptable tide) − (−.3) (the height of the low tide), 1.6 feet. To find the earliest time at which the ship may pass, the tide cycle from the 0600 high to the 1217 low tide is considered. The range of tide during this cycle is 5.1 feet, and the duration of the fall is 6 hours 17 minutes. Working backward through Table 3, the nearest tabulated correction corresponding to the computed maximum allowable height of 1.6 feet is first found, using as entering argument the tabulated range of tide nearest the calculated value. The calculated range is 5.1 feet, so the tabulated range of 5.0 is used as the entering argument. Since the computed 1.6 correction lies exactly between two tabulated values—1.5 and 1.7—the smaller of the two tabulated corrections is selected, as this will produce a narrower safe time frame than would the selection of the larger value, thus introducing some margin for error. The next step is to follow the column containing the 1.5 correction up to the top half of the table to find the time from the nearest tide. The time opposite the tabulated duration of fall closest to the calculated duration of fall is selected; in this case, it is 2 hours 19 minutes. To complete the calculation, this time of 2 hours 19 minutes is subtracted from 1217, to yield 0958.

To find the later time, the same procedure is followed with the tide cycle following the 1217 low. The resultant latest time is 1429. Thus, the ship could physically pass beneath Brooklyn Bridge at any time between 0958 and 1429 on 9 November, as indicated in Figure 11–12.

If the ship's passage under the bridge could be scheduled with more latitude, it would also be possible to pass underneath during the time periods roughly centered around the last low tide on the night of 8 November, or the first low tide in the early morning of 10 November. Hence, if all possible passage time frames were required for 9 November, it would be necessary to repeat the above procedure two more times, to determine all possible safe-passage periods.

The Shoal Problem

An analogous situation to the bridge clearance problem occurs when a ship is required to pass over a shallow bar or shoal. In this type of problem, it is necessary to compute the times on each side of a high tide between which the tide will be above a certain level. In the shoal problem, the actual depth of water must be equal to or greater than the sum of the ship's draft, measured from the waterline to the keel, plus an appropriate safety factor. The physical relationships existing

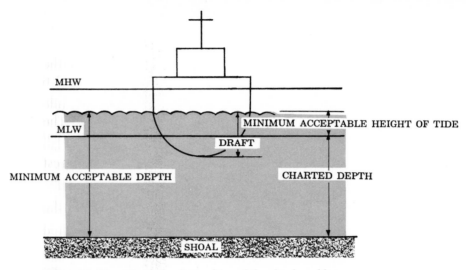

Figure 11–13. Physical relationships of the shoal problem.

in the shoal problem are represented in Figure 11–13. Once the minimum allowable height of tide is determined, the time frame is found by methods identical to those described above for the bridge problem, using Table 3 of the *Tide Tables*.

Effect of Unusual Meteorological Conditions

The navigator should always bear in mind that the tide time and height predictions given in the *Tide Tables* are based on the assumption that normal weather conditions will prevail. Although the predictions in the *Tables* are usually quite accurate, they can be greatly affected by unusual conditions of wind and barometric pressure, both at sea and in coastal waters. Heights of tide, for example, are based on a normal barometric pressure of 29.92 inches of mercury. If the pressure falls by one inch to 28.92 inches, sea level in the area may rise by as much as a foot. In practice, therefore, the navigator faced with transiting either an overhead obstruction such as a cable or bridge, or a shallow bar or shoal, will usually recommend scheduling the passage of the obstacle as near as possible to the computed times of high or low tide, as applicable, to allow the largest possible safety margin.

Summary

This chapter has examined the causes and effects of tide, as well as the more common types of problems associated with tide predictions. The navigator must be continually aware of the effects of tide, especially when passage under a low overhead obstruction or over

shallow ground is scheduled. Water depths may occasionally be less than those indicated on a chart, and overhead clearances may sometimes be lower, especially when the less extreme reference planes such as mean high and mean low water are used as the basis for charted depths and clearances. To be certain of passing an obstacle safely, the navigator should always include a suitable safety margin in his calculations of computed clearances and depths. If possible, he should try to schedule passage when extreme tide conditions exist that will provide the greatest possible margin for safety in the given situation. Finally, the navigator should bear in mind that, although the predictions in the *Tide Tables* will be accurate most of the time, there exists the possibility that actual tide levels may differ from the predictions, particularly under unusual conditions of wind and barometric pressure.

12

Current

In the preceding chapter, tide was defined as a vertical rise or fall in the surface of a body of water. This chapter will examine a related phenomenon known as *current*, which, in contrast to tide, is strictly defined as a horizontal movement of water. As will be explained, this horizontal movement is caused in large measure by tide, but other physical factors such as wind, rain, and the Coriolis effect also come into play. Like tide, current is also of great interest and importance to the navigator, both at sea and especially when piloting in coastal waters, because it continually affects the movements of his ship as it proceeds through the water.

There are two main types of current with which the surface navigator is concerned—ocean and tidal. The first part of this chapter will examine the causes, effects, and methods of prediction of ocean currents, and the second part will discuss similar aspects of tidal current.

Ocean Current

The oceans and their currents have long been an enigma to man. Even though the oceans cover over 70 percent of the earth's surface, it has only been in comparatively recent years that the mysterious forces affecting the oceans and their currents have begun to be understood. This section will relate some of the more important facts that have been established concerning the major ocean currents.

An ocean is never in a state of equilibrium. It is, rather, always in a state of motion, attempting to attain a forever unattainable balance. The primary reason for this imbalance is the rise and fall of the ocean level caused by the gravitational forces of the sun and moon, but there are several other factors that also have an effect. The waters of the oceans are continually being heated and cooled by the earth's atmosphere, blown by its winds, and salted and diluted by evaporation and rain. All these factors combine to produce flows of both surface and subsurface waters from higher to lower levels, colder to warmer areas, and higher- to lower-density regions. Once set in motion, the flows of water in the oceans are affected by the Coriolis force associated

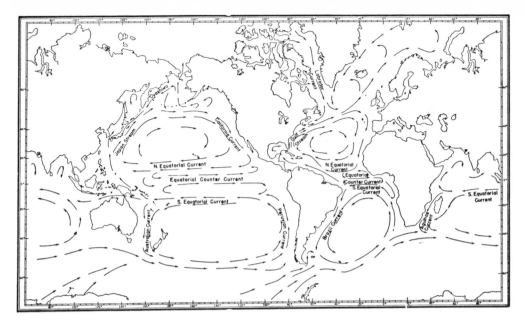

Figure 12–1. *Important ocean currents of the world.*

with the earth's rotation, so that they form giant patterns of rotation called "gyres" in each of the major ocean basins. These gyres rotate in a basically clockwise direction in the northern hemisphere, and in a counterclockwise direction in the southern hemisphere. Moreover, it is these gyres, or portions of them, that form the more well-known and well-defined of the world's great *ocean currents*, some of which are pictured in Figure 12–1.

The major ocean currents are like rivers in the oceans, but they far surpass any continental river in size and strength. The Gulf Stream, for example, is spawned in the western Caribbean and Gulf of Mexico, surges north through the Straits of Florida, and follows a meandering course off the U.S. Atlantic seaboard to Newfoundland, where it turns eastward, crosses the North Atlantic, and finally dissipates off northern Europe. Where it is strongest off Florida and South Carolina, it is 40 miles wide and 2,000 feet deep, and it carries 100 billion tons of water at a velocity sometimes approaching five knots. The fastest known current is the Somali, located in the Indian Ocean; during the summer season of the monsoon wind, it has been measured at speeds up to seven knots.

Even more interesting than the surface currents are the *deep ocean currents*, caused primarily by water-density differences. Although little was known about these currents until recent years, it is generally

accepted today that the deep-ocean current system is equally as extensive as the surface system. It is theorized that these currents have been instrumental in shaping the ocean floor in many areas of the world, particularly near the continental slopes. In the western end of the Mediterranean basin off Gibraltar, there is an incoming surface current that flows at speeds of three to five knots. At the other end of the basin, the water grows more dense because of evaporation and the influx of salts from the land, and sinks. It forms a deep ocean current that flows westward underneath the incoming surface current, thus maintaining roughly the same volume of water in the Mediterranean. Another amazing deep-sea current is the Cromwell Current, discovered in 1951. It flows at a top speed of five knots about 300 feet below and in the opposite direction to a surface current that flows westward south of the equator in the Pacific.

Research into currents is being actively pursued by oceanographers today, using both traditional methods, such as anchored and free-flowing buoys, installed current meters, dyes, and the drift of survey ships, and to an ever-increasing extent, new techniques fostered by space-age technology. Satellites equipped with infrared sensors are being used to chart currents by mapping heat flow patterns, and others have been used to track oceanographic buoys, and on at least one occasion, even an iceberg fitted with a transmitter. The ocean current system plays a large part in the world's weather patterns, primarily through transfer of heat from one place to another. As more and more becomes known about the ocean currents, better predictions of their effects will be made available to the mariner for his use in navigating on the open sea.

Predicting Ocean Currents

Although the *Planning Guides* of the *Sailing Directions* contain some information on normal locations and strengths of currents encountered on the open sea, the navigator should find the pilot chart described in Chapter 5 (see Figure 5–10 page 72) of most value in predicting the direction and velocity of ocean currents.

The direction in which a current flows is called its *set*, and its velocity is referred to as its *drift*. The navigator determines set and drift from the pilot chart by referring to the green color-coded arrows on the chart designated for that purpose. The direction of the arrows indicates the average set of the current at that location during the time period covered by the chart, and the figures printed nearby represent the average drift.

When using current data from the appropriate pilot chart, the navigator should remember that these data represent average con-

ditions based on historical records. Actual ocean currents encountered will often vary somewhat from these norms, depending on the meteorological activity taking place at the time.

Tidal Current

The major ocean currents must be reckoned with by the surface navigator while at sea; once he has entered coastal waters, however, he becomes primarily concerned with another type of current known as *tidal current*. This current is so called because it is caused primarily by the rise and fall of the tide; coastal waters affected by tidal currents are often referred to as *tidewater* areas. A tidal current that flows toward shore as a result of the approach of a high tide is called a *flood current*, and that which flows away from shore because of a low tide is an *ebb current*. During each tide cycle, there is a moment corresponding to the stand of a tide, when no horizontal movement of the water takes place as the current changes direction; this moment, which may in reality be several minutes long, is called *slack water*.

At first consideration, it might seem that the times of minimum velocity of a flood or an ebb current should coincide with the times of high and low tide stands, but in general this is not the case. The change of direction of a tidal current always lags behind the turning of the tide by a time interval of varying length, which depends on the geographic characteristics of the shoreline. Along a relatively straight coast with few indentations, the interval might be quite small. On the other hand, if a large harbor were connected to the sea by a narrow inlet, the tide and current might be out of phase by as much as three hours. In this latter kind of situation, the velocity of current in the connecting channel is usually at or near a maximum when the tide is at its extreme high or low level.

In most inshore waters, the direction of flow, or set, of an incoming flood current is not exactly opposite the direction of the outgoing ebb current, because of irregularities in the shape of the shore. Where the two currents are approximately opposite from one another, they are referred to as a *reversing current*. Offshore, where the direction of flow is not as restricted, it often happens that the effect of semidiurnal tides is to create a current that flows continuously with no clearly defined ebb or flood; its set moves completely around the compass during each tide cycle. A tidal current of this type is called a *rotary current*. As a result of the Coriolis effect, its direction of rotation is usually clockwise in the northern hemisphere, and counterclockwise in the southern hemisphere. In U.S. waters, rotary currents are fairly common off the Atlantic coast, but are rarely observed off the Gulf or Pacific coasts where the diurnal tide pattern prevails.

Prediction of Tidal Current

Prediction of tidal current is of great importance to the navigator during the practice of piloting in coastal tidewater areas. Not only must he take current into account when transiting a channel, in order to remain within its bounds, but he must also realize that some approaches into harbors and turning basins are virtually impossible for a deep-draft vessel to negotiate when a current of any appreciable velocity is flowing. A carrier or cruiser, for example, normally enters the U.S. Navy base at Mayport, Florida, at or very near the time of slack water, because of the strong tidal current flowing in the mouth of the St. Johns River at other times.

In order to make various predictions concerning tidal currents, the navigator has several means at his disposal, depending on the areas of the world in which he is operating. In the coastal waters of the United States, a set of two publications called the *Tidal Current Tables* is extensively used for tidal current predictions; in addition, a series of tidal current diagrams and charts are also available for many of the major U.S. ports. British Admiralty-produced coastal approach charts often contain a graphic known as a *current diamond*, which is used in conjunction with charted tables and the applicable volume of the *Tide Tables* to yield current estimates.

The *Tidal Current Tables*

The *Tidal Current Tables*, often referred to simply as the *Current Tables*, consist of two volumes, one titled the *Atlantic Coast of North America*, and the other the *Pacific Coast of North America and Asia*. They are drawn up for each year by the National Ocean Service, and distributed to U.S. Navy users by the Defense Mapping Agency Office of Distribution Services. In appearance, they are much like the *Tide Tables*. Each volume is divided up into several numbered tables; the West Coast edition has four tables, and the East Coast edition five, the extra table being for rotary current predictions. Like the *Tide Tables*, the *Current Tables* contain daily predictions of the times of slack water and maximum flood and ebb current velocities for each of several reference stations, and they list time and velocity difference ratios for each of several hundred subordinate locations. The navigator can apply these differences to a specified set of reference station data to obtain a complete set of current predictions for a given subordinate location for a day of interest.

Table 1 of the *Current Tables* lists the times and maximum velocities of each flood and ebb current and the time of each slack water in the chronological order of appearance for all reference stations used in that volume. Figure 12–2 opposite is a page from Table 1 of an

CHESAPEAKE BAY ENTRANCE, VIRGINIA

F-Flood, Dir. 305° True   E-Ebb, Dir. 125° True

NOVEMBER

Day	Slack Water Time h.m.	Maximum Current Time h.m.	Vel. knots	Day	Slack Water Time h.m.	Maximum Current Time h.m.	Vel. knots
1 Su	0103	0510	1.3E	16 M	0202	0600	1.7E
	0840	1120	0.9F		0923	1217	1.4F
	1422	1800	1.1E		1535	1858	1.4E
	2134	2332	0.5F		2224		
2 M	0137	0547	1.2E	17 Tu		0037	0.7F
	0926	1204	0.9F		0259	0701	1.6E
	1507	1851	1.0E		1026	1317	1.2F
	2227				1636	2004	1.3E
					2332		
3 Tu		0019	0.4F	18 W		0141	0.6F
	0218	0636	1.1E		0406	0810	1.4E
	1018	1259	0.8F		1134	1425	1.0F
	1558	1946	0.9E		1738	2108	1.3E
	2325						
4 W		0118	0.3F	19 Th	0040	0256	0.5F
	0311	0741	1.1E		0522	0921	1.3E
	1117	1356	0.8F		1244	1534	0.9F
	1657	2051	1.0E		1839	2211	1.3E
5 Th	0025	0216	0.4F	20 F	0142	0404	0.6F
	0420	0851	1.1E		0641	1029	1.3E
	1220	1459	0.8F		1352	1637	0.8F
	1758	2149	1.0E		1934	2306	1.3E
6 F	0120	0325	0.5F	21 Sa	0236	0505	0.7F
	0540	0957	1.2E		0751	1130	1.3E
	1321	1601	0.8F		1453	1733	0.8F
	1857	2245	1.2E		2021	2355	1.4E
7 Sa	0209	0427	0.6F	22 Su	0322	0606	0.8F
	0658	1058	1.3E		0851	1221	1.4E
	1419	1657	0.9F		1546	1820	0.8F
	1951	2336	1.4E		2102		
8 Su	0254	0523	0.9F	23 M		0041	1.4E
	0808	1155	1.5E		0402	0647	0.9F
	1513	1750	1.0F		0942	1306	1.4E
	2040				1633	1903	0.7F
					2138		
9 M		0020	1.6E	24 Tu		0120	1.5E
	0337	0617	1.1F		0439	0726	1.0F
	0910	1246	1.6E		1025	1351	1.4E
	1605	1840	1.1F		1714	1938	0.7F
	2126				2209		
10 Tu		0106	1.8E	25 W		0155	1.5E
	0421	0707	1.4F		0514	0801	1.0F
	1007	1337	1.8E		1104	1430	1.4E
	1654	1929	1.2F		1752	2009	0.7F
	2211				2238		
11 W		0149	1.9E	26 Th		0230	1.5E
	0506	0757	1.6F		0548	0835	1.1F
	1101	1427	1.8E		1141	1507	1.3E
	1744	2014	1.2F		1829	2041	0.6F
	2254				2305		
12 Th		0235	2.0E	27 F		0302	1.4E
	0552	0846	1.7F		0622	0908	1.1F
	1154	1516	1.9E		1215	1545	1.3E
	1834	2102	1.2F		1906	2112	0.6F
	2339				2332		
13 F		0321	2.0E	28 Sa		0334	1.4E
	0640	0935	1.7F		0656	0940	1.1F
	1247	1607	1.8E		1249	1619	1.2E
	1926	2151	1.1F		1943	2147	0.6F
14 Sa	0024	0412	2.0E	29 Su	0002	0406	1.4E
	0731	1027	1.7F		0733	1017	1.1F
	1341	1701	1.7E		1325	1657	1.2E
	2021	2240	1.0F		2023	2224	0.5F
15 Su	0111	0502	1.9E	30 M	0034	0443	1.3E
	0825	1120	1.5F		0812	1058	1.0F
	1437	1758	1.6E		1402	1735	1.1E
	2120	2335	0.8F		2106	2307	0.5F

DECEMBER

Day	Slack Water Time h.m.	Maximum Current Time h.m.	Vel. knots	Day	Slack Water Time h.m.	Maximum Current Time h.m.	Vel. knots
1 Tu	0111	0522	1.3E	16 W	0245	0016	0.7F
	0856	1139	1.0F		1004	0642	1.6E
	1443	1824	1.1E		1606	1255	1.2F
	2153	2348	0.5F		2300	1934	1.4E
2 W	0156	0611	1.2E	17 Th	0348	0115	0.7F
	0945	1228	0.9F		1107	0744	1.4E
	1529	1911	1.0E		1658	1353	1.0F
	2244					2033	1.3E
3 Th	0250	0044	0.5F	18 F	0457	0220	0.6F
	1040	0705	1.2E		1212	0849	1.3E
	1619	1321	0.9F		1751	1452	0.8F
	2338	2006	1.1E			2132	1.3E
4 F	0357	0143	0.5F	19 Sa	0102	0328	0.6F
	1141	0812	1.2E		0609	0954	1.2E
	1713	1421	0.8F		1318	1551	0.7F
		2103	1.1E		1841	2228	1.3E
5 Sa	0033	0248	0.6F	20 Su	0157	0432	0.7F
	0514	0917	1.2E		0720	1055	1.2E
	1244	1517	0.8F		1421	1648	0.6F
	1809	2159	1.3E		1929	2317	1.3E
6 Su	0126	0352	0.8F	21 M	0246	0528	0.7F
	0633	1026	1.3E		0822	1152	1.2E
	1346	1619	0.9F		1518	1743	0.6F
	1904	2255	1.4E		2013		
7 M	0218	0455	1.0F	22 Tu		0004	1.3E
	0747	1127	1.4E		0331	0615	0.8F
	1445	1717	0.9F		0916	1240	1.2E
	1958	2349	1.6E		1607	1824	0.6F
					2053		
8 Tu	0307	0553	1.2F	23 W		0047	1.4E
	0854	1224	1.6E		0411	0700	0.9F
	1541	1812	1.0F		1003	1327	1.2E
	2049				1651	1907	0.6F
					2129		
9 W		0037	1.8E	24 Th		0126	1.4E
	0356	0647	1.4F		0449	0735	1.0F
	0955	1319	1.7E		1044	1408	1.2E
	1636	1903	1.1F		1731	1942	0.6F
	2139				2203		
10 Th		0126	2.0E	25 F		0205	1.4E
	0445	0742	1.6F		0525	0811	1.0F
	1052	1412	1.7E		1122	1445	1.3E
	1728	1954	1.1F		1809	2017	0.6F
	2228				2235		
11 F		0215	2.1E	26 Sa		0239	1.5E
	0535	0830	1.7F		0600	0846	1.1F
	1146	1504	1.8E		1158	1524	1.3E
	1820	2045	1.1F		1845	2052	0.6F
	2317				2308		
12 Sa		0304	2.1E	27 Su		0312	1.5E
	0625	0923	1.7F		0636	0921	1.1F
	1239	1555	1.8E		1233	1601	1.2E
	1913	2136	1.0F		1921	2127	0.6F
					2342		
13 Su	0006	0357	2.0E	28 M		0346	1.5E
	0716	1014	1.7F		0712	1000	1.1F
	1331	1648	1.7E		1307	1635	1.2E
	2006	2227	1.0F		1957	2204	0.6F
14 M	0056	0447	1.9E	29 Tu	0018	0421	1.4E
	0810	1106	1.6F		0751	1037	1.1F
	1422	1741	1.6E		1343	1712	1.2E
	2102	2320	0.9F		2036	2243	0.6F
15 Tu	0149	0544	1.8E	30 W	0058	0502	1.4E
	0905	1158	1.4F		0833	1118	1.1F
	1514	1839	1.5E		1420	1751	1.2E
	2159				2117	2328	0.6F
				31 Th	0145	0547	1.4E
					0919	1200	1.0F
					1459	1837	1.2E
					2204		

Time meridian 75° W.   0000 is midnight.   1200 is noon.

Figure 12–2.   A reference station page, Atlantic Coast Tidal Current Tables, *Table 1.*

NO.	PLACE	METER DEPTH	POSITION Lat.	POSITION Long.	TIME DIFF. Min. before Flood	Flood	TIME DIFF. Min. before Ebb	Ebb	SPEED RATIO Flood	Ebb	Minimum before Flood knots	deg.	Maximum Flood knots	deg.	Minimum before Ebb knots	deg.	Maximum Ebb knots	deg.
		ft	° ' N	° ' W	h. m.	h. m.	h. m.	h. m.										
	HAMPTON ROADS Time meridian, 75°W			on CHESAPEAKE BAY ENTRANCE, p.64														
3505	Thimble Shoal Lt., 1.4 miles ESE of <32> Fort Wool		37 00.3	76 12.8	-1 01	-0 23	+0 03	-0 25	0.9	0.7	0.0	--	0.9	298	0.0	--	1.0	117
3510	0.9 mile northeast of----------------		36 59.8	76 17.2	-1 26	-2 03	-1 42	-1 47	1.0	1.2	0.0	--	1.0	265	0.0	--	1.8	080
3515	0.4 mile northeast of----------------		36 59.5	76 17.8	-1 41	-1 46	-1 25	-1 52	1.0	0.9	0.0	--	1.0	258	0.0	--	1.4	066
3520	0.7 mile southwest of----------------		36 58.85	76 18.95	-2 09	-2 39	-2 24	-2 17	0.6	0.9	0.0	--	0.6	250	0.0	--	1.3	045
3525	0.2 mile northwest of----------------		36 59.30	76 18.42	-1 47	-2 07	-1 39	-1 54	1.3	1.3	0.0	--	1.3	240	0.0	--	2.0	050
	Old Point Comfort																	
3530	midchannel <33>---------------------		36 59.3	76 19.3	-1 15	-1 07	-0 34	-1 09	1.5	1.0	0.0	--	1.5	260	0.0	--	1.5	055
3535	0.4 mile east of--------------------		37 00.2	76 18.0	-2 15	-2 02	-1 15	-2 12	1.3	0.8	0.0	--	1.3	235	0.0	--	1.2	045
3540	0.2 mile south of-------------------		36 59.77	76 18.88	-1 15	-1 20	-1 22	-1 56	1.7	0.9	0.0	--	1.7	240	0.0	--	1.4	075
3545	0.9 mile southwest of---------------		36 59.33	76 19.57	-1 31	-1 09	-0 30	-1 42	1.7	1.0	0.0	--	1.7	240	0.0	--	1.5	050
3550	Willoughby Spit, 0.8 mile northwest of--		36 58.6	76 18.4	-2 10	-2 25	-2 10	-2 25	0.7	0.7	0.0	--	0.7	260	0.0	--	1.0	040
3555	Willoughby Spit, 0.7 mile north of------		36 58.8	76 17.3	-2 58	-2 05	-1 32	-3 05	1.0	0.5	0.0	--	1.0	285	0.0	--	0.8	080
3560	Willoughby Bay entrance-----------------		36 57.7	76 17.9	-2 50	-2 50	-2 50	-2 50	0.3	0.3	0.0	--	0.3	135	0.0	--	0.4	330
3565	Sewells Point, channel west of----------		36 57.5	76 20.4	-1 19	-1 42	-1 52	-1 42	0.9	0.8	0.0	--	0.9	195	0.0	--	1.2	000
3567	Sewells Point, 1 mile south of----------	15	36 56.22	76 20.40	-2 06	-1 33	-1 40	-2 19	0.5	0.5	0.0	--	0.5	172	0.0	--	0.7	359
3570	Sewells Point, pierhead----------------	7	36 56.8	76 20.1	-1 30	-1 35	-1 30	-1 35	0.6	0.5	0.0	--	0.6	195	0.0	--	0.8	010
	Newport News																	
3575	Channel, east end--------------------	15	36 57.47	76 21.37	-1 36	-0 55	-0 22	-1 33	1.0	0.6	0.0	--	1.0	229	0.0	--	1.0	355
3580	Channel, middle---------------------	15	36 57.3	76 22.9	-1 21	-1 18	-0 41	-1 33	1.1	0.7	0.0	--	1.1	244	0.0	--	1.1	076
3585	Channel, west end--------------------	15	36 57.20	76 24.80	-0 54	-1 15	-0 26	-0 40	0.7	0.4	0.0	--	0.7	280	0.1	010	0.6	092
3590	Middle Ground, 1 mile south of-------	7	36 56.0	76 23.2	-0 05	-0 05	-0 05	-0 05	1.1	0.8	0.0	--	1.1	270	0.0	--	1.2	100
	ELIZABETH RIVER																	
3595	Craney Island-----------------------	15	36 53.68	76 20.15	-1 55	-2 10	-2 22	-2 19	0.7	0.6	0.1	098	0.7	177	0.2	270	0.9	001
3600	Lambert Point------------------------	15	36 52.50	76 19.95	-2 41	-2 16	-2 23	-2 21	0.5	0.5	0.0	--	0.5	143	0.0	--	0.7	328
3605	West Norfolk Bridge, Western Branch-----		36 51.5	76 20.6	-2 35	-2 35	-2 35	-2 35	0.5	0.5	0.0	--	0.6	260	0.0	--	0.7	080
3610	Southern RR. wharves, Pinner Point------		36 51.6	76 19.0	-2 46	-2 30	-2 00	-2 40	0.4	0.3	0.0	--	0.4	140	0.0	--	0.4	290
3620	Berkley Bridge, Eastern Branch---------		36 50.5	76 17.0	- -	- -	- -	- -	- -	- -	- -	- -	0.3	120	0.0	--	0.4	295
3625	Virginian RR. bridge, Eastern Branch----		36 50.2	76 14.7	-2 10	-2 10	-2 10	-2 10	0.4	0.4	0.0	--	0.4	100	0.0	--	0.6	280
3630	Berkley, Southern Branch-------------		36 50.0	76 17.8	- -	- -	- -	- -	- -	- -	- -	- -	0.3	215	0.0	--	0.3	330
3635	Chesapeake, Southern Branch-----------		36 48.5	76 17.4	-2 36	-2 11	-1 59	-2 24	0.7	0.4	0.0	--	0.7	180	0.0	--	0.6	000
3640	Gilmerton Hwy. bridge, Southern Branch--		36 46.5	76 17.7	-2 46	-2 14	-2 12	-2 34	0.6	0.5	0.0	--	0.6	180	0.0	--	0.7	000
	NANSEMOND RIVER																	
3645	Pig Point, 1.8 miles northeast of------		36 55.4	76 25.1	-1 26	-1 02	-0 24	-1 12	0.8	0.7	0.0	--	0.8	285	0.0	--	1.0	070
3650	Town Point Bridge, 0.5 mile east of-----		36 53.3	76 29.0	-2 03	-1 54	-1 20	-1 38	0.9	0.5	0.0	--	0.9	265	0.0	--	0.8	070
3655	Dumpling Island----------------------		36 48.5	76 33.5	-1 55	-1 55	-1 55	-1 55	1.0	0.7	0.0	--	1.0	175	0.0	--	1.0	345
	JAMES RIVER																	
	Newport News																	
3660	0.1 mile off shipbuilding plant------	8	36 58.8	76 26.5	+0 09	-0 18	+0 10	-0 08	0.9	0.9	0.0	--	0.9	325	0.0	--	1.3	145
3665	0.8 mile SW of shipbuilding plant----		36 58.5	76 27.3	-0 35	-0 37	-0 16	-0 27	1.0	0.8	0.0	--	1.0	325	0.0	--	1.2	140

Endnotes can be found at the end of Table 2.

Figure 12–3. A subordinate location data page, Atlantic Coast Tidal Current Tables, *Table 2.*

Atlantic Coast Current Tables volume. Inasmuch as the tidal currents are primarily caused by the action of the tides, there are usually four maximum currents for every tidal day period of 24 hours and 50 minutes, with a slack water between each one. As was the case with tide, however, a vacancy does occasionally appear; in these instances, the slack or maximum current that seems to be missing will be found to occur either just before midnight on the preceding day or just after midnight on the following day. For ease of reference, successive slack-water times are placed in one column, and maximum current velocities are arranged in consecutive order in a second column. As in the *Tide Tables*, times given in Table 1 are based on the standard time zone in which the reference station is located.

Table 2 consists of time differences for flood and ebb currents, and for the minimum currents preceding them (usually but not always

NO.	PLACE	METER DEPTH (ft)	POSITION Lat. (°' N)	POSITION Long. (°' W)	TIME DIFF. Min. before Flood (h.m.)	TIME DIFF. Flood (h.m.)	TIME DIFF. Min. before Ebb (h.m.)	TIME DIFF. Ebb (h.m.)	SPEED RATIOS Flood	SPEED RATIOS Ebb	Minimum before Flood (knots deg.)	Maximum Flood (knots deg.)	Minimum before Ebb (knots deg.)	Maximum Ebb (knots deg.)
	KENNEBEC RIVER Time meridian, 75°W			on PORTSMOUTH HARBOR ENTRANCE, p.10										
180	Bluff Head, west of--------------------		43 51.3	69 47.8	+0 33	+0 53	+0 26	+0 24	1.9	1.9	0.0 - -	2.3 014	0.0 - -	3.4 184
185	Fiddler Ledge, north of-----------------		43 52.8	69 47.8	+0 47	+1 12	+0 22	+0 48	1.6	1.4	0.0 - -	1.9 267	0.0 - -	2.6 113
190	Doubling Point, south of----------------		43 52.8	69 48.4	+0 28	+0 49	+0 23	+0 53	2.2	1.7	0.0 - -	2.6 300	0.0 - -	3.0 127
195	Lincoln Ledge, east of------------------		43 53.8	69 48.6	+0 32	+0 45	+0 23	+0 34	1.6	1.6	0.0 - -	1.9 359	0.0 - -	2.8 174
200	Bath, 0.2 mile south of bridge <3>------		43 54.5	69 48.5	+0 29	+1 28	+0 43	+0 23	0.8	0.8	0.0 - -	1.0 003	0.0 - -	1.5 177
	CASCO BAY													
205	Broad Sound, west of Eagle Island------		43 42.7	70 03.8	-1 16	-1 05	-1 27	-0 59	0.8	0.7	0.0 - -	0.9 010	0.0 - -	1.3 168
210	Hussey Sound, SW of Overset Island------	15	43 40.27	70 10.52	-1 28	-1 18	-0 58	-1 30	0.9	0.6	0.0 - -	1.1 316	0.3 189	1.2 153
	---do.-----------------------------------	25	43 40.27	70 10.52	-1 39	-1 19	-1 06	-1 32	0.9	0.6	0.0 - -	1.1 318	0.3 211	1.1 155
	---do.-----------------------------------	40	43 40.27	70 10.52	-1 58	-1 16	-1 05	-1 32	0.9	0.5	0.1 228	1.1 314	0.3 200	1.0 154
212	Hussey Sound, SE of Pumpkin Nob---------		43 40.45	70 10.78	-2 21	-1 29	-1 32	-1 14	1.0	0.5	0.1 068	1.2 346	0.1 066	0.9 168
215	Hussey Sound, east of Crow Island-------	40	43 41.33	70 10.79	-2 18	-0 42	-0 55	-1 24	0.7	0.4	0.1 114	0.9 016	0.0 - -	0.8 197
220	Portland Hbr. ent., SW of Cushing I.----		43 37.9	70 12.7	-1 43	-1 11	-1 20	-0 58	0.8	0.6	0.0 - -	1.0 322	0.0 - -	1.1 154
225	Diamond I. Ledge, midchannel SW. of-----		43 39.6	70 13.5	-1 26	-1 12	-1 11	-1 06	0.8	0.5	0.0 - -	0.9 300	0.0 - -	0.9 150
230	Portland Breakwater Light 0.3 mi. NW of <1> <4>---------------		43 39.5	70 14.5	- - -	-0 47	- - -	-1 07	0.3	0.3	0.0 - -	0.4	0.0 - -	0.5 048

TABLE 2. - CURRENT DIFFERENCES AND OTHER CONSTANTS

Figure 12–4. Tabulated data for multiple meter depths are given for Hussey Sound in the Atlantic Coast Current Tables.

slack waters), speed ratios and average speed and true direction of flood and ebb currents, and other useful information for each of the subordinate locations located within the area of coverage of that volume. As in the *Tide Tables*, the subordinate location data is arranged in geographic sequence, with an alphabetic index to Table 2 located in the back of each volume. A sample page from Table 2 of an *Atlantic Coast Current Tables* appears in Figure 12–3 opposite. Notice that at some locations the water depth for which the data were determined is given, as in the case of the Hampton Roads Newport News stations in Figure 12–3. If no water depths appear, the data presented represent average data for all depths. At a few locations precise data are supplied for multiple meter depths, as in the case of Hussey Sound, whose data are partially reproduced in Figure 12–4 above.

The differences in velocity of maximum currents between a subordinate location and its reference station in the *Current Tables* are always expressed in the form of arithmetic ratios, rather than as algebraic differences predominantly used in the *Tide Tables*. As will be seen in the following section, this leads to the use of multiplication rather than algebraic addition when converting reference station current velocities in order to construct a complete current table for a given location of interest.

Table 3 of the *Current Tables* is used to find the velocity of current for a given time at a reference station or subordinate location, after a complete current table has been constructed by use of the first two tables as applicable. This table is reproduced in Figure 12–5.

TABLE 3.—VELOCITY OF CURRENT AT ANY TIME

TABLE A

Interval between slack and desired time	Interval between slack and maximum current													
h. m.	h. m. 1 20	h. m. 1 40	h. m. 2 00	h. m. 2 20	h. m. 2 40	h. m. 3 00	h. m. 3 20	h. m. 3 40	h. m. 4 00	h. m. 4 20	h. m. 4 40	h. m. 5 00	h. m. 5 20	h. m. 5 40
0 20	0.4	0.3	0.3	0.2	0.2	0.2	0.2	0.1	0.1	0.1	0.1	0.1	0.1	0.1
0 40	0.7	0.6	0.5	0.4	0.4	0.3	0.3	0.3	0.3	0.2	0.2	0.2	0.2	0.2
1 00	0.9	0.8	0.7	0.6	0.6	0.5	0.5	0.4	0.4	0.4	0.3	0.3	0.3	0.3
1 20	1.0	1.0	0.9	0.8	0.7	0.6	0.6	0.5	0.5	0.5	0.4	0.4	0.4	0.4
1 40	------	1.0	1.0	0.9	0.8	0.8	0.7	0.7	0.6	0.6	0.5	0.5	0.5	0.4
2 00	------	------	1.0	1.0	0.9	0.9	0.8	0.8	0.7	0.7	0.6	0.6	0.6	0.5
2 20	------	------	------	1.0	1.0	0.9	0.9	0.8	0.8	0.7	0.7	0.7	0.6	0.6
2 40	------	------	------	------	1.0	1.0	1.0	0.9	0.9	0.8	0.8	0.7	0.7	0.7
3 00	------	------	------	------	------	1.0	1.0	1.0	0.9	0.9	0.8	0.8	0.8	0.7
3 20	------	------	------	------	------	------	1.0	1.0	1.0	0.9	0.9	0.9	0.8	0.8
3 40	------	------	------	------	------	------	------	1.0	1.0	1.0	0.9	0.9	0.9	0.9
4 00	------	------	------	------	------	------	------	------	1.0	1.0	1.0	1.0	0.9	0.9
4 20	------	------	------	------	------	------	------	------	------	1.0	1.0	1.0	1.0	0.9
4 40	------	------	------	------	------	------	------	------	------	------	1.0	1.0	1.0	1.0
5 00	------	------	------	------	------	------	------	------	------	------	------	1.0	1.0	1.0
5 20	------	------	------	------	------	------	------	------	------	------	------	------	1.0	1.0
5 40	------	------	------	------	------	------	------	------	------	------	------	------	------	1.0

TABLE B

Interval between slack and desired time	Interval between slack and maximum current													
h. m.	h. m. 1 20	h. m. 1 40	h. m. 2 00	h. m. 2 20	h. m. 2 40	h. m. 3 00	h. m. 3 20	h. m. 3 40	h. m. 4 00	h. m. 4 20	h. m. 4 40	h. m. 5 00	h. m. 5 20	h. m. 5 40
0 20	0.5	0.4	0.4	0.3	0.3	0.3	0.3	0.3	0.2	0.2	0.2	0.2	0.2	0.2
0 40	0.8	0.7	0.6	0.5	0.5	0.5	0.4	0.4	0.4	0.4	0.3	0.3	0.3	0.3
1 00	0.9	0.8	0.8	0.7	0.7	0.6	0.6	0.5	0.5	0.5	0.4	0.4	0.4	0.4
1 20	1.0	1.0	0.9	0.8	0.8	0.7	0.7	0.6	0.6	0.6	0.5	0.5	0.5	0.5
1 40	------	1.0	1.0	0.9	0.9	0.8	0.8	0.7	0.7	0.7	0.6	0.6	0.6	0.6
2 00	------	------	1.0	1.0	0.9	0.9	0.9	0.8	0.8	0.7	0.7	0.7	0.7	0.6
2 20	------	------	------	1.0	1.0	1.0	0.9	0.9	0.8	0.8	0.8	0.7	0.7	0.7
2 40	------	------	------	------	1.0	1.0	1.0	0.9	0.9	0.9	0.8	0.8	0.8	0.7
3 00	------	------	------	------	------	1.0	1.0	1.0	0.9	0.9	0.9	0.9	0.8	0.8
3 20	------	------	------	------	------	------	1.0	1.0	1.0	1.0	0.9	0.9	0.9	0.8
3 40	------	------	------	------	------	------	------	1.0	1.0	1.0	1.0	0.9	0.9	0.9
4 00	------	------	------	------	------	------	------	------	1.0	1.0	1.0	1.0	0.9	0.9
4 20	------	------	------	------	------	------	------	------	------	1.0	1.0	1.0	1.0	0.9
4 40	------	------	------	------	------	------	------	------	------	------	1.0	1.0	1.0	1.0
5 00	------	------	------	------	------	------	------	------	------	------	------	1.0	1.0	1.0
5 20	------	------	------	------	------	------	------	------	------	------	------	------	1.0	1.0
5 40	------	------	------	------	------	------	------	------	------	------	------	------	------	1.0

Use table A for all places except those listed below for table B.
Use table B for Cape Cod Canal, Hell Gate, Chesapeake and Delaware Canal and all stations in table 2 which are referred to them.

1. From predictions find the time of slack water and the time and velocity of maximum current (flood or ebb), one of which is immediately before and the other after the time for which the velocity is desired.
2. Find the interval of time between the above slack and maximum current, and enter the top of table A or B with the interval which most nearly agrees with this value.
3. Find the interval of time between the above slack and the time desired, and enter the side of table A or B with the interval which most nearly agrees with this value.
4. Find, in the table, the factor corresponding to the above two intervals, and multiply the maximum velocity by this factor. The result will be the approximate velocity at the time desired.

Figure 12–5. Atlantic Coast Tidal Current Tables, *Table 3.*

Note that Table 3 actually consists of two similarly constructed but separate tables. Table 3A is used for most computations, with Table 3B being reserved for use only with those locations listed in Table 2 that use Cape Cod Canal, Hell Gate, or the Chesapeake and Delaware Canal as their reference stations.

Table 4, pictured in Figure 12–6 next page, is used to find the time frame around a slack-water or minimum current within which the current velocity will be below a desired maximum. Like Table 3, it is made up of two parts, A and B, with B used only for subordinate locations referred to Cape Cod Canal, Hell Gate, and the Chesapeake and Delaware Canal.

Table 5, as mentioned earlier, appears only in the *Atlantic Coast* volume of the *Current Tables*. It is designed for use in predicting the set and drift of an offshore rotary current at 46 locations off the Atlantic seaboard. A page from this table appears in Figure 12–7 on page 217. For each location listed in Table 5, the average direction and velocity of the rotary current are given for each hour after the occurrence of maximum flood current at a specified reference station. The velocities given should be increased by about 15 to 20 percent at the time of a new or full moon, and decreased by the same amount when the moon is at or near quadrature. In practice, this table is not extensively used, inasmuch as the rotary current velocities listed for most locations are so small for the most part as to be considered negligible.

Use of the *Tidal Current Tables*

In solving tidal current prediction problems involving the use of the *Current Tables*, the navigator normally uses a standard current form such as that shown in Figure 12–8 on page 218. The form has three sections: the first is for constructing a complete current table for a given location of interest in conjunction with Tables 1 and 2; the second is for computing the velocity of current for a given time, using Table 3; and the third is for use with Table 4 in finding the earliest and latest times on each side of a slack water within which the velocity of current will be below a specified limit.

As an example of the use of the *Current Tables*, suppose that the navigator desired to compute the set and drift of the current at mid-channel off Old Point Comfort, Hampton Roads, Virginia, at 0900 on 6 December. To accomplish this, the navigator must first construct a complete current table for Old Point Comfort midchannel for 6 December, and then use this computed table in conjunction with Table 3 to find the set and drift of the current at 0900.

To construct the complete current table, the Old Point Comfort

TABLE 4.—DURATION OF SLACK

The predicted times of slack water given in this publication indicate the instant of zero velocity, which is only momentary. There is a period each side of slack water, however, during which the current is so weak that for practical purposes it may be considered as negligible.

The following tables give, for various maximum currents, the approximate period of time during which weak currents not exceeding 0.1 to 0.5 knot will be encountered. This duration includes the last of the flood or ebb and the beginning of the following ebb or flood, that is, half of the duration will be before and half after the time of slack water.

Table A should be used for all places *except* those listed below for Table B.

Table B should be used for **Cape Cod Canal, Hell Gate, Chesapeake and Delaware Canal,** and all stations in Table 2 which are referred to them.

Duration of weak current near time of slack water

TABLE A

Maximum current	Period with a velocity not more than—				
	0.1 knot	0.2 knot	0.3 knot	0.4 knot	0.5 knot
Knots	*Minutes*	*Minutes*	*Minutes*	*Minutes*	*Minutes*
1.0	23	46	70	94	120
1.5	15	31	46	62	78
2.0	11	23	35	46	58
3.0	8	15	23	31	38
4.0	6	11	17	23	29
5.0	5	9	14	18	23
6.0	4	8	11	15	19
7.0	3	7	10	13	16
8.0	3	6	9	11	14
9.0	3	5	8	10	13
10.0	2	5	7	9	11

TABLE B

Maximum current	Period with a velocity not more than—				
	0.1 knot	0.2 knot	0.3 knot	0.4 knot	0.5 knot
Knots	*Minutes*	*Minutes*	*Minutes*	*Minutes*	*Minutes*
1.0	13	28	46	66	89
1.5	8	18	28	39	52
2.0	6	13	20	28	36
3.0	4	8	13	18	22
4.0	3	6	9	13	17
5.0	3	5	8	10	13

When there is a difference between the velocities of the maximum flood and ebb preceding and following the slack for which the duration is desired, it will be sufficiently accurate for practical purposes to find a separate duration for each maximum velocity and take the average of the two as the duration of the weak current.

Figure 12–6. Atlantic Coast Tidal Current Tables, *Table 4.*

TABLE 5.—ROTARY TIDAL CURRENTS 193

Georges Bank — Lat. 41°50′ N., long. 66°37′ W.

Time	Direction (true) Degrees	Velocity Knots
0	285	0.9
1	304	1.1
2	324	1.2
3	341	1.1
4	10	1.0
5	43	0.9
6	89	1.0
7	127	1.3
8	147	1.6
9	172	1.4
10	197	0.9
11	232	0.8

(Time column labeled: Hours after maximum flood at Pollock Rip Channel, see page 28)

Georges Bank — Lat. 41°54′ N., long. 67°08′ W.

Time	Direction (true) Degrees	Velocity Knots
0	298	1.1
1	325	1.4
2	344	1.5
3	0	1.2
4	33	0.7
5	82	0.8
6	118	1.1
7	138	1.5
8	153	1.2
9	178	1.1
10	208	0.9
11	236	0.8

Georges Bank — Lat. 41°48′ N., long. 67°34′ W.

Time	Direction (true) Degrees	Velocity Knots
0	325	1.5
1	332	2.1
2	342	2.0
3	358	1.3
4	35	0.7
5	99	0.8
6	126	1.3
7	150	2.0
8	159	1.9
9	169	1.7
10	197	1.2
11	275	0.9

Georges Bank — Lat. 41°42′ N., long. 67°37′ W.

Time	Direction (true) Degrees	Velocity Knots
0	316	1.1
1	341	1.3
2	356	1.0
3	16	0.8
4	43	0.6
5	92	0.8
6	122	1.0
7	146	1.1
8	170	1.1
9	195	1.0
10	215	1.0
11	272	0.9

Georges Bank — Lat. 41°41′ N., long. 67°49′ W.

Time	Direction (true) Degrees	Velocity Knots
0	318	1.6
1	320	1.8
2	325	1.4
3	330	0.8
4	67	0.3
5	111	0.9
6	117	1.5
7	126	1.7
8	144	1.7
9	160	1.1
10	242	0.8
11	292	1.2

Georges Bank — Lat. 41°30′ N., long. 68°07′ W.

Time	Direction (true) Degrees	Velocity Knots
0	312	1.5
1	338	1.7
2	346	1.5
3	14	1.1
4	59	0.9
5	99	0.9
6	123	1.3
7	144	1.7
8	160	1.6
9	187	1.3
10	244	1.0
11	274	1.1

Georges Bank — Lat. 41°29′ N., long. 67°04′ W.

Time	Direction (true) Degrees	Velocity Knots
0	277	1.0
1	302	1.2
2	329	1.4
3	348	1.3
4	15	1.2
5	48	1.1
6	85	1.2
7	122	1.4
8	145	1.5
9	166	1.3
10	194	1.2
11	223	1.1

Georges Bank — Lat. 41°14′ N., long. 67°38′ W.

Time	Direction (true) Degrees	Velocity Knots
0	305	1.4
1	332	1.6
2	355	1.6
3	15	1.4
4	38	1.1
5	77	0.9
6	112	1.2
7	141	1.6
8	162	1.6
9	187	1.5
10	214	1.4
11	252	1.2

Georges Bank — Lat. 41°13′ N., long. 68°20′ W.

Time	Direction (true) Degrees	Velocity Knots
0	319	1.5
1	332	2.0
2	345	1.4
3	9	0.8
4	42	0.6
5	80	0.7
6	118	1.0
7	138	1.3
8	154	1.4
9	169	1.5
10	188	1.3
11	236	0.9

Georges Bank — Lat. 40°48′ N., long. 67°40′ W.

Time	Direction (true) Degrees	Velocity Knots
0	304	0.9
1	340	0.9
2	353	0.8
3	29	0.6
4	56	0.6
5	83	0.6
6	107	0.9
7	140	1.0
8	156	1.0
9	175	0.9
10	202	0.8
11	245	0.8

Georges Bank — Lat. 40°49′ N., long. 68°34′ W.

Time	Direction (true) Degrees	Velocity Knots
0	301	1.2
1	326	1.5
2	345	1.4
3	8	1.1
4	36	0.8
5	69	0.8
6	106	1.0
7	139	1.4
8	153	1.5
9	175	1.4
10	201	1.1
11	237	0.9

Great South Channel, Georges Bank — Lat. 40°31′ N., long. 68°47′ W.

Time	Direction (true) Degrees	Velocity Knots
0	320	0.7
1	331	0.9
2	342	1.1
3	3	1.0
4	23	0.8
5	63	0.4
6	129	0.7
7	140	0.9
8	164	1.0
9	179	1.0
10	190	0.8
11	221	0.6

Figure 12–7. Atlantic Coast Tidal Current Tables, *Table 5.*

COMPLETE CURRENT TABLE

Locality: _____   Date: _____

Reference Station: _____

Time Difference: Min Bef Flood: _____   Flood: _____
                     Min Bef Ebb: _____   Ebb: _____
     Speed Ratio: Flood: _____
                          Ebb: _____

Maximum Flood Direction: _____
Maximum Ebb Direction: _____

Reference Station: _____ Locality: _____

_____ _____   _____ _____
_____ _____   _____ _____
_____ _____   _____ _____
_____ _____   _____ _____
_____ _____   _____ _____
_____ _____   _____ _____
_____ _____   _____ _____
_____ _____   _____ _____
_____ _____   _____ _____
_____ _____   _____ _____

VELOCITY OF CURRENT AT ANY TIME

Int. between slack and desired time: _____
Int. between slack and maximum current: _____   (Ebb) (Flood)
Maximum current: _____
Factor, Table 3   _____
Velocity: _____
Direction: _____

DURATION OF SLACK

Times of maximum current: _____   _____
Maximum current: _____   _____
Desired maximum: _____   _____
Period — Table 4:   _____   _____
Sum of periods: _____
Average period: _____
Time of slack: _____
Duration of slack: From: _____ To: _____

Figure 12–8. U.S. Naval Academy Current Form.

Figure 12–9.   A portion of the Index to Table 2, Atlantic Coast Tidal Current Tables.

difference data are first located by referring to the index to Table 2 located in the back of the Atlantic Coast volume of the *Tidal Current Tables.* A portion of the index appears in Figure 12–9. From the index, the sequential numbers of the subordinate locations at Old Point Comfort are found to begin with 3530. Turning to this number in Table 2, illustrated in Figure 12–3, the "midchannel" data of interest to the navigator are noted and recorded on the top of the current form. The reference station for this and other nearby locations appears in bold-faced type above the various difference data; it is Chesapeake Bay Entrance, located in Table 1 starting on page 65.

Opening Table 1 to the proper page, the daily predictions for 6 December appearing in Figure 12–2 are found and recorded on the left side of the current form in chronological order. In addition, the first current event on the following day is also recorded, for reasons to be explained later. At this point, the partially completed current form now appears as in Figure 12–10A on the following page.

As the final step in the production of a complete current table for Old Point Comfort midchannel for 6 December, the time differences for slack-water minimums and maximum currents are added algebraically to the times at Chesapeake Bay Entrance, and the reference station maximum velocities are multiplied by the appropriate velocity ratios. Care must be taken to use the flood ratio for flood-current velocity conversions, and the ebb ratio for ebb-current conversions. The converted velocities are always rounded to the nearest tenth. At

COMPLETE CURRENT TABLE

Locality: __OLD POINT COMFORT MIDCHANNEL__   Date: ___6 DECEMBER___

Reference Station: __CHESAPEAKE BAY ENTRANCE__

Time Difference: Min Bef Flood: ___−1:15___   Flood: __−1:07__
                    Min Bef Ebb: ___−:34___   Ebb: __−1:09__
   Speed Ratio: Flood: ___1.5___
                    Ebb: ___1.0___

Maximum Flood Direction: ___260°___
Maximum Ebb Direction: ___055°___

Reference Station: __CHESAPEAKE BAY ENTRANCE__   Locality: __OLD PT. C. M/C__

0126	0		
0352	.8 F		
0633	0		
1026	1.3 E		
1346	0		
1619	.9 F		
1904	0		
2255	1.4 E		
0218	0 (7 DEC)		

Figure 12–10A. A partially completed current table.

this time, the time difference between minimums before flood currents at Chesapeake Bay Entrance and at Old Point Comfort midchannel—1 hour 15 minutes—is applied to the first current event on the following day—the slack before flood at 0218 7 December—to determine whether that event will occur on the day in question at the subordinate location. It will not, so the 0218 slack on 7 December at Chesapeake Bay Entrance is dropped from further consideration. The reason for recording the first current event on the succeeding day at the reference station should now be obvious. If the subordinate location time differences are negative, they may result in a following day's current event being moved forward into the day in question at the subordinate location. On the other hand, if the time differences are positive, the result may be to move a prior day's current event into the day under consideration. Hence the following rule: when the time corrections are *positive*, the last event on the *preceding day* at

the reference station should always be recorded; when the corrections are *negative*, the first event on the *following day* should be recorded.

After all current time and velocity corrections have been made, the complete current table for 6 December at Old Point Comfort mid-channel appears as illustrated in Figure 12–10B.

Now the velocity of current at 0900 is ready to be computed. First, the quantities indicated on the second part of the form must be figured and recorded, for use with Table 3. The interval between the nearest slack water—the 0559 slack—and the desired time is 3 hours and 1 minute; the interval between this slack and the next maximum current, the 0917 ebb, is 3 hours and 18 minutes; and the maximum current is the 0917 ebb, 1.3 knots in velocity. Table A of Table 3 is entered using as entering arguments the tabulated values closest to the computed values for the interval between slack and desired time,

---

NAVIGATION DEPARTMENT

COMPLETE CURRENT TABLE

Locality: __OLD POINT COMFORT MIDCHANNEL__   Date: ____6 DECEMBER____

Reference Station: __CHESAPEAKE BAY ENTRANCE__

Time Difference: Min Bef Flood: ____−1:15__ Flood: __−1:07__
                          Min Bef Ebb: ____− :34__ Ebb: __−1:09__
   Speed Ratio: Flood: ____1.5_____
                               Ebb: ____1.0_____

Maximum Flood Direction: _____260°_____
Maximum Ebb Direction: _____055°_____

Reference Station: __CHESAPEAKE BAY ENTRANCE__   Locality: __OLD PT. C , M/C__

0126	0		0011	0
0352	.8 F		0245	1.2 F
0633	0		0559	0
1026	1.3 E		0917	1.3 E
1346	0		1231	0
1619	.9 F		1512	1.4 F
1904	0		1830	0
2255	1.4 E		2146	1.4 E
0218	0 (7 DEC)			

Figure 12–10B. Complete current table for Old Point Comfort Midchannel for 6 December.

COMPLETE CURRENT TABLE

Locality: __OLD POINT COMFORT MIDCHANNEL__   Date: ___6 DECEMBER___

Reference Station: __CHESAPEAKE BAY ENTRANCE__

Time Difference:	Min Bef Flood:	__−1:15__ Flood: __−1:07__
	Min Bef Ebb:	__−:34__ Ebb: __−1:09__
Speed Ratio:	Flood:	__1.5__
	Ebb:	__1.0__

Maximum Flood Direction: __260°__
Maximum Ebb Direction: __055°__

Reference Station: __CHESAPEAKE BAY ENTRANCE__   Locality: __OLD PT. C. M/C__

0126	0	0011	0
0352	.8 F	0245	1.2 F
0633	0	0559	0
1026	1.3 E	0917	1.3 E
1346	0	1231	0
1619	.9 F	1512	1.4 F
1904	0	1830	0
2255	1.4 E	2146	1.4 E
0218	0 (7 DEC)		

VELOCITY OF CURRENT AT ANY TIME

Int. between slack and desired time:	__3:01__	
Int. between slack and maximum current:	__3:18__	(Ebb) (Flood)
Maximum current:	__1.3__	
Factor, Table 3	__1.0__	
Velocity:	__1.3__	
Direction:	__055°__	

Figure 12–10C. Computed current at 0900 6 December at Old Point Comfort Midchannel.

and the interval between slack and maximum current. For this problem, the table yields a velocity factor of 1.0. This factor is then multiplied by the velocity of the 0917 ebb, resulting in a velocity of current of 1.3 knots at 0900; the product of this multiplication is always rounded to the nearest tenth. Since this is an ebb current, its direction from Table 2 is 055°T. The form now appears as in Figure 12–10C.

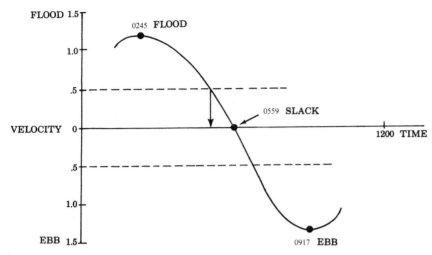

Figure 12–11.   Velocity of tidal current versus time.

Now, suppose that the captain desired to pass Old Point Comfort midchannel on 6 December when the current was below .5 knot. Referring to the calculated complete current table for this location, it is apparent that there would be at least four time periods on this date within which the current would be below the .5 knot velocity limit, since there are four slack waters on the sixth. If no time of day were indicated for the passage, the navigator would have to compute all four time frames. Let it be further stipulated, therefore, that the passage must be made near the time of the 0559 slack water. The problem can be visualized by a plot of the velocity of current versus time, which appears in Figure 12–11. For the current to be below the .5 knot maximum, the location must be passed between the times represented by the arrows on the time scale.

To work this problem, the bottom third of the current form is used in conjunction with Table 4 of the *Current Tables*. The maximum currents on either side of the desired slack are the 0245 flood and the 0917 ebb; their times and maximum velocities are recorded on the form, along with the desired maximum velocity of .5 knot. Next, Table A of Table 4 is entered twice, using as entering arguments the tabulated maximum currents closest to the predicted values and the desired maximum velocity. For the 0245 flood current, a total period of 120 minutes is obtained, and for the 0917 ebb current, a period of 78 minutes is extracted. The period in each case represents the total duration of the desired maximum velocity, *assuming the maximum velocities of the current on each side of the slack are identical.* Since consecutive flood and ebb currents seldom have the same velocity, it is necessary to find the *average* period; it is this period that

the navigator will use to compute the desired time frame. To find the average period, the length of each tabulated period is entered on the form, added together, and divided by 2. In this example, the average period is:

$$\frac{120 + 78}{2} \text{ or 99 minutes in duration.}$$

To complete the problem, it is further assumed that the slack water will occur in the middle of the average period of low velocity. Thus, the times within which the velocity of the current will be below .5 knot are given by the expression

$$0559 \pm \frac{99}{2} = 0509.5, \ 0648.5$$

Rounding off to the nearest whole minute, the resultant times are 0510 and 0649. The completed current form appears in Figure 12–12.

If daylight savings time is in effect, the navigator first must convert all reference station daily current predictions to daylight savings times by adding one hour. The complete current table, current at a given time, and duration of slack are then computed in the usual manner.

Current Diagrams and Charts

In addition to the tabulated data in the *Tidal Current Tables*, there are several types of graphic aids available to help the navigator to predict the set and drift of tidal current at certain heavily traveled locations in U.S. waters. Two of the more widely used of these are a set of current diagrams included in the *Atlantic Coast Tidal Current Tables*, and a series of 13 *Tidal Current Charts* issued by NOS covering eleven of the more important bays, harbors, and sounds on the East and West Coasts of the continental United States.

One example of the current diagrams contained in the *Atlantic Coast Tidal Current Tables*, the one for Chesapeake Bay, appears in Figure 12–13 on page 226. These diagrams are based on the occurrence of specified maximum current or slack water events at a designated reference station. The diagrams use as their entering arguments the number of hours preceding or following the designated event, and the ship's speed. The diagram in Figure 12–13 is entered by setting one edge of a parallel ruler along the applicable speed line of the nomogram to the right of the diagram, and then expanding the ruler until the opposite edge lies on the vertical line associated with the proper time interval adjacent to the location of interest listed along the left border. The numbers on the horizontal lines indicate current velocities. Those figures that appear alongside the expanded

COMPLETE CURRENT TABLE

Locality: __OLD POINT COMFORT MIDCHANNEL__   Date: ___6 DECEMBER___

Reference Station: _CHESAPEAKE BAY ENTRANCE_

Time Difference: Min Bef Flood: ___−1:15___ Flood: __−1:07__
                     Min Bef Ebb: ___− :34___ Ebb: __−1:09__
  Speed Ratio: Flood: ___1.5___
                     Ebb: ___1.0___

Maximum Flood Direction: ___260°___
Maximum Ebb Direction: ___055°___

Reference Station: _CHESAPEAKE BAY ENTRANCE_   Locality: _OLD PT. C. M/C_

0126	0	0011	0
0352	.8 F	0245	1.2 F
0633	0	0559	0
1026	1.3 E	0917	1.3 E
1346	0	1231	0
1619	.9 F	1512	1.4 F
1904	0	1830	0
2255	1.4 E	2146	1.4 E
0218	0 (7 DEC)		

VELOCITY OF CURRENT AT ANY TIME

Int. between slack and desired time:	3:01
Int. between slack and maximum current:	3:18 (Ebb) (Flood)
Maximum current:	1.3
Factor, Table 3	1.0
Velocity:	1.3
Direction:	055°

DURATION OF SLACK

Times of maximum current:	0245	0917
Maximum current:	1.2	1.3
Desired maximum:	.5	.5
Period – Table 4:	120	78
Sum of periods:		198
Average period:		99
Time of slack:		0559

Duration of slack: From: __0510__ To: __0649__

Figure 12–12.   *Completed current form for Old Point Comfort Midchannel, showing computed duration of slack water.*

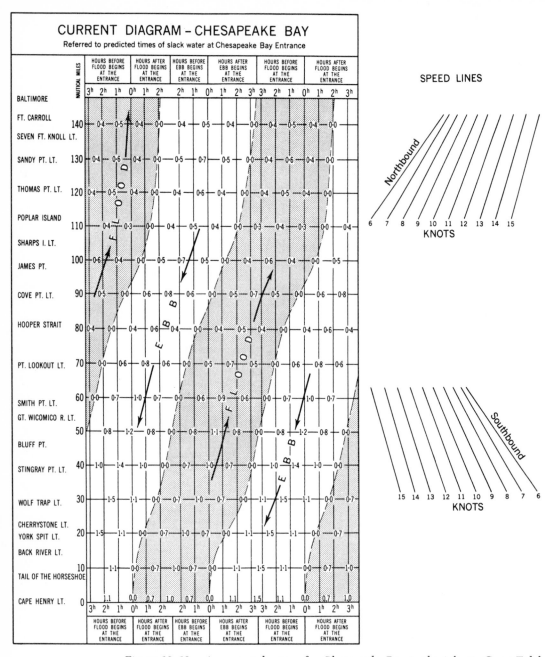

Figure 12–13. A current diagram for Chesapeake Bay in the Atlantic Coast Tidal Current Tables.

ruler edge on the diagram indicate the current that should be encountered at each successive location as the ship proceeds either northbound or southbound at the given speed. Set of the current is northerly during flood currents and southerly during ebb currents;

in the diagram it is determined by whether the edge of the ruler cuts a horizontal line in a shaded flood-current area or an unshaded ebb-current area.

Tidal Current Charts are sets of small-scale chartlets of a particular bay, harbor, or sound, which depict the normal set and drift of the tidal current for each hour after occurrence of a particular current event at a designated reference station. Figure 12–14 next page shows a portion of one of a set of eleven such chartlets that show the predicted current at various locations within Narragansett Bay for each hour after high water at Newport, Rhode Island. The direction of the arrows on the chartlet represents the set of the tidal current at that place and time, and the numbers near the arrows represent the drift.

In addition to the current diagrams and charts described above, NOS has recently begun annual publication of a series of four graphic aids called *Tidal Current Diagrams* designed for use with the *Tidal Current Charts* described above. Graphs in these *Diagrams* indicate directly the current flow for a particular date and time for Long Island Sound, Boston Harbor, Upper Chesapeake Bay, and New York Harbor.

Tidal Current Diamonds

For predictions of the set and drift of tidal currents in foreign coastal areas, the navigator can use the applicable *Enroute* volume of the *Sailing Directions*, which contains tidal current diagrams and chartlets for selected ports, foreign-produced equivalents of the *Tidal Current Tables* and *Charts*, or charted current information. On charts produced by the British Admiralty, lettered symbols called *tidal current diamonds* are printed at various locations along major channels of interest to the navigator. The letters within the diamonds refer to small tables printed on the chart, which give hourly predictions of set and drift based on the occurrence of a certain tide event at a designated reference station, usually high water. A portion of a British Admiralty chart showing a current diamond and its associated charted information appears in Figure 12–15. The table shown contains two velocity columns, one for current velocities at spring tides, and the other for velocities at neap tides.

To use this type of charted current information, the navigator must initially determine whether the date in question is at spring tides or neap tides, or somewhere in between. To do this, the applicable volume of the *Tide Tables* is used to determine the heights of high and low water on the day under consideration at the reference station referred to by the charted current diamond table. These heights are

Figure 12–14. A portion of a tidal current chart for Narragansett Bay.

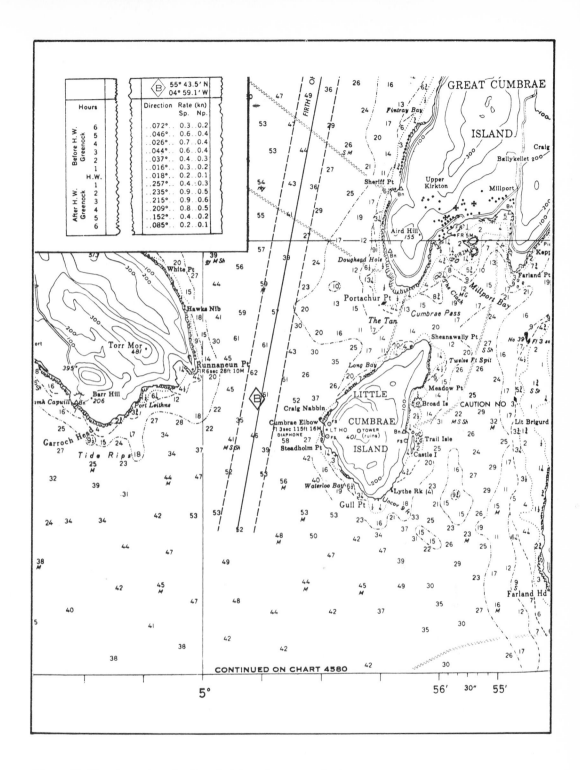

Figure 12–15. A portion of a British Admiralty approach chart.

then compared to the corresponding heights of spring and neap tides at the reference station; this latter information is usually given on the chart. To predict the tidal current at the site of the current diamond for a given time on the date in question, it is necessary to interpolate first for current strength values for that day of the month, and then for current strength and direction for times between the whole hours listed.

As an example, suppose that the high and low water levels at Greenock on a particular day were computed by use of a *Tide Table* to be 9.9 feet and 1.7 feet, respectively. From charted information (not shown in Figure 12–15), it is found that high water at spring tide is 10.8 feet, and a neap tide, 9.0 feet. Low waters at these times are .8 and 2.9 feet, respectively. By comparing the computed heights with these charted values, it appears that this day is about midway between the spring tide and neap tide times of the month. Returning to the table for diamond "B" in Figure 12–15, current velocities for each whole hour are thus computed by interpolating halfway between the velocities given for spring and neap tides.

After computing the current velocity for each hour, it is then necessary to complete the set and drift determination by interpolating in both these computed strengths and in the direction column for the values corresponding to the exact time of interest. If, for example, it were 4 hours 30 minutes before high water at Greenock, the interpolated direction would lie halfway between the tabulated directions 026° and 046° , or 036° . The computed current 5 hours before high water would be .5 knots (halfway between 0.6 and 0.4), and for 4 hours before, .55 knots. Thus, for 4 hours 30 minutes, a drift of .525 knots is obtained; it would be rounded off to .5 knots in practice.

Wind-Driven Currents

When using current predictions from any source, the navigator should always bear in mind that the predictions are subject to error during unusual conditions of wind or river discharge. Figure 12–16 is an excerpt from the *Current Tables* showing the average current velocity that will normally result when the surface of a body of water

Wind velocity (miles per hour)	10	20	30	40	50
Average current velocity (knots) due to wind at following lightship stations:					
Boston and Barnegat	0.1	0.1	0.2	0.3	0.3
Diamond Shoal and Cape Lookout Shoals	0.5	0.6	0.7	0.8	1.0
All other locations	0.2	0.3	0.4	0.5	0.6

Figure 12–16.   Velocities of wind-driven currents.

is subjected to continuing winds of various strength. The set of wind-driven currents is normally slightly to the right of the direction in which the wind is blowing in the northern hemisphere, and slightly to the left in the southern hemisphere. The navigator may wish to combine estimated wind-driven current vectors with predicted ocean or tidal current vectors in order to obtain a more accurate prediction of the actual currents existing at a particular location and time.

Summary

This chapter has described the two major types of surface currents normally encountered by the seagoing surface navigator, and has examined in some detail the methods used to predict their effects at a given place and time. Although the predictions contained in the various reference publications discussed in this chapter are usually accurate most of the time, the navigator must always be alert for those occasions when the actual currents encountered will differ greatly from those predicted as a result of unusual wind or storm conditions.

The following chapter on current sailing will explain how the navigator takes the estimated currents that he predicts by methods described in this chapter into account in directing the movements of his vessel safely from one point to another.

13

Current Sailing

Chapter 8, "Dead Reckoning," states that a vessel's calculated dead reckoning course and speed line is seldom if ever identical to her actual track over the surface of the earth. It also mentions that the difference between the two is caused by the effect of current. Used in this context, current refers not only to the horizontal movement of the water through which the ship is proceeding, but also to all other factors which, operating singly or combined, might have caused the ship's projected position to differ from her true position at any time. Among these additional factors are the following:

Wind
Steering inaccuracy
Undertermined compass or gyro error
Error in engine or shaft RPM indications
Excessively fouled bottom
Unusual conditions of loading or trim

Current, in the strict oceanographic sense of horizontal water movement, however, usually accounts for the major portion of the discrepancy between a vessel's DR and fix positions.

Current sailing refers to the methods used by the navigator to take the effects of current into account in directing the movements of his vessel. There are two distinct phases of current sailing. During the first phase, called the *pre-sailing* or *planning* phase, the navigator uses methods described in the previous chapter to obtain a predicted or estimated current. He then applies this estimate to his intended track to find the optimum course and speed to order. In the second phase, referred to as *post-sailing*, the navigator computes the "actual" current that has acted on his ship during a transit between two points on the track. He then uses this computed current as an estimate for the next leg of the track, if the situation is such that the current may be expected to remain unchanged. During pre-sailing procedures, the current is considered to consist entirely of predicted horizontal water movements; however, the current used during the post-sailing

phase consists not only of water movements, but also all other factors mentioned above that may have affected the ship's movement during the time of the plot.

To work both pre-sailing problems involving an estimated current or post-sailing problems with an actual current, the navigator uses a graphic drawing known as a *current triangle*. This is a geometric, scaled drawing that uses vectors to represent the relationships between the ship's ordered course and speed, the current, and the resulting ship's track. There are two types of current triangles. The one employed during the pre-sailing or planning phase of current sailing is called the *estimated current triangle*. The other, used during the post-sailing phase, is known as the *actual current triangle*.

The Estimated Current Triangle

In the estimated current triangle, the three legs represent the ordered course (C) and speed (S), the estimated set (S) and drift (D) of the current, and the resultant track (TR) and speed of advance (SOA). An estimated current triangle appears in Figure 13–1. As can be seen from Figure 13–1, the vector representing the track and speed of advance of the ship is the vectorial sum of the vectors representing the ship's ordered course and speed and the current. Furthermore, if the triangle were drawn to scale using some convenient speed scale and compass rose, it should be apparent that any one of the vectors constituting the triangle could be found if the other two were known. The navigator makes use of the geometric solution of the estimated current triangle to solve the following two basic types of problems. The predetermined set and drift of the estimated current are considered given quantities in each case:

1. To find the expected track and speed of advance when the vessel proceeds at a given course and speed; and
2. To find the course and speed a vessel should order to achieve an intended track and speed of advance.

As a medium on which to work the current triangle, most navigators prefer the maneuvering board. This device incorporates several speed scales, and by locating the foot of the track and ordered course vectors at the center of the maneuvering board, direction measurements are

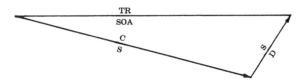

Figure 13–1. The estimated current triangle.

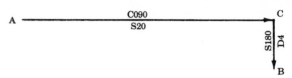

Figure 13–2A. Determining Track and SOA by a current triangle.

greatly facilitated. The estimated current triangle may also be worked on a portion of the chart isolated from the plot, or within a convenient chart compass rose. It is also possible to construct the triangle directly on the dead reckoning plot, using the DR course line as the ordered course and speed vector for purposes of the triangle. Inasmuch as this practice leads to a proliferation of extra lines in the region of the DR plot, however, it is usually best to work the triangle away from the plot if at all possible.

Solving the Estimated Current Triangle

As an example of the use of the current triangle to solve the first type of estimated current problem mentioned above, suppose that a navigator wished to determine the track and SOA that his ship was making while proceeding on an ordered course of 090°T at an ordered speed of 20 knots. By use of the *Current Tables,* he has estimated that a current having a set of 180°T and a drift of 4 knots is flowing in the area.

To find the solution, the navigator selects a blank portion of his chart away from his dead reckoning plot for the site of the current triangle. Using the distance scale of the chart as a speed scale for vector lengths, a vector 20 units long in the direction 090°T is first drawn to represent the ordered course and speed. From the tip or head of this vector, a second vector 4 units long in the direction 180°T, representing the estimated current, is laid down and labeled. At this stage, the semicompleted current triangle appears as shown in Figure 13–2A. To complete the triangle, a vector representing the expected track and speed of advance is drawn, as in Figure 13–2B. The track is obtained by measuring the direction of vector AB; it is 101°T. Measuring the length of the vector against the chosen speed scale yields the speed of advance, 21 knots.

To solve the second type of estimated current problem, in which

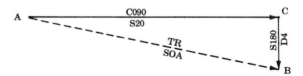

Figure 13–2B. Completed estimated current triangle.

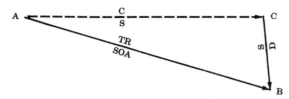

Figure 13–3. Solving for ordered course and speed using the current triangle.

the direction of the intended track and the speed of advance are either given or calculated, and the proper ordered course and speed are required, the solution would be found in a similar fashion, except that the vector representing the track and speed of advance would be drawn first, followed by the current vector. The required course and speed would then be the missing vector. Figure 13–3 illustrates the estimated current triangle as it appears in this type of problem. The direction of the intended track vector, AB in the figure, is implicitly determined as soon as the navigator decides to proceed from a point A to a point B across the chart. The speed of advance, or length of vector AB, can be specified in a number of ways. It may simply be given by directive; a certain distance may be required to be traversed in a given amount of time; or it may be required to arrive at a point B from a point A at a given time, as when proceeding to rendezvous. SOAs in the latter two cases must be calculated by means of the speed-time-distance formula.

Chapter 8, "Dead Reckoning," states that the navigator normally lays down an intended track on his chart prior to sailing through a given area, to represent the planned path and speed of the ship. Each leg of the track is labeled with its direction and the intended SOA. It should be apparent that to achieve this planned track and SOA, the navigator must solve a problem similar to the one just discussed. In practice, the navigator will often simply make an intelligent guess as to the proper ordered course and speed, taking into consideration the estimated set and drift, rather than draw an estimated current triangle. He will recommend a course a few degrees to the right or left of the intended track, and a speed close to the SOA; his recommendations are then adjusted based on the trend of the subsequent fix information.

The Estimated Position Allowing for Current

Returning for a moment to the first estimated current problem discussed in the last section, suppose that the navigator desired to obtain an estimate of his position 30 minutes after proceeding from a 1200 fix position on an ordered course of 090°T and at an ordered speed of 20 knots. Referring to his plot, he could simply locate a DR

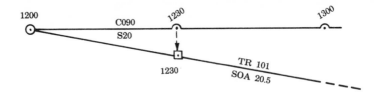

Figure 13–4. Determining an estimated position (EP) with allowance for current.

position for 1230 on the 090°T DR course line. But suppose further that the navigator desired to be more accurate in the determination of this position, and it was not possible to obtain a fix or running fix. If he applied the estimated current to the DR position, a more accurate *estimated position* previously described in Chapter 8 could well result, assuming that the actual current encountered was close to the predicted estimated current.

To obtain an estimated position (EP) for a given time taking current into account, a DR position is first plotted for the time of interest. Then the DR position is "adjusted" in the direction of the set of the current the amount of distance the drift of the current would have carried the ship in the time since the last fix. If done properly, the adjusted DR should fall on the resultant track previously computed. The estimated position is labeled by the box symbol for an EP, as shown in Figure 13–4. In the figure, the 1230 DR is advanced 2 miles, the distance the current would carry the ship in 30 minutes, in the direction of the set of the current, 180°T. An estimated position can be plotted at whatever time intervals the navigator desires in this way; all EPs should lie on the intended track. This fact leads to an alternative method of locating estimated positions by extending the intended track out from the last good fix, and calculating the distance traveled along the track by using the speed of advance. In this way EPs can be plotted on the intended track similarly to the way in which DR positions are plotted on the DR course line.

As pointed out in Chapter 8, the navigator should refrain from placing an unwarranted amount of confidence in any estimated position, including one derived by the method described above. The reliability of a position obtained by any means other than crossing two or more simultaneous LOPs is always open to doubt, and must be decided by the navigator based on the circumstances of each case.

Determining an EP from a Running Fix

During the discussion of the running fix in Chapter 8, "Dead Reckoning," it was explained that in the piloting environment a line of position, or LOP, was never advanced more than 30 minutes in order to obtain a running fix. Furthermore, in that chapter the effects

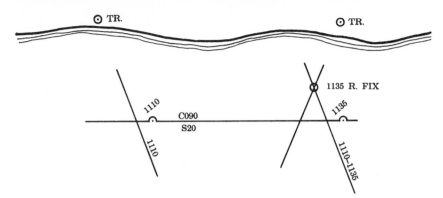

Figure 13–5A. A 25-minute running fix.

of any possible current were not considered in advancing the LOP. In certain circumstances, particularly when the navigator has observed that a strong current is flowing, he may wish to improve on the potential accuracy of his running fix by forming an estimated position based on it, that takes into account the estimated current.

To do this, the navigator simply adjusts the final position of his advanced LOP by moving it in the direction of the set a distance the drift of the current would have moved it in the time interval through which the LOP was advanced. Consider the situation in Figure 13–5A, in which an LOP was advanced 25 minutes for a running fix at 1135.

Suppose that the navigator has estimated a current to be flowing with a set of 230°T and a drift of 5 knots. In order to obtain an estimated position incorporating this estimated current, the 1110–1135 LOP must be adjusted for the estimated current by moving it in the direction 230°T a distance of $\frac{25}{60}$ × 5 or 2.1 miles. This is done by advancing a point on the line and then reconstructing the remainder of the LOP through the point, just as was originally done to advance

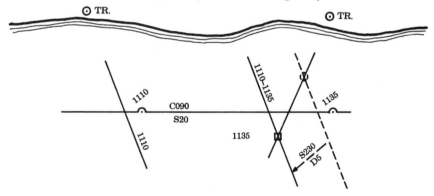

Figure 13–5B. Obtaining an EP from a running fix.

the 1110 LOP. The repositioned LOP and resulting estimated position appear in Figure 13–5B.

As is the case with all estimated positions, the navigator should be wary in placing complete confidence in an estimated position derived from a running fix, which is itself an artificial position.

The Actual Current Triangle

In the actual current triangle, the three legs represent the ordered course (C) and speed (S), the course over the ground (COG) and speed over the ground (SOG), and the resultant set (S) and drift (D) of the actual current experienced. An illustration of an actual current triangle appears in Figure 13–6.

Note that the intended track and speed of advance of the estimated current triangle have become the course and speed over the ground in the actual current triangle. In practice, this type of current triangle is used mainly to find the actual current vector, with the other two sides of the triangle given or determined from the navigation plot.

As an example, suppose that a ship had been steaming on a course of 270°T at a speed of 15 knots for 60 minutes from a 1200 fix position. At the end of the 60 minutes, a new fix was obtained, which placed the ship 3 miles south of its 1300 DR position, as shown in Figure 13–7A. Why didn't the 1300 fix show the ship to be at the 1300 DR position? The answer to this question, as mentioned at the beginning of this chapter, is that some "current" acted on the ship during its 60 minutes of travel, causing its actual course and speed over the ground to be different from the DR course line, representing the ordered course and speed.

To visualize the situation, the navigator could draw an actual cur-

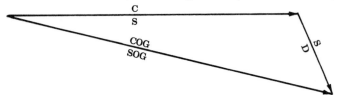

Figure 13–6. *The actual current triangle.*

Figure 13–7A. *A 60-minute DR track.*

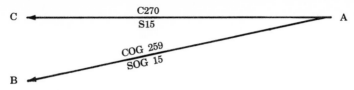

Figure 13–7B. The known legs of the actual current triangle.

rent triangle. A vector representing the ordered course and speed from the 1200 fix is first drawn and labeled, followed by a second vector representing the actual path the ship must have followed between the 1200 and 1300 fixes. At this point, the triangle appears as in Figure 13–7B. The direction of the COG-SOG vector is the direction of the 1300 fix from the 1200 fix. Its length represents the speed at which the ship must have traveled to traverse the distance between the two fixes in the given 60 minutes of time—the SOG. Since the path of the ship with respect to the ground is the resultant of its ordered course and speed and the actual current, the missing vector in Figure 13–7B must be the actual current vector. Furthermore, it must lie in the direction from C to B in the figure, by the laws of vector mechanics. The completed actual current triangle appears in Figure 13–7C. The set and drift of the actual current experienced between 1200 and 1300 can be obtained by measuring the direction and length of vector CB; they are 180°T and 3 knots. The navigator could then use these values as the estimated current during the planning phase for the next part of the transit.

In practice, it not usually required to draw a separate current triangle to determine the actual current. A ship can always be thought of as being displaced from its DR course line by the actual current acting on it in the time interval since the last good fix. Moreover, the ship's course and speed do not alter the effect of the current, because the whole body of water in which the ship floats is moving with the set and drift of the current. Thus, the ship's fix position at any time will always be offset from the corresponding DR position in the direction of the set of the current by the distance the drift would have carried the ship during the time period.

It follows, therefore, that the actual set can be determined simply by measuring the direction of a fix from the corresponding DR position

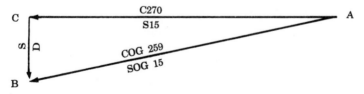

Figure 13–7C. The completed actual current triangle.

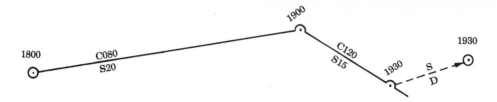

Figure 13–8. Determining actual set and drift after a course change.

at any time; the actual drift can be determined by measuring the distance from the DR to the fix position, and then dividing this distance by the number of hours elapsed since the last good fix. In the plot of Figure 13–7A, the set could have been obtained by measuring the direction of the 1300 fix from the 1300 DR position, 180°T. The drift could have been figured by dividing the distance, 3 miles, by the number of hours, 1 in this case, for a drift of 3 knots. As a second example, consider the plot in Figure 13–8. Even though there was a course and speed change at 1900, the effect of the current on the vessel is everywhere the same during the interval from 1800 to 1930. Thus, the set of the actual current can be measured from the 1930 DR to the 1930 fix; it is 070°T. The drift is found to be 4 knots by dividing the distance from the 1930 DR to the 1930 fix, 6 miles, by the number of hours, 1½.

Whenever actual current is to be determined as described above, care must be taken to use only the last two good fixes as a basis for the determination. If one or more running fixes were plotted in the interim between the two fixes, they must be disregarded, since the effect of any current was not taken into account while advancing the earlier LOPs to form the running fix(es). In this situation, the DR course line must be replotted from the time of the first fix to the time of the second, ignoring any intervening running fixes.

For purposes of comparison, the estimated and actual current triangles have been superimposed on each other in Figure 13–9.

As can be seen, the ordered course and speed vector is common to both triangles. In the estimated current triangle, the estimated current vector combines with the ordered course and speed vector to form the resultant track and speed of advance. In the actual current triangle, the actual current vector causes the ship to proceed along the vector representing its actual course and speed over the ground.

As a final cautionary note, it should be pointed out that current set and drift often undergo rapid change with both time and location in relatively constricted bodies of water such as straits, bays, rivers, and river mouths. When piloting in such waters, the navigator should not

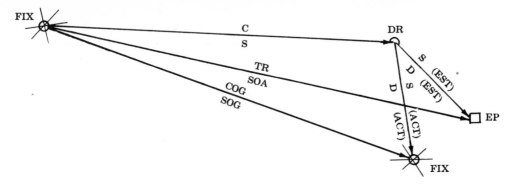

Figure 13–9. Relationship between the estimated and actual current triangles.

consider the actual current previously determined to be valid as a future estimate for more than a reasonably short time period, usually about 15 minutes.

Summary

This chapter has explained the procedures used by the navigator to take the estimated current into account as he directs the movements of his vessel from one point to another. The technique of determining the actual current, which may have been the result of many other factors in addition to water movement, has also been examined in some detail.

The navigator must be cognizant of the fact that current predictions from whatever source are always subject to error, and may not represent the actual conditions existing at a future time. Because of this, the navigator should always regard any estimated position with suspicion, tempered by experience. In the last analysis, the fix is the only certain way to ascertain the ship's position with 100 percent accuracy.

14

Precise Piloting and Anchoring

In Chapter 8 of this text dealing with dead reckoning, it was assumed that a vessel executed a change of course or speed instantaneously as it proceeded along the intended track laid down on the chart by the navigator. This is not the case, of course, in reality. Generally speaking, the larger the vessel, the more time and distance are required for either a course or speed change to be effected. On a comparatively small-scale chart of a large coastal approach region or ocean area, this time and distance can be disregarded for the most part, and any errors in the dead reckoning track so introduced are insignificant.

During piloting operations in more constricted waters and in narrow channels, however, the increased degree of accuracy required makes it impossible to ignore these factors. The practice of taking the turning diameter, time to turn, and acceleration and deceleration data into account when plotting and directing the movements of a vessel is termed *precise piloting*. The principles of precise piloting come into play whenever the vessel is engaged in any type of maneuvering requiring high precision, especially when following a narrow channel or when anchoring.

The first part of this chapter will explain the techniques involved in precise piloting, and the second part will examine in some detail the procedures used in selecting, plotting, and executing an anchorage.

Ship's Handling Characteristics

The attributes of a particular vessel relating to her performance in making turns at various rudder angles and speeds, and in accelerating and decelerating from one speed to another, are collectively called the vessel's *handling characteristics*. In the case of naval warships, these characteristics are referred to by the more specific term *tactical characteristics*, since these handling qualities bear directly upon the tactics that may be employed by the ship and others of her type and class. Every ship has a set of handling characteristics peculiar to herself; even warships within the same class often differ to some extent in the manner in which they respond to a given rudder angle

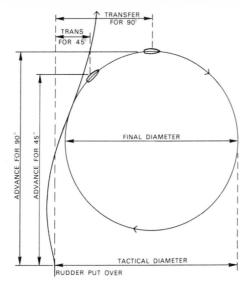

Figure 14–1. A ship's turning characteristics.

or engine speed change, although nominally they are designed to have identical tactical characteristics.

The handling or tactical characteristics pertaining to the ship's performance during turns are called her *turning* characteristics. The following terms are extensively used in describing these characteristics. They are illustrated in Figure 14–1.

Advance is the distance gained in the direction of the original course until the ship steadies on her final course. It is measured from the point at which the rudder is put over, and will be a maximum for a turn of 90°.

Transfer is the distance gained at right angles to the direction of the original course until the ship steadies on her final course.

The *turning circle* is the path followed by the point about which the ship seems to pivot—the pivoting point—as she executes a 360° turn. Every rudder angle and speed combination will normally result in a different turning circle.

Tactical diameter is a naval term referring to the distance gained at a right angle to the left or right of the original course in executing a single turn of 180°. Tactical diameter can be thought of as the transfer for a turn of 180°; it will be different for each rudder angle and speed combination.

Final diameter is the diameter of the turning circle the ship would describe if she were allowed to continue a particular turn indefinitely. For all but a few small ships, the final diameter will

always be less than the tactical diameter, due to the initial "kick" of the ship away from the direction of the turn, shown in a somewhat exaggerated manner in Figure 14–1.

Standard tactical diameter is a predetermined tactical diameter established by various tactical publications, most notably *ATP 1-B Volume 1*, for each ship type. It is used to standardize the tactical diameters for all ships by ship type, and finds its most extensive application when maneuvering in formation.

Standard rudder is the amount of rudder necessary to turn a ship in her standard tactical diameter at standard speed. It varies with the ship type, and also with the class of ship within a particular type.

Angle of turn is the horizontal angle through which the ship swings in executing a turn, measured from a ship's original course to her final course.

Each ship will normally have on board a complete list of all turning and acceleration/deceleration data pertaining to the ship. Merchant vessels operating in U.S. waters are required by Coast Guard regulations to have handling characteristics tables readily available to the conning officer. On Navy ships, the ship's tactical data are maintained in the form of a file called the *tactical data folder*. This folder contains a wealth of information in tabular form about the ship and her tactical turning and acceleration/deceleration characteristics for different rudder angle and speed combinations, and engine speed changes. All conning officers should be intimately familiar with their ship's handling and tactical characteristics and data.

One part of these tables of particular interest to the navigator is the turning characteristics for the ship for various rudder and speed combinations. This information usually appears in tabulated form, with the advance and transfer listed for various angles of turn at various rudder angles and speeds. Following is a typical table of this kind, showing the advance and transfer for every 15° of angle of turn using 15° rudder at 15 knots of speed.

Angle of Turn	Advance	Transfer	Angle of Turn	Advance	Transfer
15°	180	18	105°	330	280
30°	230	30	120°	310	335
45°	270	60	135°	270	380
60°	310	110	150°	230	418
75°	330	170	165°	180	470
90°	335	220	180°	100	500

In the case of Navy ships, tables of this type are normally based on the tactical characteristics of the first ship of the class within the type, as the time and expense of making the large number of runs required to generate a similar table for each ship would be prohibitive. When extensive alterations are made on a ship, however, a new set of tactical data should be experimentally determined by procedures set forth in the U.S. Navy publication *NavShips Technical Manual.* Unfortunately, in practice it is not uncommon to find tactical data folders on older ships that have never been updated, even though many extreme modifications may have been made since she was commissioned. A newly reported navigator faced with this situation should recommend that a period of the ship's operating schedule be set aside to update her tactical data folder at the earliest opportunity, and he should regard any advance, transfer, and acceleration data currently being used with extreme suspicion, until such a time as new data are obtained.

Merchant ships' handling characteristics are usually determined for different conditions of loading at the builder's trials. As is the case for older Navy ships, the data is often outdated or incomplete on many older vessels.

Use of Advance and Transfer During Piloting

As has been previously stated, the navigator should always take the turning characteristics of his ship into consideration during piloting operations in which any degree of precision is necessary. In practice, the professional navigator will use these characteristics whenever the scale of the chart on which he is plotting allows, in order to increase the accuracy of his dead reckoning plot. Consider the following example.

Suppose that a navigator of a medium-sized vessel has laid down an intended track on his chart to negotiate a 50° bend in a narrow river channel, as illustrated in Figure 14–2A on the next page. If the ship's rudder were not put over until she reached the intersection of the old and new track directions, point A in Figure 14–2A, the turning diameter of the ship might cause her to go aground on the left side of the channel as she was making the turn; at the very least, it would be far to the left of the intended track. Obviously, the navigator should recommend a point on the old track, called the "turning point," at which the ship should put her rudder over, taking her turning circle into account, so that the ship will come out of the turn on the new track leg.

To do this, the navigator uses the proper advance and transfer table for the amount of rudder and speed to be used in making the turn.

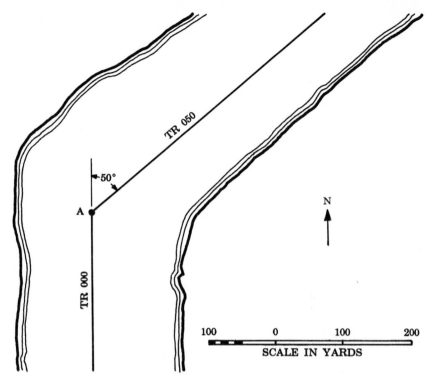

Figure 14–2A. A track laid down to negotiate a narrow channel.

For this example, the table just discussed will be used. First, the amount of advance and transfer for an angle of turn of 50° must be calculated by interpolation in the table. The advance for a turn of 45° is 270 yards, and for a turn of 60°, 310 yards. Interpolating between these values, the following expression is set up to yield the advance for a 50° turn:

$$270 + \frac{5}{15}(310 - 270) = 283 \text{ yards.}$$

In similar fashion, the transfer for this turn is calculated to be 77 yards.

After the advance and transfer have been calculated for the given angle of turn, the next step is to apply their values to the track to determine the point at which the rudder should be put over. Initially, there is no point upon which to base the advance, so the transfer is laid off first. The transfer is indicated by drawing a dashed construction line parallel to the original track, at a distance from it equal to the amount of transfer, as shown in Figure 14–2B. The intersection of the dashed transfer line with the new track, labeled point B in Figure

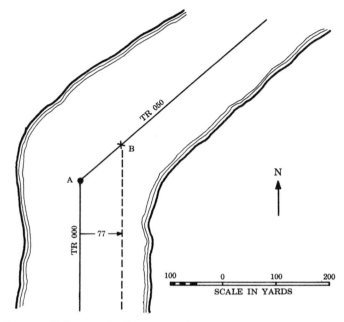

Figure 14–2B. Laying off the transfer for a turn of 50°.

14–2B, defines the point on the new track at which the turn should be completed, taking transfer into account. Next, advance is included by laying its value off from point B back along the dashed transfer line. This locates a second point on the transfer line, labeled point C in Figure 14–2C on the next page. To locate the turning point on the original track, a perpendicular is dropped to it from point C. The resulting point, labeled point D, is also shown in Figure 14–2C.

To complete the plot, the navigator locates a suitable object along the shoreline to use as the basis for a predicted line of position referred to as a *turn bearing*. It is drawn from the turning point toward but not through the object, and it is labeled with the true bearing of the object from the turning point, as indicated in Figure 14–2C. After determining the turn bearing, the navigator informs the conning officer of its value so that he may put the ship's rudder over when this bearing is sighted on the gyro repeater.

When selecting an object for a turn bearing, the ideal object should be located as nearly as possible at a right angle from the turning point on the original track. The effect of any unknown gyro error on a bearing to an object in this position is minimized, and the rate of change of bearing is maximized, thus enhancing the probability of starting the turn at the proper moment. The object used for a turn bearing in Figure 14–2C is so located. Many navigators prefer to select an object on the same side of the ship as the direction of the

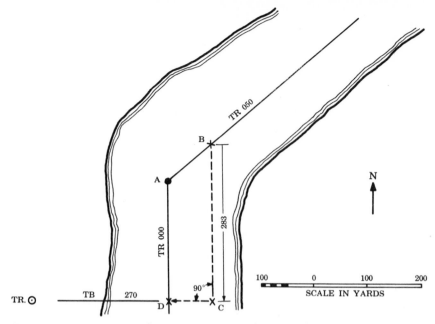

Figure 14–2C. Locating the turning point on the original track.

turn if available, especially if the conning officer who will be giving greatest attention to that side is to personally sight the turn bearing. In practice, however, there are usually no objects in the ideal position, and in coastal piloting, there may often be only one or two distinguishable landmarks available anywhere along the shore, so the navigator must often select a less than optimum turn bearing.

It should be reiterated that in the example discussed above, the advance and transfer table used to determine the turning point was compiled for a rudder angle of 15 degrees and a speed of 15 knots. If either a different rudder angle or different speed had been used, the advance and transfer, and hence the turning point and turn bearing, would vary. An appropriate advance and transfer table consistent with the planned rudder angle and speed must always be used to determine the proper advance and transfer values. When using any advance and transfer tables, the navigator must also bear in mind that the data in the tables were compiled under zero wind and current conditions. When an appreciable wind or current exists, the actual response of a ship in a turn may differ markedly from the tabulated data. Ships with high freeboard such as a container ship or aircraft carrier usually turn into a wind much more rapidly than they would turn under zero wind conditions, and turn out of the wind much more slowly. Ships with a deep draft may be much affected by even a slight current. Thus, the navigator must always be ready to adjust his rec-

ommendations as to the time to turn and the amount of rudder to use, based on the physical conditions existing at the given location and time.

Anchoring

Among the more critical of all precision piloting evolutions carried out by the navigator is anchoring in the exact center of a predetermined anchorage. It is one of the few instances in which all of the piloting skills of the navigator are brought into play, including plotting, computing tide and current, and precise piloting. Moreover, it is not enough that the navigator ready his own piloting team for the evolution. He must also brief the anchor detail personnel who will handle the ground tackle, and the commanding officer or master and sea detail deck watch officers who will have the responsibility for ship control. All other personnel who will be concerned with the anchorage must also be apprised of the overall plan and the part they are to play in it.

From the navigator's point of view, much of the effort in anchoring actually is expended before the evolution ever takes place. There are four stages in any successful anchoring, although they may not be formally recognized as such—selection, plotting, execution, and post-anchoring procedures. They will each be examined in the following sections of this chapter.

Selection of an Anchorage

An anchorage position in most cases is specified by higher authority. Anchorages for most ports are assigned by the local port authority in response to individual or joint requests for docking or visit. Naval ships submit a Port Visit (PVST) Request letter or Logistic Requirement (LOGREQ) message well in advance of the ship's scheduled arrival date. Operational anchorages in areas outside of the jurisdiction of an established port authority are normally assigned by the Senior Officer Present Afloat (SOPA) for ships under his tactical command.

If a ship is steaming independently and is required to anchor in other than an established port, the selection of an anchorage is usually made by the navigator and then approved by the commanding officer or master. In all cases, however, regardless of whether the anchorage is selected by higher authority or by the navigator himself, the following conditions should always apply insofar as possible:

The anchorage should be at a position sheltered from the effects of strong winds and currents.

The bottom should be good holding ground, such as mud or sand, rather than rocks or reefs.

The water depth should be neither too shallow, hazarding the ship, nor too deep, facilitating the dragging of the anchor.

The position should be free from such hazards to the anchor cable as fish traps, buoys, and submarine cables.

The position should be free from such hazards to navigation as shoals and sand bars.

There should be a suitable number of landmarks, daymarks, and lighted navigation aids available for fixing the ship's position both by day and by night.

If boat runs to shore are to be made, the anchorage chosen should be in close proximity to the intended landing.

Even when an anchorage has been specified by higher authority, the commanding officer or master, inasmuch as he is ultimately responsible for the safety of his ship, has the prerogative of refusing to anchor at the location assigned if he judges it to be unsafe. In these circumstances, he should request an alternate location less exposed to hazard.

Many of the coastal charts of the United States and its possessions drawn up by the National Ocean Survey contain colored anchorage circles of various sizes for different types of ships, located on the chart in those areas best suited for anchoring, taking into account the factors listed above. These circles are lettered and numbered, allowing a particular berth to be specified. Foreign charts often have anchorage areas specified as well. Amplifying information on possible anchorage sites can be obtained from the applicable volume of the *Coast Pilots* for U.S. waters, from the proper volume of the *Enroute Sailing Directions* for foreign waters, and from the *Fleet Guide* for ports in both foreign and domestic waters frequented by U.S. Navy ships.

When it is desired to anchor at a location other than that shown as an anchorage berth on a chart, the anchorage is normally specified by giving the range and bearing to it from some charted reference point, along with the radius of the berth.

Plotting the Anchorage

After the anchorage position has been determined, the navigator is ready to begin plotting the anchorage. In so doing, reference is often made to the following terms:

The approach track　This is the track along which the ship must proceed in order to arrive at the center of the anchorage. Its length will vary from 2,000 yards or more for a large ship, to 1,000 yards for a ship the size of a Navy destroyer or smaller.

Under most circumstances, it should never be shorter than 1,000 yards.

The head bearing   If at all possible, the navigator selects an approach track such that a charted navigational aid will lie directly on the approach track if it were extended up to the aid selected. The bearing to the aid thus described is termed the "head" bearing; it should remain constant if the ship is on track during the approach.

The letting-go circle   This is a circle drawn around the intended position of the anchor at the center of the berth with a radius equal to the horizontal distance from the hawsepipe to the pelorus.

The letting-go bearing   Sometimes referred to as the "drop" bearing, this is a predetermined bearing drawn from the intersection of the letting-go circle with the approach track to a convenient landmark or navigation aid, generally selected near the beam.

Range circles   These are preplotted semicircles of varying radii centered on the center of the anchorage, drawn so that the arcs are centered on the approach track. Each is labeled with the distance from that arc to the letting-go circle.

Swing circle   This is a circle centered at the position of the anchor, with a radius equal to the sum of the ship's length plus the length of chain let out.

Drag circle   This is a circle centered at the final calculated position of the anchor, with a radius equal to the sum of the hawsepipe to pelorus distance and the final length of chain let out. All subsequent fixes should fall within the limits of the drag circle.

The actual radii of both the swing and drag circles will in reality be less than the values used by the navigator in plotting them on the chart, because the catenary of the chain from the hawsepipe to the bottom is disregarded. Thus, a built-in safety factor is always included in the navigator's plot.

Prior to commencing the anchorage plot, it is always wise to draw a swing circle of estimated radius around the designated anchorage site to check whether any charted hazards will be in close proximity to the ship at any time as she swings about her anchor. If any such known hazards are located either within or near the swing circle, an alternate berth should normally be requested.

If the anchorage appears safe, the navigator begins the anchorage plot by selecting the approach track. During this process, due regard

must always be given to the direction of the predicted wind and current expected in the vicinity of the anchorage. Insofar as possible, the approach should always be made directly into whichever of these two forces is predicted to be strongest at the approximate time at which the anchorage is to be made.

The letting-go circle is drawn with a radius equal to the horizontal distance between the anchor-hawsepipe and the pelorus from which bearings will be observed. If the anchor were not let go until the pelorus were over the center of the assigned berth, the anchor would miss the center by the length of the ship from the hawsepipe to the pelorus. Thus, when the letting-go bearing, measured from the intersection of the letting-go circle and the approach track, is observed on the pelorus, the anchor will be in position directly over the center of the assigned berth.

The anchorage plot is completed by laying down the remainder of the intended track leading up to the approach track, and then swinging the range circles across the track. These arcs are normally drawn at 100-yard intervals measured outward from the letting-go circle to 1,000 yards, and at ranges of 1,200, 1,500, and 2,000 yards thereafter. After the anchor has been let go and the chain let out to its final length, a second swing circle is plotted, followed by the drag circle.

The use of these various quantities is best illustrated by an example. Suppose that a ship having a total length of 300 feet (100 yards) and a hawsepipe to pelorus distance of 150 feet (50 yards) has been directed to anchor at the position specified in the bay pictured in Figure 14–3A.

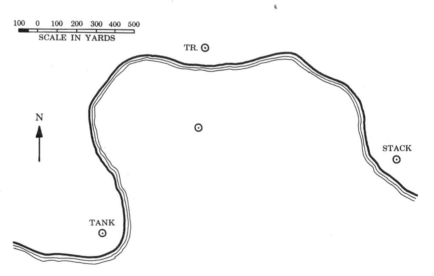

Figure 14–3A. An anchorage assignment.

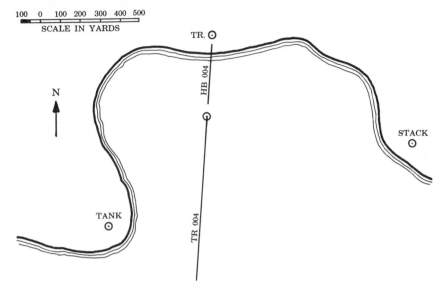

Figure 14–3B. Laying down the approach track.

After an estimated swing circle has been plotted and the anchorage has been determined to be safe, the navigator is ready to begin the construction of the anchorage plot. As the first step, the approach track is selected by considering the different objects available for a head bearing, taking into account the expected winds and current in the bay. Assuming negligible current and a northerly wind, the tower in Figure 14–3A is a good choice for a head bearing, especially since it is doubtful that an approach track of sufficient length could be constructed using any other navigation aid shown. The approach track is then laid off from it, as in Figure 14–3B.

As the next step, the intended track leading to the final approach track is laid down, with care being taken to allow for the proper length for the approach track. It is assumed here that the ship will be approaching the bay from the southwest. The advance and transfer for the 60° left turn onto the approach track are obtained, and the turning point is located. A turn bearing is drawn from this point and labeled. At this stage, the plot appears as in Figure 14–3C on the next page.

To complete the plot, the letting-go circle is drawn, with a radius equal to the 50-yard hawsepipe to pelorus distance. The letting-go bearing is then constructed using the stack, as it is nearly at a right angle to the approach track. Finally, range circle arcs are drawn and labeled, centered on the middle of the anchorage, with radii measured in 100-yard increments outward from the letting-go circle to 1,000 yards. Arcs are also swung for 1,200-, 1,500-, and 2,000-yard distances to the letting-go circle; the last of these is not shown because of space

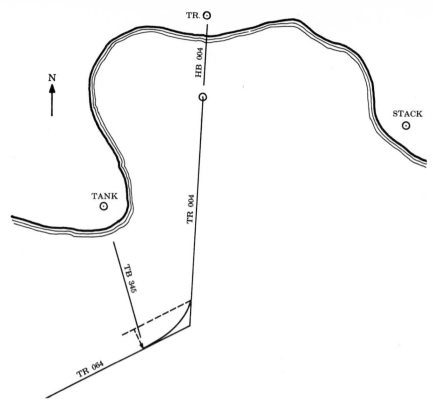

Figure 14–3C. Laying down the intended track and turn bearing.

limitations. The anchorage plot is now complete, and appears as shown in Figure 14–3D.

The various arcs, bearings, and tracks are drawn as shown to make the reading of information from the plot fast and easy for the navigator as the ship is approaching its anchorage. He will already have briefed the captain, conning officer, and pilot if aboard as to the intended head bearing and letting-go bearing; they could then drop anchor in the proper location without any further word from the navigator, by keeping the head bearing always constant and dropping anchor when the letting-go bearing reached the proper value. The navigator does not remain silent during this time, however. Quite to the contrary, he obtains fixes as often as possible and maintains a running commentary to inform all concerned as to the distance to go to the drop circle and whether the ship is to the right, to the left, or on the approach track as it proceeds to the anchorage.

Executing the Anchorage

When executing the actual anchorage, the navigator's dual objective is to keep the ship as nearly as possible on its preplanned approach

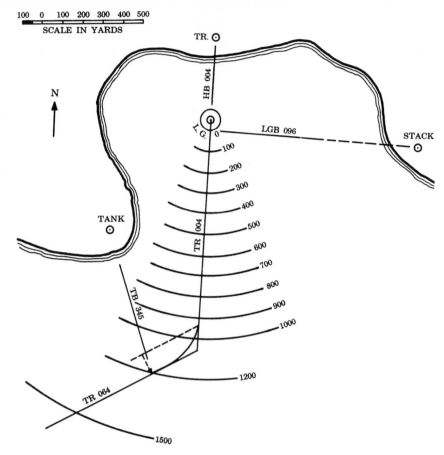

Figure 14–3D.   The completed anchorage plot.

track, and to have all headway off the ship when the hawsepipe is directly over the center of the anchorage. As mentioned above, he obtains frequent fixes as the ship proceeds along its track, and he keeps the bridge continually informed as to the position of the ship in relation to the track and the letting-go circle. He recommends courses to get back onto track if necessary. Since every ship has her own handling characteristics, speeds that should be ordered as the ship proceeds along the track are difficult to specify. In general, however, with 1,000 yards to go, most ships are usually slowed to a speed of five to seven knots. Depending on wind and current, engines should be stopped when about 300 yards from the letting-go circle, and the anchor detail should be instructed to "Stand by." As the vessel draws near the drop circle, engines are normally reversed so as to have all remaining headway off the ship as she passes over the letting-go circle. When the pelorus is exactly at the letting-go bearing,

the word "Let go the anchor" is passed to the anchor detail, and the anchor is dropped.

As the anchor is let go, the navigator should immediately call for a round of bearings to be taken, and he should record the ship's head. After the resulting fix is plotted, a line is extended from it in the direction of the ship's head, and the hawsepipe to pelorus distance is laid off along the line, thus plotting the position of the anchor at the moment that it was let go. If all has gone well, the anchor should have been placed within 50 yards of the center of the anchorage.

Post-Anchoring Procedures

After the anchor has been let go, chain is let out or "veered" until a length or "scope" of chain five to seven times the depth of water is reached. At this point, the chain is secured and the engines are backed, causing the flukes of the anchor to dig into the bottom, thereby "setting" the anchor. When the navigator receives the word that the chain has been let out to its full precomputed length and that the anchor appears to be holding with a moderate strain on the chain, he again records a round of bearings and the ship's head, as well as the direction in which the chain is tending. With this information, he plots another fix, and he recomputes the position of the anchor by laying off the sum of the hawsepipe to pelorus distance plus the scope of chain in the direction in which the chain is tending. This second calculation of the position of the anchor is necessary because it may have been dragged some distance from its initial position during the process of setting the anchor.

After the final position of the anchor has been thus determined, the navigator then draws a second swing circle, this time using the computed position of the anchor as the center, and the sum of the ship's length plus the actual scope of chain let out as the radius. If any previously undertemined obstruction, such as a fish net buoy or the swing circle of another ship anchored nearby, is found to lie within this circle, the ship may have to weigh anchor and move away from the hazard. If the ship is anchored in a designated anchorage area, due care should be taken to avoid fouling the area of any adjacent berths, even though they might presently be unoccupied. If the swing circle intersects another berth, it may be necessary to take in some chain to decrease the swing radius; if this is not possible, a move to a larger berth may be advisable.

If he is satisfied that no danger lies within the swing circle, the navigator then draws the drag circle concentric with the swing circle, using as a radius the sum of the hawsepipe to pelorus distance plus the scope of chain. All fixes subsequently obtained should fall within

the drag circle; if they do not, the anchor should be considered to be dragging. Both the swing circle and drag circle are shown in Figure 14–4, assuming that a scope of chain of 50 fathoms to the hawsepipe has been let out.

After plotting the drag circle, the navigator then selects several lighted navigation aids suitable for use in obtaining fixes by day or night, and he enters them in the *Bearing Book* for use by the anchor-bearing watch. The anchor-bearing watch is charged with obtaining and recording in the *Bearing Book* a round of bearings to the objects designated by the navigator at least once every 15 minutes, and plotting the resulting fix on the chart each time. Should any fix fall outside the drag circle, another round of bearings is immediately obtained and plotted. If this second fix also lies outside the drag circle, the ship is considered to be dragging anchor, and all essential personnel are alerted.

In practice, if the ship is to be anchored for any length of time, the navigator will usually cover the area of the chart containing the drag circle with a sheet of semi-clear plastic, so that the chart will not be damaged by the repeated plotting and erasures of fixes within the drag circle.

The importance of maintaining a competent and alert anchor-bearing watch cannot be overemphasized, as many recorded groundings have occurred because the duties of this watch were improperly performed. When a ship is dragging anchor, especially in high wind

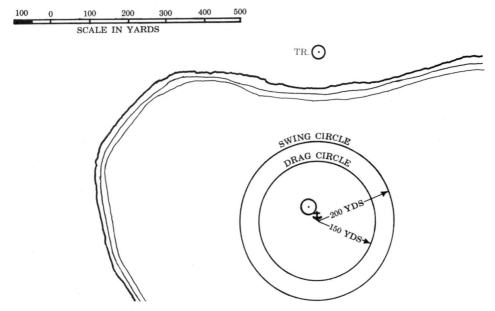

Figure 14–4. *An expanded view of the swing and drag circles.*

conditions, there is often no unusual sensation of ship's motion or other readily apparent indication of the fact. The safety of the ship depends on the ability of the anchor-bearing watch to accurately plot frequent fixes and to alert all concerned if they begin to fall outside the drag circle. If conditions warrant, the ship may have to get under way. As interim measures to be taken while the ship is preparing to do this, more chain may be veered to increase the total weight and catenary of chain in the water, and a second anchor may be dropped if the ship is so equipped.

In situations in which high winds are forecast, the ship should assume an increased degree of readiness, with a qualified conning officer stationed on the bridge, and a skeleton engineering watch standing by to engage the engines if necessary. As an example, during a Mediterranean deployment a U.S. Navy destroyer was anchored off Cannes, France, in calm waters with less than five knots of wind blowing. Because high winds had been forecast for later in the day, the officer of the deck was stationed on the bridge, and a skeleton engineering watch was charged with keeping the engines in a five-minute standby condition. Two hours after anchoring, after two liberty sections had gone ashore, the wind began to increase. In the next 45 minutes, wind force increased to the point where 75-knot gusts were being recorded. The ship got under way and steamed throughout the night until the storm abated the next day.

Summary

This chapter has described the procedures used by the navigator during precise piloting in constricted waters when the accuracy required necessitates the incorporation of his vessel's handling characteristics data in the dead reckoning plot. The navigator should take the effects of wind and current into consideration when using handling data, and he should also be aware that subsequent alterations might have changed the ship's actual characteristics somewhat from the listed figures.

Precision anchoring involves almost every skill the navigator possesses in order to drop the anchor successfully within 50 yards of the center of the designated anchorage. Seventy-five percent of the navigator's work in anchoring is accomplished before the actual evolution takes place. The navigator must ensure that all members of the ship's company who will be involved are prepared to play their roles.

After the anchor has been set and the chain secured at its final length, the navigator must make sure that a properly trained and alert anchor-bearing watch is stationed to plot and record bearings for a

fix at least once every 15 minutes to insure that the ship is not dragging anchor. In the event that the anchor is discovered to be dragging, the ship must be prepared to get under way to reset the anchor if necessary.

15

Voyage Planning

The previous chapters in this text have each dealt in detail with some aspect of either the marine navigator's environment or his duties during the practice of piloting. This chapter will examine how these different factors are brought together and employed by the navigator in planning an extended voyage.

Since a knowledge of time and how it is determined is essential during voyage planning for estimating departure and arrival times and dates, for planning time zone adjustments en route, and for many other purposes, an introductory section on time is included at the beginning of this chapter. It will provide a brief overview of the determination of time, time zones, and their use. Subsequent sections of this chapter will examine various aspects of the planning activities of a Navy navigator as he proceeds through the steps in preparation for a typical Mediterranean deployment of a Navy ship with the U.S. Sixth Fleet, with the main emphasis being placed on the planning of a voyage from Norfolk, Virginia, to Gibraltar, and thence to Naples, Italy.

Although the particular planning process discussed herein as an example is oriented toward a Navy ship, most aspects of the process will have direct parallels in the case of a navigator planning a similar transoceanic voyage on a commercial ship or private vessel. In fact, the same concerns and general procedures apply to almost any major voyage to be undertaken on almost any type of surface vessel.

Time

The consideration of time is always of major importance in every voyage planning process. Almost every planning action of the navigator is concerned in some way with the timely arrival of the ship at her destination and at intermediate points en route. In the case of both commercial and naval ships, the time and date of arrival are normally specified by higher authority in the form of an estimated time of arrival (ETA) that must be met. In order to arrive on time, the navigator must compute the estimated time and date of departure

(ETD) as well as the ETAs at several points along the track to check his progress. As the ship moves from one time zone into another as she proceeds along her route, the navigator must recommend when to reset the ship's clocks and in what direction, to keep the ship's time in accord with the local zone time. As will be explained in *Marine Navigation 2: Celestial and Electronic*, the consideration of time is also very important in the performance of celestial navigation practiced after the ship has passed beyond the limits of piloting waters.

The Development of Zone Time

To understand the concept of time, it is necessary to have a basic knowledge of how time is derived. Until the nineteenth century, man had been accustomed to reckoning time according to the apparent motion of the sun across the sky. In the late 1800s, however, the development of comparatively rapid transit systems such as the railroad and the steamship made keeping time according to the motions of the "apparent" sun impractical, inasmuch as timepieces had to be adjusted every time one's position on the earth changed. Moreover, because of the elliptic path of the earth around the sun, the rate at which the apparent sun moves across the sky is not constant, but varies from day to day. To avoid this latter difficulty, the concept of a theoretical "mean" sun passing completely around the earth at the equator once every 24 hours came to be widely used for marking the passage of time. One "mean" solar day is 24 hours in length, with each hour consisting of 60 minutes and each minute, 60 seconds. Since the mean sun completes one circuit of the earth every 24 hours, it follows that it moves at the rate of 15° of arc as measured at the equator—or 15° of longitude—every hour. This fact will assume great significance in *Marine Navigation 2* when the procedures to use the sun to obtain celestial lines of position are discussed.

At first, this "mean" solar time was reckoned according to the travel of the mean sun relative to the observer's meridian; this was called *local mean time* (LMT). Shortly thereafter, however, to eliminate the necessity for frequent resetting of timepieces, the earth's surface was divided into 24 vertical sectors called *time zones*, each of which is 15° of longitude in width. Time within each zone is reckoned according to the position of the mean sun in relation to the central meridian of the zone. Thus, clocks are changed by one hour increments only when transiting from one fiteen-degree longitude zone to another, and every location within the same time zone keeps the same "standard" time. Time reckoned in this manner is called *zone time* (ZT).

Figure 15–1 depicts a standard time zone chart of the world. Because it is based on a Mercator projection, each sector appears as a

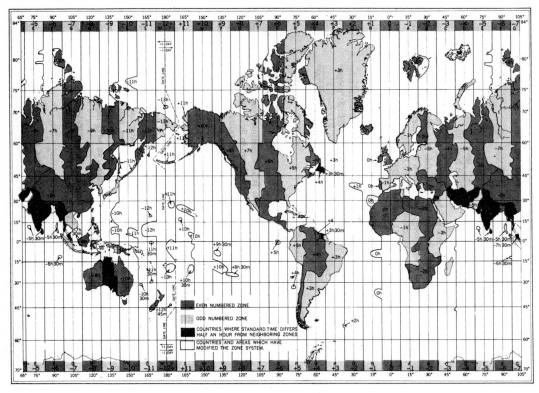

Figure 15–1. Standard time zone chart of the world.

vertical band 15° of longitude in width. As can be seen by an inspection of the time zone chart, each zone is defined by the number of hours of difference existing between the time kept within that zone and the time kept within the zone centered on the prime (0°) meridian passing through Greenwich, England. Each zone is also labeled with letters called *time zone indicators* that assist in identification of the zone.

Time based upon the relationship of the mean sun with the prime meridian is called *Greenwich Mean Time*, abbreviated GMT; it is often referred to as ZULU time, because of its time zone indicator letter. The farther to the west of Greenwich that a time zone lies, the earlier will the time kept in that zone be in relation to GMT. This is indicated by placing a plus (+) sign in front of the hourly difference figure, to indicate the number of hours that must be added to the local zone time to convert it to GMT. By the same token, the farther to the east of Greenwich a time zone is located, the later its time will be relative to Greenwich Mean Time. This fact is indicated by minus (−) signs, indicating the number of hours that must be subtracted from local zone time to obtain GMT. The Greenwich zone is centered

at the prime meridian, and it extends 7½ degrees to either side. A new time zone boundary lies every 15 degrees thereafter across both the eastern and western hemispheres, resulting in the twenty-fourth zone being split into two halves by the 180th meridian. The half on the west side of this meridian keeps time 12 hours behind GMT, making its difference + 12 hours, while the half on the east side keeps time 12 hours ahead, resulting in a difference of − 12 hours. Thus, there are in reality 25 diffenent standard time zones, numbered from + 1 through + 12 to the west of the Greenwich zone, and − 1 through − 12 to the east. These differences are usually referred to as the *zone description* or *zone difference*, abbreviated ZD.

From the preceding discussion and from an inspection of the time zone chart, it follows that as a general rule the standard time zone in which any particular position on earth is located can be found simply by dividing its longitude by 15. If the remainder from such division is less than 7½° , the quotient represents the number of the zone; if the remainder is greater than 7½° , the location is in the next zone away from the Greenwich meridian. If there is no remainder, the location lies exactly on the central meridian of a time zone. The sign of the zone is determined by the hemisphere in which the position is located. In the western hemisphere the sign is positive, and in the eastern hemisphere the sign is negative. Applying the proper sign to the number found by division yields the zone description (ZD). As an example, if the standard time zone in which Norfolk, Virginia, is located were required, its longitude, 76° 18.0′ West, would be divided by 15 to yield a quotient of 5 with a remainder of 1° 18′. Thus, it is located in the + 5 time zone, which has the time zone indicator letter *R* assigned.

For the navigator sailing the world's oceans the 15 degree-wide standard time zones are convenient to use, but in practice many places do not always adhere precisely to the time of the zone in which they are physically located, as indicated in Figure 15–1; to do so would in many cases cause a great deal of confusion in conducting business and travel. As a result, time zone boundaries follow state and country boundaries in many cases, while some regions establish their own local "standard" times different from any of the 24 accepted zones. The South American country of Guyana, for example, maintains time 3 hours and 30 minutes behind GMT. In Antarctica, where all 24 time zones converge at the pole, GMT is used throughout.

Daylight savings time is a device adopted by some countries, including the United States, to extend the hours of daylight in the evening during summer. Locations keeping daylight savings time keep the time of the next zone to the east of the time zone in which

they are located. When daylight savings time is being kept in Norfolk, for example, it keeps +4Q time rather than its standard +5R time. For purposes of the voyage described in this chapter from Norfolk to Naples, daylight savings time has been disregarded, even though it probably would be in effect during the time of the year in question. The time kept at any particular location and time of year can normally be found in the applicable volume of the *Sailing Directions* for foreign ports or the *Coast Pilots* for U.S. ports.

Time Conversions

Because of the difficulties inherent in working with the different zone times in calculating times of arrival and departure on long voyages such as the Norfolk to Naples transit of this chapter, the navigator usually first converts all times to GMT prior to the initial planning stages of a voyage. After all ETAs and ETDs have been computed in GMT, certain times of interest can then be converted to local zone time. To do this, extensive use is made of the formula

$$ZT + ZD = GMT,$$

in which ZT is the local zone time for the location of interest, and ZD is the zone description previously described. When correcting from zone time to GMT, the zone difference is added algebraically to the zone time; but when converting to zone time from GMT, the sign opposite the zone difference must be used.

As an example, suppose that the navigator wished to convert 0800 local zone time on 30 June at Naples, Italy, to Greenwich Mean Time. Since the longitude of Naples is 14° 16′ East, it lies in the −1A time zone, and the local zone time to be converted could be written as 0800A 30 June. Time zone indicators are normally placed after all four-digit time figures in the manner shown here to avoid confusion during time conversion. To convert 0800A 30 June to GMT, the zone difference of −1 hour is added algebraically to yield 0700Z 30 June.

As a second example, suppose that it is desired to convert 1000Z 18 June at Norfolk, Virginia, to local zone time. To accomplish this, the formula above is rearranged to the following form:

$$ZT = GMT - ZD;$$

thus it is necessary to subtract the +5 zone difference from GMT, yielding 0500R 18 June.

When making time conversions, the navigator must be alert for situations in which the date changes as a result of the conversion process. For instance, suppose in the first example discussed above that the time at Naples to be converted were 0030A 30 June. Sub-

tracting the -1 zone difference yields 2330Z 29 June. Likewise, if the time in the second example at Norfolk had been 0245Z 18 June, conversion to local zone time would result in 2145R *17* June.

Further examples of time conversions are given in the following table:

Location/Position	Local Zone Time (ZT)	Zone Description (ZD)	Greenwich Mean Time (GMT)
Norfolk, Va.	0700R 30 Jun	+5R	1200Z 30 Jun
Greenwich, Eng.	1200Z 15 Jul	+0Z	1200Z 15 Jul
Tokyo, Japan	1200I 15 Apr	−9I	0300Z 15 Apr
160°W	1200X 15 May	+11X	2300Z 15 May
165°W	1400X 15 Jun	+11X	0100Z 16 Jun
173°W	1200Y 17 Jul	+12Y	2400Z 17 Jul
160°E	1200L 1 Jul	−11L	0100Z 1 Jul
165°E	0800L 1 Jul	−11L	2100Z 30 Jun
173°E	1200M 2 Sep	−12M	0000Z 2 Sep

To assist in conversions of times from one zone to another, particularly those involving date changes, an aid called the Time Comparison Table can be used; it is printed on most standard time zone charts. A sample table of this type appears in Figure 15–2. The table

ZONE	+12	+11	+10	+9	+8	+7	+6	+5	+4	+3	+2	+1	0	−1	−2	−3	−4	−5	−6	−7	−8	−9	−10	−11	−12
00	00	01	02	03	04	05	06	07	08	09	10	11	12	13	14	15	16	17	18	19	20	21	22	23	/
01	01	02	03	04	05	06	07	08	09	10	11	12	13	14	15	16	17	18	19	20	21	22	23	/	01
02	02	03	04	05	06	07	08	09	10	11	12	13	14	15	16	17	18	19	20	21	22	23	/	01	02
03	03	04	05	06	07	08	09	10	11	12	13	14	15	16	17	18	19	20	21	22	23	/	01	02	03
04	04	05	06	07	08	09	10	11	12	13	14	15	16	17	18	19	20	21	22	23	/	01	02	03	04
05	05	06	07	08	09	10	11	12	13	14	15	16	17	18	19	20	21	22	23	/	01	02	03	04	05
06	06	07	08	09	10	11	12	13	14	15	16	17	18	19	20	21	22	23	/	01	02	03	04	05	06
07	07	08	09	10	11	12	13	14	15	16	17	18	19	20	21	22	23	E	01	02	03	04	05	06	07
08	08	09	10	11	12	13	14	15	16	17	18	19	20	21	22	23	G	01	02	03	04	05	06	07	08
09	09	10	11	12	13	14	15	16	17	18	19	20	21	22	23	N	01	02	03	04	05	06	07	08	09
10	10	11	12	13	14	15	16	17	18	19	20	21	22	23	A	01	02	03	04	05	06	07	08	09	10
11	11	12	13	14	15	16	17	18	19	20	21	22	23	H	01	02	03	04	05	06	07	08	09	10	11
12	12	13	14	15	16	17	18	19	20	21	22	23	C	01	02	03	04	05	06	07	08	09	10	11	12
13	13	14	15	16	17	18	19	20	21	22	23		01	02	03	04	05	06	07	08	09	10	11	12	13
14	14	15	16	17	18	19	20	21	22	23	E	01	02	03	04	05	06	07	08	09	10	11	12	13	14
15	15	16	17	18	19	20	21	22	23	T	01	02	03	04	05	06	07	08	09	10	11	12	13	14	15
16	16	17	18	19	20	21	22	23	A	01	02	03	04	05	06	07	08	09	10	11	12	13	14	15	16
17	17	18	19	20	21	22	23	D	01	02	03	04	05	06	07	08	09	10	11	12	13	14	15	16	17
18	18	19	20	21	22	23	/	01	02	03	04	05	06	07	08	09	10	11	12	13	14	15	16	17	18
19	19	20	21	22	23	/	01	02	03	04	05	06	07	08	09	10	11	12	13	14	15	16	17	18	19
20	20	21	22	23	/	01	02	03	04	05	06	07	08	09	10	11	12	13	14	15	16	17	18	19	20
21	21	22	23	/	01	02	03	04	05	06	07	08	09	10	11	12	13	14	15	16	17	18	19	20	21
22	22	23	/	01	02	03	04	05	06	07	08	09	10	11	12	13	14	15	16	17	18	19	20	21	22
23	23	/	01	02	03	04	05	06	07	08	09	10	11	12	13	14	15	16	17	18	19	20	21	22	23
ZONE	+12	+11	+10	+9	+8	+7	+6	+5	+4	+3	+2	+1	0	−1	−2	−3	−4	−5	−6	−7	−8	−9	−10	−11	−12

(Left margin column labeled TIME)

Figure 15–2. Time Comparison Table.

can be used to determine the zone time and date of any zone from a given zone time and date. When crossing the date change line shown in the table from left to right, one day is added to the date; when crossing from right to left, one day is subtracted. As an example of its use, if the time and date at Washington, D.C. (+5R) were 1900R 1 February, the corresponding time and date at Sydney, Australia (−10K) would be 1000K 2 February. As a second example, if the time and date at Tokyo, Japan (−9I) were 0900I 1 February, the time and date in San Francisco (+8U) would be 1600U 31 January. The circled figures in the diagram refer to these two examples.

In labeling the track, as well as in writing messages, times and dates are usually written in an alphanumeric format called a *date-time group*, as illustrated in Figure 15–3. If all dates fall within the same month and year, the month and year abbreviations can be omitted when labeling a track. Several examples of the use of a date-time group in labeling the track are given later in this chapter.

1 7	1 0 0 0	R	J U N	8 2
DAY OF MONTH	4-DIGIT TIME	TIME ZONE INDICATOR	3-LETTER MONTH ABB	LAST TWO DIGITS OF YEAR

Figure 15–3. *Format of a date-time group.*

Changing Time Zones and Dates En Route

To avoid undue psychological stress to crew members, ship's clocks are normally set to conform to the time zone in which the ship is operating. The navigator, therefore, should include the schedule of time zone changes in his planning efforts.

When steaming in an easterly direction, as in traveling from Norfolk to Naples, it is necessary to advance ship's clocks by one hour periodically in order to conform to the proper zone time. When proceeding in a westerly direction, the clocks must be continually retarded. In cases requiring clocks to be advanced, the normal procedure in the Navy is to advance them during the morning hours, generally at either 0100 or 0200, so as not to disrupt the normal working day. When clocks are set back, the period during the second dog watch from 1800 to 2000 is usually selected, to avoid the possibility of a five-hour watch in a four-hour watch rotation cycle. To assist in determining when the ship's clocks should be reset, the navigator should label his track at intervals with the estimated times of arrival

(ETAs) at these points. The times are normally first calculated in Greenwich time and then converted to local zone time. Legs of the track that will require adjustment of the ship's clocks can then be determined by inspection.

The 180th meridian is designated as the International Date Line, because the time kept in the 7½-degree wide zones on either side of it differs by 24 hours or one complete day. When crossing the date line on a *westerly* heading, the zone description changes from + 12 to − 12, so ship's clocks must in effect be *advanced* 24 hours, thereby "losing" (ie, skipping ahead) one calendar day. Conversely, when crossing the date line on an *easterly* heading, the ship's clocks must in effect be *retarded* 24 hours, since the zone description changes from − 12 on the western side of the line to + 12 on the eastern side. In this instance, the ship is said to have "gained" a day, since one calendar day is repeated. For convenience, ships crossing the date line will usually effect the date change at night, normally at midnight. In the Navy, if the date change would cause a Sunday or holiday to be repeated or lost, the ship may operate for a time using a zone description of + 13 or − 13, to allow the date change to be made on either the preceding or the following working day.

The Voyage-Planning Process

The key to successful planning of either a single voyage or an extended deployment is early advance preparation by the navigator. In the case of Navy ships, the first notification of a deployment and its associated transits is normally received when the *Yearly Employment Schedule* for the fleet of which the ship is normally a part is published by the cognizant fleet commander. This publication is normally received by the ship's operations officer, who extracts all pertinent information on the ship's schedule, compiles it, and then disseminates it to all officers and personnel concerned. If a deployment is scheduled during the first quarter of the year covered by the *Yearly Employment Schedule*, advance notice is usually received by other means three to six months before. Confirmed deployment dates will be indicated in subsequently published *Quarterly Employment Schedules* or by message.

As soon as the navigator receives word as to the area of the world to which the ship will proceed, he should commence the voyage planning process. In order not to overlook any of the many facets of an extended voyage and deployment, the navigator should normally lay out all of his anticipated requirements in the form of a check-off list. The list below, although compiled for a sample deployment from

Norfolk, Virginia, to the Sixth Fleet port of Naples, Italy, is representative of the general items such a list should contain.

1. Publications and Charts
    a. Is the normal operating allowance of publications and charts on board?
    b. Are all charts and publications in normal use corrected and up to date with the latest *Notice to Mariners?*
    c. Which chart portfolios and publications not in normal use, on allowance and on board, will be required for the voyage and which must be brought up to date by the use of chart and publication correction cards listing *Notice to Mariners* corrections? Are any out of date or superseded?
    d. Which charts and publications not normally on allowance have to be ordered?
    e. Do the pre-deployment check-off lists and Sixth Fleet operation orders and Fleet instructions require any additional charts and publications?
    f. Do the operation orders require any classified combat charts for gunfire support?
    g. Are the latest pilot charts on board for use in the initial transit?

2. Equipment
    a. Are magnetic compasses within acceptable limits or is a period at sea to swing ship required prior to departure?
    b. Is the degaussing system operating properly? When was the last degaussing run completed? Will the ship be required to run the degaussing range prior to departure?
    c. Does the ship have a sufficient number of spare PMP parts on board for use by both the navigator and CIC over a six-month deployment?
    d. Is there a three-arm protractor on board for emergency use?
    e. Are the Loran/Omega/satellite navigation sets working properly? If not, is the electronics material officer aware they need repair? Are spare parts for these receivers on board?
    f. Are all chronometers, stop watches and comparing watches operating properly and up to allowance? Are

the ship's clocks up to allowance and properly calibrated?

    g. Are the echo sounder and the pit log operating prop-
erly?

3. Personnel
    a. Is the department/division up to allowance in key petty
officers?
    b. Will all key personnel remain on board throughout the
cruise?
    c. Are sufficient numbers of qualified strikers on board to
stand long periods of underway watches?

4. Operations
    a. What is the total distance from Norfolk to Naples?
    b. What will be the ETD and ETA?
    c. Is the voyage at least 1,500 miles so that Optimum Track
Ship Routing must be requested from the appropriate
Naval Oceanography Center?
    d. If the voyage is less than 1,500 miles, is there a pos-
sibility of hurricanes or typhoons en route so that OTSR
can be requested?
    e. Should the track originally recommended be changed
based on OTSR recommendations? Will the ETD and
ETA change?
    f. Have climatological summaries been requested from
the appropriate Oceanography Center to give frequency of various wind speeds and their direction, wave
heights, currents, and the probability of rain and storms
en route?
    g. Has the operations officer been provided with the final
ETD from Norfolk and the ETA at Naples? Has he been
given the date-time group (DTG) in Greenwich Mean
Time (GMT) and the latitude and longitude of all major
junction points of the coastal, rhumbline, and great
circle tracks to be employed?
    h. Has the operations officer been advised to request
weather en route (WEAX) in the movement report?

As can be seen by an inspection of this check-off list, the navigator
is faced with many areas of concern in planning a voyage of any
duration. As the days and weeks pass, and the departure date draws
ever closer, the intensity of his planning steadily increases until several days before the start of the journey, when virtually all his working

hours will probably be filled checking the innumerable and inevitable last-minute details.

The remaining sections of this chapter will focus on the key aspects of planning the transit from Norfolk, Virginia, to Naples, Italy, and will conclude with a discussion of some general points applicable to both this transit and the subsequent deployment.

Obtaining and Updating Charts and Publications

Among the first tasks of the navigator after he has been informed of a scheduled extended voyage or deployment is to ascertain which charts and publications will be required, and of these, which are on board and which are not. To find this information, he consults the applicable allowance lists contained within the *Allowance List* part of *Pub. 1–N*, described in Chapter 5. Amplifying information on requirements for charts and publications can also be found in applicable standing fleet operation orders and instructions. Any portfolios and charts either not on board or outdated are ordered from the Defense Mapping Agency Office of Distribution Services (DMAODS).

Having ascertained his requirements, the navigator then consults the DMAHTC *Catalog of Nautical Charts, Publication 1–N*, and the NOS *Nautical Chart Catalog 1* to find the numbers of all charts and *Coast Pilot* and *Sailing Directions* volumes that will be of use in planning and executing the initial transit from Norfolk to Naples. He directs the chart petty officer to pull all those charts and publications currently on board from their storage locations, along with their correction cards, and to apply to them all the necessary corrections from the file copies of the *Notice to Mariners*.

When the ordered charts and publications are received from DMA and corrected up to date, the navigator then has on hand all the information he needs to successfully and intelligently plan for the forthcoming transit and deployment. The navigator does not wait for any missing material to arrive, however, before he begins planning. Instead, he proceeds with the determination of the departure and if necessary the arrival dates and times, as set forth in the following section.

Determination of the Departure and Arrival Dates

When a Navy ship is to deploy, one of two possible situations will exist. She will either deploy in company with other ships in a group referred to as a "task group," or she will transit independently to her assigned fleet, joining it on arrival. When a ship is to deploy as part of a task group, an operation order is usually made up by the task group commander and his staff to cover the transit from the home

port to the destination. After completion of the deployment, a second operation order is compiled to govern the return trip. In situations of this type, the navigator of the flagship is normally assigned additional duties as staff navigator, except in those rare instances where a staff is so large as to have a permanently assigned navigator attached to it. In any event, the staff navigator will assume the responsibility of planning the voyage for the entire task group, with possible assistance rendered by navigators from the various ships comprising the group. When a ship is to transit independently, on the other hand, all the work of preplanning must be accomplished by the ship's navigator and his assigned personnel. For the Norfolk to Naples voyage under discussion, it will be assumed that the ship is to conduct an independent transit.

Regardless of whether the ship is transiting in company or independently, the arrival time at the destination is usually specified by higher authority. For purposes of the mock voyage described herein, it will be assumed that the ship has been directed to arrive at Naples at 0800A 30 June. Should the navigator later conclude that this arrival time is unsafe for some reason, such as a low tide or high current velocity, a request can be initiated to change the ETA; such instances, however, are rare. The navigator, then, is usually concerned mainly with the determination of an estimated time of departure that will ensure the timely arrival of the ship at the destination.

In the case of a ship conducting an independent transit, the date and time of departure is normally specified in the form of a message from the ship's squadron commander, but it is based on the ETD which the ship's navigator has previously recommended as best fitting the ship's needs for the transit. To determine the estimated time and date at which the ship should depart, the navigator must first determine the approximate distance between Norfolk and Naples along the most suitable route at the given time of year. The total number of hours required for the journey at the approximate speed of advance (SOA) to be used is then computed by the use of the speed-time-distance formula. The distance between Norfolk and Naples can be found easily by use of *Publication No. 151, Distance Between Ports*. The navigator can sail a great circle route across the Atlantic to Gibraltar, but sailing a great circle from Gibraltar to Naples is impossible due to the geography of the Mediterranean. Thus, *Publication No. 151* is entered first for the great circle distance from Norfolk to Gibraltar, 3,335 miles. It is entered a second time for the distance from Gibraltar to Messina, on the toe of Italy, 1,049 miles, and a third time for the distance between Messina and Naples, 175 miles. Adding the three together yields a total distance from Norfolk to Naples of 4,559

miles. The maximum SOA is normally specified by higher authority; for most conventionally powered destroyers it is usually set at about 16 knots, the most economical cruising speed for this type of ship. By dividing the 4,559 mile total by 16 knots, a total time in transit of 285 hours, or 11 days and 21 hours is computed.

At the beginning of this chapter in the section on time, it was shown that Norfolk lies in the +5R time zone, while Naples is in the −1A zone. Thus, the ship will "lose" 6 hours of time en route because of time zone adjustments; this figure must also be added to the initial time requirement figure. The ship must depart Norfolk, therefore, at least 12 days and 3 hours before her scheduled arrival time at Naples. Subtracting this time from 0800A 30 June yields an estimated time of departure from Norfolk of 0500R on 18 June. An alternative and in most cases preferred method of obtaining the same result would be to first convert the designated arrival time at Naples to GMT. The computed transit time, 285 hours, is then subtracted, and finally the resulting time of departure from Norfolk expressed in GMT is converted to Romeo time by subtracting the +5 zone difference. To facilitate departure ceremonies, an ETD for deployments between the hours of 0800 and 1000 local time is usually specified by local authorities. The navigator in this case, therefore, would recommend to his commanding officer that the departure be scheduled for 1000R 17 June. The extra time would be compensated for by specifying a slower SOA on some legs of the transit.

The navigator should also consider the predicted tides and currents at Norfolk on the day of departure and at Naples on the day of arrival. Depending on the size of the ship and the location of the port, this consideration may be the overriding factor in deciding on departure time or entry time. A carrier, for example, will only enter or depart when the tide is high and the current is determined to be slack at the Navy piers in Norfolk. The state of the tide and current in Norfolk present no undue difficulties to a destroyer, except possibly during peak ebb or flood and during storm conditions, as all channels are wide, deep, and easily negotiated. In any event, either the navigator or one of his quartermasters will compute complete tide and current tables for the restricted water piloting portion of the transit on the day of departure. They will also compute a complete tide table for Naples for 30 June using the "Europe" volume of the *Tide Tables*, and obtain tidal current information for this date from charted information such as current diamonds and from the *Enroute* volume of the *Sailing Directions* for the Western Mediterranean. If information from these sources indicates that it is unsafe to enter port at the designated time, a decision may be made to request permission to

adjust the arrival time so as to enter port when tide and current conditions are more favorable.

Plotting the Intended Track

Concurrent with the determination of the ETD, the navigator will usually begin to plot his initial estimate of the ship's track for the transit. Before he actually begins to plot, however, he first gathers all available information applicable to a transit of the ocean areas under consideration during the given time of year. A most valuable aid at this stage is a climatological summary for the time of year and area under consideration, which is furnished upon request to Navy navigators by the Naval Oceanography Command Center serving his area. It will provide the normal wind speeds and directions, wave heights, currents, and the probability of rain and storms en route. Other sources of the same type of information are the appropriate editions of the *Sailing Direction Planning Guides*, the *Coast Pilots*, and the pilot chart. For the voyage under consideration, the navigator would need the *Planning Guides* for the Western Mediterranean and for the North Atlantic, the Atlantic Coast volume of the *Coast Pilots* and the North Atlantic pilot chart for the month of June. His objective in using all these references is to lay down the optimum track from Norfolk to Naples, taking advantage of great circle routes to the fullest extent possible consistent with the meteorological conditions normally prevailing during the month of June. In this regard, it will sometimes be faster when contemplating an extended voyage to choose a longer route, if in so doing known areas of major weather disturbance can be avoided. In addition to these on-board references and the Naval Oceanography Center climatological summary, the Navy navigator should also request another service available to Navy ships known as Optimum Track Ship Routing, or OTSR. In contrast to the aids discussed above, which are based on historical meteorological data, the OTSR is an optimum track for the specific requesting ship based on the actual climatological and hydrographic forecasts covering the time of the voyage. Optimum Track Ship Routing will be described in greater detail later in this chapter.

Since the shortest distance between Norfolk and the Strait of Gibraltar lies along a great circle drawn between them, the navigator begins his plot by constructing the great circle track as a straight line on a gnomonic projection. In this case, the gnomonic projection with a point of tangency in the North Atlantic is chosen; it is shown in Figure 15–4A next page. Since this chart cannot be used for navigational purposes, it is then necessary to transfer the great circle route onto a Mercator projection on which the great circle track is

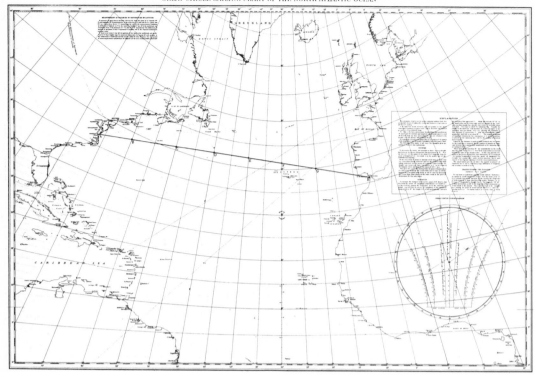

Figure 15–4A. Great circle route, Norfolk to Gibraltar.

approximated by a series of rhumb lines. To transfer the track, the navigator selects convenient points along the great circle on the gnomonic chart, about 300 to 500 miles apart, and he transfers these points to the Mercator chart by replotting them at their correct latitude and longitude. Each consecutive point is labeled either with letters or a sequential alphanumeric designation such as R–1, R–2, etc. Rhumb lines connecting these points are then drawn. At this stage, the track appears as shown in Figure 15–4B. Since the small scale of the North Atlantic chart shown in Figure 15–4B makes its use for actual navigation impractical, the points along the track and the rhumb lines connecting them are also laid off on larger scale "working" charts and plotting sheets of smaller portions of the North Atlantic, which are suitable for plotting celestial and electronic fixes.

It should be mentioned here that it is also possible to determine a series of rhumb lines approximating a great circle track by use of a modern electronic calculator or a ship's computer to solve a computational algorithm developed for this purpose. More sophisticated calculators, and calculators designed especially for marine navigation, come equipped with this algorithm "built in" to their navigation pro-

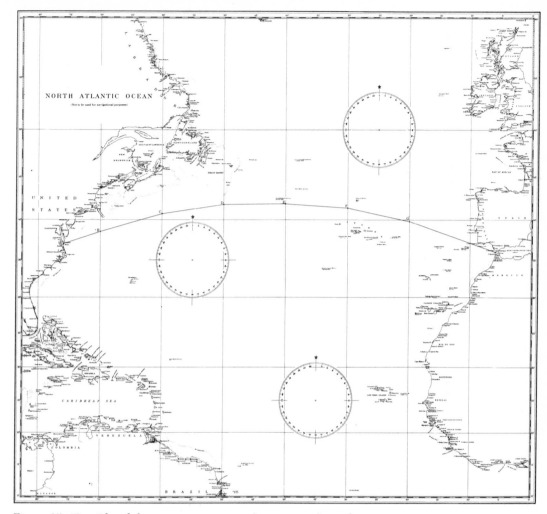

Figure 15–4B. Rhumb line approximation to the great circle track.

gram packages or internal memories. Other applications for the electronic calculator in marine navigation are discussed in Appendix A of *Marine Navigation 2*.

Completing the Track

After the track from Point A off Hampton Roads to the entrance to Naples has been laid down, the exit track from Norfolk and the entrance track into Naples are plotted in accordance with procedures set forth in Chapter 8. Figure 8–16 on page 146 shows a portion of a typical exit track from Norfolk. If the ship is to anchor at a predesignated anchorage at Naples, the anchorage plot may also be com-

pleted at this time using procedures described in the preceding chapter.

To complete the plot, distances along the rhumb line segments of the track are measured and totaled to obtain the actual distance measured along the track from Norfolk to Naples. This figure should agree fairly well with the distance found previously from the *Distance Between Ports* publication, but differences of 10 or 20 miles are not uncommon due to slight variations in the tracks upon which they were based. After computing the speeds of advance for the various legs of the track as explained in the following section, most navigators label each junction point on the large-scale chart with the distance remaining to the destination and the computed ETA at each point, expressed in both Greenwich mean and local zone time. Other navigators may record this information in a separate table, while a few do both. In any case, the information should be readily available for use by all concerned. As a minimum, each rhumb line segment on the large-scale plot should be labeled with the track direction and

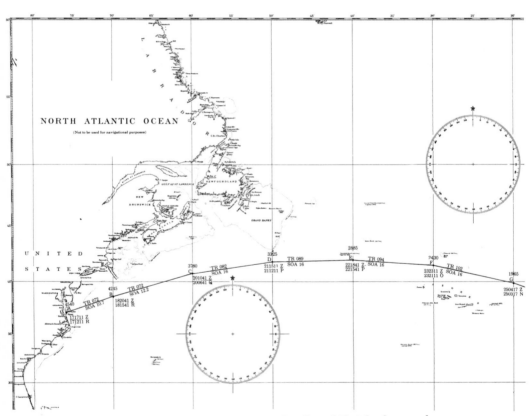

Figure 15–5. Segment of a completed Norfolk-Gibraltar track.

speed of advance, and each junction point should be labeled with the distance remaining and the ETA in local zone time. A portion of the completed track appears in Figure 15–5, and a sample table supporting it follows.

Track Description Norfolk to Gibraltar

Point	Latitude	Longitude	Course to Next Point	Distance to Next Point	ETA	Distance to Naples
Pier	36–57.0N	76–20.0W	Various	19	———	4599
A	36–56.5N	75–56.0W	072	295	171711Z	4540
B	38–46.0N	70–00.0W	075	465	182041Z	4245
C	40–40.0N	60–00.0W	082	455	201041Z	3780
D	41–50.0N	50–00.0W	089	440	211511Z	3325
E	42–00.0N	40–00.0W	094	455	221841Z	2885
F	41–30.0N	30–00.0W	102	465	232311Z	2430
G	39–50.0N	20–00.0W	109	690	250417Z	1965
H	36–00.0N	06–00.0W	090	51	262323Z	1275
Gibraltar	35–57.0N	05–45.0W	———	———	270235Z	1225

If the ship is to steam within sight of land at night during the transit, as when passing Gibraltar or approaching Naples, the applicable volume of the *List of Lights* should be consulted, and a table should be prepared listing the computed visibility of all lights that should be seen. After the track has been drawn, the visibility arcs for all these lights should be laid down on the large-scale coastal charts of the area in accordance with procedures set forth in Chapter 6.

Determining SOAs of the Track Legs

It was shown previously that the ship must depart Norfolk no later than 0500R on 18 June (180500R JUN) to arrive at Naples on time using an average SOA over the entire track of 16 knots. It was further stated, however, that the navigator would probably recommend that the ship actually depart at 1000R on the seventeenth, to facilitate her departure. Thus, if the commanding officer agrees with this recommendation, the ship will have 19 hours of slack time to expend en route. The navigator could account for this slack time in a number of ways. After consulting with the ship's planning board for training, he might plan for the extra time to be expended in the course of emergency drills, standard training exercises, and shiphandling drills, or he could simply lower the overall SOA somewhat in order to proceed at a slower speed. In any event, he must carefully plan for

expending the time because once Navy ships are under way, they are normally required to maintain position within 50 miles or four hours of steaming time, whichever is less, from their prefiled planned positions at any time. As can be seen by an inspection of the SOA figures in Figure 15–5 and the accompanying table, in this case the navigator has planned to use up all the slack time during the ship's transit to Point C on the track. Because the ship must proceed at slow speeds while exiting Hampton Roads, he has allowed for an overall SOA from the pier to Point A of only 8.7 knots, which accounts for one hour of the extra time. He has planned to use 9 of the remaining 18 hours between Points A and B, and the other nine hours between Points B and C, thus in effect reducing the required SOAs for these legs from 16 knots to 10.7 knots and 12.2 knots, respectively. In reality, the ship may actually slow down or stop several times during her travel from Point A to Point C while conducting drills and exercises, and then proceed at 20 or 25 knots for a short time to maintain the planned SOAs.

The theoretical position of the ship on the intended track at any time is sometimes referred to as her *position of intended movement*, abbreviated PIM. The PIM moves along the track at the SOA, and the actual ship's position can be described in relation to it. If the ship arrives two hours ahead of the planned ETA at point C, for example, it might be said that she is "two hours ahead of PIM" at that point. At the start of a transit, it is the usual practice to proceed at an ordered speed slightly faster than the SOA, in order to build up some additional slack time in case some unanticipated delay occurs en route, such as a storm or mechanical breakdown. In general, it is always best to proceed slightly ahead of PIM at first, and to decrease speed to account for any time thus gained toward the end of the voyage, rather than being late, after having been delayed en route by some unexpected occurrence.

Optimum Track Ship Routing

Optimum Track Ship Routing, or OTSR, is a forecasting service provided by the Naval Oceanography Command, which seeks to provide an optimum track for naval vessels making transoceanic voyages. This track is selected by applying ocean wave forecasts and climatological information to individual ship loading and performance data. The use of the OTSR recommendations does not guarantee smooth seas and following winds, but it does promise a high probability of sailing the safest and most rapid route consistent with the requirements of the requesting ship. Since the service became fully operational in 1958, ships of the Navy have been able to stay on schedule

during transits to a degree seldom seen before the institution of the service, and ship and cargo damage have been significantly reduced.

OTSR is an advisory service only, and use of its recommended track does not relieve the commanding officer and navigator of Navy ships from the responsibility to properly plan the voyage using all other means available. Neither does it seek to limit the prerogative of the commanding officer in deciding upon the route his vessel will follow. If the ship does deviate to any great extent from the OTSR route, the Naval Oceanography Command Center that performs the service should be informed by message.

Using OTSR Services

Requests for the OTSR service should be submitted for receipt by the Naval Eastern Oceanography Center (NEOC) Norfolk, Virginia, (Atlantic area) or by the Naval Western Oceanography Center (NWOC), Pearl Harbor, Hawaii, (Pacific area) at least 72 hours prior to the estimated time of departure (ETD). In practice, most navigators submit their requests several weeks before this deadline, and they include a request for a route recommendation based on long-range climatological data. The OTSR route recommendation based on the actual short-range weather forecast is normally received by message about 36 hours prior to the ETD.

Information on the OTSR request includes the following:

Name and type of ship
Point of departure and ETD
Destination, including the latest acceptable arrival time
Intended SOA
Draft
Any unusual conditions of loading

Specific instructions for requesting OTSR services can be found in the U.S. Navy *NAVOCEANCOMINST 3140.1* series. After receipt, the navigator should compare the OTSR recommendations with his own preplanned track. Usually they will not vary a great deal if there is no major storm activity along the ship's intended track. On those occasions when the two tracks do diverge significantly, the navigator will normally recommend altering his track to conform to the OTSR track, adjusting the SOA if necessary so as to arrive on time at the destination. Should the ship not adhere to the OTSR recommendations and in so doing encounter a storm en route which she would have avoided had she sailed the OTSR route, the commanding officer may be held responsible for any storm damage incurred.

While en route the navigator is required to send a short OTSR weather report at 0800 local zone time each day to NEOC or NWOC. Using this information, the cognizant Center continuously surveys and reevaluates the route. The message includes the following information:

Position in latitude and longitude
Course and speed being made good
Wind direction and speed
Wave direction, period, and height
Swell direction, period, and height

Miscellaneous Considerations

The preceding sections of this chapter have each dealt with some aspect of the procedures used by the Navy navigator in deciding upon and laying down the ship's track in preparation for an extended voyage. Although preparation of the track is probably the major concern of the navigator during voyage planning, there are many other areas to which he must devote some attention both before and after getting under way on the deployment. Several of the more important of these concerns are highlighted in the following paragraphs.

The Ship's Position Report

By both custom and formal regulations, Navy navigators are required to submit a ship's position report in writing to the commanding officer three times daily when the ship is under way, at 0800, 1200, and 2000 local zone time. This report is submitted on a standard form called a *ship's position report*, shown in Figure 15–6. For the transit from Norfolk to Naples, each report should have entered on it, in addition to the other required data, the distance to and ETA at the next track junction point and, under the remarks section, the amount of time by which the ship is ahead or behind the intended track. All position reports should be signed by the navigator.

Routine Messages

The navigator of a ship such as the destroyer cited in this chapter that does not have a separate meterological division on board is responsible for the transmission of synoptic weather observation reports four times each day when under way. The reports are transmitted to the Naval Oceanography Command Center in whose area of responsibility the ship is located during its transit. The content of these messages is discussed in Chapter 2 (see Figure 2–3 page 8).

Several days prior to entering any port at which a U.S. naval facility

SHIP'S POSITION
NAVSHIPS-1111 (REV. 5-62)

TO:

COMMANDING OFFICER, USS

AT *(Time of day)*	DATE

LATITUDE	LONGITUDE	DETERMINED AT

BY *(Indicate by check in box)*

☐ CELESTIAL   ☐ D. R.   ☐ LORAN   ☐ RADAR   ☐ VISUAL

SET	DRIFT	DISTANCE MADE GOOD SINCE *(time)* *(mi.)*

DISTANCE TO	MILES	ETA

TRUE HDG. °	ERROR	° GYRO	° GYRO	VARIATION °

MAGNETIC COMPASS HEADING *(Check one)*   °

☐ STD   ☐ STEER-ING   ☐ REMOTE IND   ☐ OTHER

DEVIATION °	1104 TABLE DEVIATION	DG: *(Indicate by check in box)*

☐ ON   ☐ OFF

REMARKS

RESPECTFULLY SUBMITTED

CC:

Figure 15–6. A ship's position report.

is located, a message apprising the cognizant naval authorities of any logistic support required on arrival is normally transmitted by Navy ships. This message, called a *LOGREQ*, for Logistics Requirements, is compiled and written in a standard format by the operations officer and transmitted by the ship's communication facility. The navigator has several items of information that he must supply, including the estimated times of arrival and departure. Like other departments, he may also request any necessary supplies he requires on arrival via the message. For foreign ports not having a U.S. naval facility, a comparable message or letter is usually sent to the appropriate civil port authorities prior to the ship's arrival.

The Captain's Night Orders

It is customary for the commanding officer or master of all ships at sea to set down in writing each night the ship is under way a complete set of instructions concerning the activities to be carried out by the watch during the night. In the Navy, these instructions are called the *Captain's Night Orders*. In practice on most Navy ships, the navigator often writes the rough draft of the night orders for further annotation and signature by the captain. The night orders should contain all the navigation instructions necessary for the OOD to safely conn the ship during the night. They are normally prepared and signed by the captain each night the ship is under way, even if the captain remains on the bridge throughout the night. Night orders may also be written when the ship is at anchorage, if unusual conditions of wind or sea warrant, for use by the in-port OOD and anchor bearing watch.

Consumable Items

In addition to the foregoing preparations, the navigator should also ensure that a large supply of all routinely used navigational materials is on board prior to departure, such as maneuvering boards, plotting sheets, spare tide and current forms and smooth deck logs, pencils, and all other miscellaneous items required. It is always preferable to have a bit too much than too little.

Summary

This chapter concludes *Marine Navigation 1: Piloting* with an overview of the procedures used by the navigator in planning an extended voyage, using as an example a theoretical transit from Norfolk, Virginia, to Naples, Italy. Throughout the process of voyage planning, virtually all of the skills of the navigator are employed in order that he may be thoroughly prepared for getting under way on the appointed date and time.

During the planning stage of a voyage, it is essential that the navigator make use of all resources available to him in deciding upon his final intended track. Climatological summaries and Optimum Track Ship Routing provided by the Naval Oceanographic Command, *Sailing Direction Planning Guides*, *Coast Pilot* volumes, and pilot charts are especially helpful during the planning process. It is often possible to maintain a faster speed of advance by deviating somewhat from the great circle route if by doing so areas of major weather disturbance can be avoided. Even after getting under way a cautious navigator can still plan for unexpected delays by traveling at a speed slightly faster than the required SOA during the first part of a voyage.

While proper and sufficient planning cannot ensure success, it nevertheless can do much to make the voyage easier. Insufficient or inattentive planning, on the other hand, can lead to disaster or serious embarrassment. As an example of the latter, consider the predicament of a Navy navigator who one day found that the vital publications and charts needed to visit a small, infrequently visited but highly desirable European port were not on board, even though they were specified as being in the ship's allowance by the *1–N*.

The subject of time, introduced in this chapter, is covered in more detail in *Marine Navigation 2: Celestial and Electronic*. A profile of a typical day's work in navigation at sea while en route on the voyage planned herein is also presented.

Chart No. 1

United States of America

Nautical Chart Symbols and Abbreviations

Eighth Edition
NOVEMBER 1984

Prepared jointly by

DEPARTMENT OF COMMERCE
National Oceanic and Atmospheric Administration
National Ocean Service

DEPARTMENT OF DEFENSE
Defense Mapping Agency
Hydrographic/Topographic Center

Published at Washington, D.C.
DEPARTMENT OF COMMERCE
National Oceanic and Atmospheric Administration
National Ocean Service
Washington, D.C. 20230

INTRODUCTION

General Remarks—This publication (Chart No. 1) contains symbols and abbreviations that have been approved for use on nautical charts published by the United States of America. A Glossary of Terms used on the charts of various nations is also included. The user should refer to DMAHTC Pub. No. 9, **American Practical Navigator** (Bowditch), Volume I, for the use of the chart in the practice of navigation and more detailed information pertaining to the chart sounding datum, tides and currents, visual and audible aids to navigation, etc.

Numbering—Terms, symbols, and abbreviations are numbered in accordance with a standard format approved by a 1952 resolution of the International Hydrographic Organization (IHO). Although the use of IHO-approved symbols and abbreviations is not mandatory, the United States has adopted many IHO-approved symbols for standard use. Style differences of the alphanumeric identifiers in the first column of the following pages show the status of symbols and abbreviations.

VERTICAL FIGURES indicate those items for which the symbols and abbreviations are in accordance with resolutions of IHO.

SLANTING FIGURES indicate those symbols for which no IHO resolution has been adopted.

SLANTING FIGURES ASTERISKED indicate IHO and U.S. symbols do not agree.

SLANTING LETTERS IN PARENTHESIS indicate that the items are in addition to those appearing in the IHO STANDARD LIST OF SYMBOLS AND ABBREVIATIONS.

Metric Charts and Feet/Fathom Charts—In January 1972 the United States began producing certain new nautical charts in meters. Since then many charts have been issued with soundings and contours in meters; however, for some time to come there will still be many charts on issue depicting sounding units in feet or fathoms. Modified reproductions of foreign charts are being produced retaining the native sounding unit value. The sounding unit is stated in bold type outside the border of every chart and in the chart title.

Chart Modernization—Chart symbols and labeling are brought into reasonable agreement with uniform international charting standards and procedures as quickly as opportunity affords. An example of this is the trend toward using vertical type for labeling items referred to the shoreline plane of reference, and slant type for all items referred to the sounding datum. This is not completely illustrated in this publication but is reflected in new charts produced by this country in accordance with international practices.

Soundings—The sounding datum reference is stated in the chart title. In all cases the unit of depth used is shown in the chart title and in the border of the chart in bold type.

Drying Heights—On rocks and banks that cover and uncover the elevations are above the sounding datum as stated in the chart title.

Shoreline—Shoreline shown on charts represents the line of contact between the land and a selected water elevation. In areas affected by tidal fluctuation, this line of contact

is usually the mean high-water line. In confined coastal waters of diminished tidal influence, a mean water level line may be used. The shoreline of interior waters (rivers, lakes) is usually a line representing a specified elevation above a selected datum. Shoreline is symbolized by a heavy line (A9).

Apparent Shoreline is used on charts to show the outer edge of marine vegetation where that limit would reasonably appear as the shoreline to the mariner or where it prevents the shoreline from being clearly defined. Apparent shoreline is symbolized by a light line (A7, C17).

Landmarks—A conspicuous feature on a building may be shown by a landmark symbol with a descriptive label. (See I 8b, 36, 44, 72.) Prominent buildings that are of assistance to the mariner may be shown by actual shape as viewed from above (See I 3a, 19, 47, 66). Legends associated with landmarks when shown in capital letters, indicate conspicuous or the landmarks may be labeled "CONSPIC" or "CONSPICUOUS."

Buoys—The buoyage systems used by other countries often vary from that used by the United States. U.S. Charts show the colors, lights and other characteristics in use for the area of the individual chart. Certain U.S. distributed modified reproduction charts of foreign waters may show shapes and other distinctive features that vary from those illustrated in this chart.

In the U.S. system, on entering a channel from seaward, buoys on the starboard side are red with even numbers, on the port side, black or green with odd numbers. Lights on buoys on the starboard side of the channel are red or white, on the port side, white or green. Mid-channel buoys have red and white or black and white vertical stripes and may be passed on either side. Junction or obstruction buoys have red and green or red and black horizontal bands, the top band color indicating the preferred side of passage. This system does not apply to foreign waters.

IALA Buoyage System—The International Association of Lighthouse Authorities (IALA) Maritime Buoyage System (combined Cardinal-Lateral System) is being implemented by nearly every maritime buoyage jurisdiction worldwide as either REGION A buoyage (red to port) or REGION B buoyage (red to starboard). The terms "REGION A" and "REGION B" will be used to determine which type of buoyage is in effect or undergoing conversion in a particular area. The major difference in the two buoyage regions will be in the lateral marks. In REGION A they will be red to port; in REGION B they will be red to starboard. Shapes of lateral marks will be the same in both REGIONS, can to port; cone (nun) to starboard. Cardinal and other marks will continue to follow current guidelines and may be found in both REGIONS. A modified lateral mark, indicating the preferred channel where a channel divides, will be introduced for use in both REGIONS. Section L and the color plates at the back of this publication illustrate the IALA buoyage system for both REGIONS A and B.

Aids to Navigation Positioning—The aids to navigation depicted on charts comprise a system consisting of fixed and floating aids with varying degrees of reliability. Therefore, prudent mariners will not rely solely on any single aid to navigation, particularly a floating aid.

The buoy symbol is used to indicate the approximate position of the buoy body and the sinker which secures the buoy to the seabed. The approximate position is used because of practical limitations in positioning and maintaining buoys and their sinkers in precise geographical locations. These limitations include, but are not limited to, inherent imprecisions in position fixing methods, prevailing atmospheric and sea conditions, the

slope of and the material making up the seabed, the fact that buoys are moored to sinkers by varying lengths of chain, and the fact that buoy body and/or sinker positions are not under continuous surveillance but are normally checked only during periodic maintenance visits which often occur more than a year apart. The position of the buoy body can be expected to shift inside and outside the charting symbol due to the forces of nature. The mariner is also cautioned that buoys are liable to be carried away, shifted, capsized, sunk, etc. Lighted buoys may be extinguished or sound signals may not function as the result of ice, running ice, other natural causes, collisions, or other accidents.

For the foregoing reasons a prudent mariner must not rely completely upon the position or operation of floating aids to navigation, but will also utilize bearings from fixed objects and aids to navigation on shore. Further, a vessel attempting to pass close aboard always risks collision with a yawing buoy or with the obstruction the buoy marks.

Colors—Colors are optional for characterizing various features and areas on the charts. For instance the land tint in this publication is gold as used on charts of the National Ocean Service; however, charts of the DMA show land tint as gray.

Heights—Heights of lights, landmarks, structures, etc. are referred to the shoreline plane of reference. Heights of small islets or offshore rocks, which due to space limitations must be placed in the water area, are bracketed. The unit of height used is shown in the chart title.

Conversion Scales—Depth conversion scales are provided on all charts to enable the user to work in meters, fathoms, or feet.

Improved Channels—Improved channels are shown by dashed limit lines with the depth and date of the latest examination placed adjacent to the channel or in a channel tabulation.

Longitudes—Longitudes are referred to the meridian of Greenwich.

Traffic Separation Schemes—Traffic separation schemes show established routes to increase safety of navigation, particularly in areas of high density shipping. These schemes were established by the International Maritime Organization (IMO) and are described in the IMO publication "Ships Routing".

Traffic separation schemes are generally shown on nautical charts at scales of 1:600,000, and larger. When possible, traffic separation schemes are plotted to scale and shown as depicted in Section P.

Names—Names on nautical charts compiled and published by the United States of America are in accordance with the principles of the Board of Geographic Names.

Correction Dates—The dates of New Editions are shown below the lower left border of the chart. These include the date of the latest Notice to Mariners applied to the charts.

U.S. Coast Pilots, Sailing Directions, Light Lists, Lists of Lights—These related publications furnish information required by the navigator that cannot be shown conveniently on the nautical charts.

U.S. Nautical Chart Catalogs and Indexes—These list nautical charts, auxiliary maps, and related publications and include general information relative to the use and ordering of charts.

Special and Foreign Symbols—Some differences may be observed between the symbols shown in Chart No. 1 and symbols shown on certain special charts and reproductions of foreign charts. A glossary of foreign terms and abbreviations is generally shown on charts on which they are used, as well as in the Sailing Directions. In addition, an extensive glossary is found at the back of this publication.

TABLE OF CONTENTS

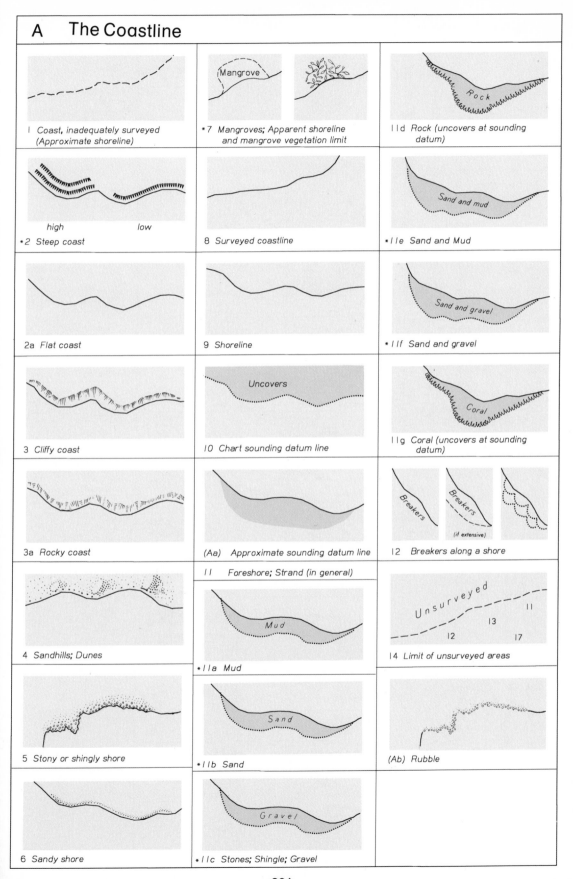

A The Coastline

1 Coast, inadequately surveyed
(Approximate shoreline)

*7 Mangroves; Apparent shoreline
and mangrove vegetation limit

11d Rock (uncovers at sounding
datum)

high   low
*2 Steep coast

8 Surveyed coastline

*11e Sand and Mud

2a Flat coast

9 Shoreline

*11f Sand and gravel

3 Cliffy coast

10 Chart sounding datum line

11g Coral (uncovers at sounding
datum)

3a Rocky coast

(Aa) Approximate sounding datum line

11 Foreshore; Strand (in general)

12 Breakers along a shore

4 Sandhills; Dunes

*11a Mud

14 Limit of unsurveyed areas

5 Stony or shingly shore

*11b Sand

(Ab) Rubble

6 Sandy shore

*11c Stones; Shingle; Gravel

291

B Coast Features

1	G	Gulf
2	B	Bay
(Ba)	B	Bayou
3	Fd	Fjord
4	L	Loch; Lough; Lake
5	Cr	Creek
5a	C	Cove
6	In	Inlet
7	Str	Strait
8	Sd	Sound
9	Pass	Passage; Pass
	Thoro	Thoroughfare
10	Chan	Channel
10a		Narrows
11	Entr	Entrance
12	Est	Estuary
12a		Delta
13	Mth	Mouth
14	Rd	Road; Roadstead
15	Anch	Anchorage
16	Hbr	Harbor
16a	Hn	Haven
17	P	Port
(Bb)	P	Pond
18	I	Island
19	It	Islet
20	Arch	Archipelago
21	Pen	Peninsula
22	C	Cape
23	Prom	Promontory
24	Hd	Head; Headland
25	Pt	Point
26	Mt	Mountain; Mount
27	Rge	Range
27a		Valley
28		Summit
29	Pk	Peak
30	Vol	Volcano
31		Hill
32	Bld	Boulder
33	Ldg	Landing
	Lndg	Landing
34		Tableland
		Plateau
35	R, Rk, Rks	Rock, Rocks
36		Isolated rock
(Bc)	Str	Stream
(Bd)	R	River
(Be)	Slu	Slough
(Bf)	Lag	Lagoon
(Bg)	Apprs	Approaches
(Bh)	Rky	Rocky
(Bi)	Is	Islands
(Bj)	Ma	Marsh
(Bk)	Mg	Mangrove
(Bl)	Sw	Swamp

C The Land

1 Contour lines (Contours)

1a Contour lines, approximate (Contours)

2 Hachures

2a Form lines, no definite interval

2b Shading

3 Glacier

4 Saltpans

5 Isolated trees

5a Deciduous; of unknown or unspecified type

5b Coniferous

5c Palm tree

5d Nipa palm

5e Filao

5f Casuarina

5g Evergreen tree (other than coniferous)

6 Cultivated fields

6a Grass fields

7 Paddy (rice) fields

7a Park; Garden

8 Bushes

8a Tree plantation in general

9 Deciduous woodland

10 Coniferous woodland

10a Woods in general

11 Tree top elevation (above shoreline datum)

12 Lava flow

13 River; Stream

14 Intermittent stream

15 Lake; Pond

16 Lagoon (Lag)

17 Marsh; Swamp

18 Slough (Slu.)

19 Rapids

20 Waterfalls

21 Spring

292

D Control Points

I	△	Triangulation point (station)	4	⊕ Obs Spot	Observation spot
Ia		Astronomic station	5	⊼ o BM	Bench mark
2	⊙	Fixed point (landmark, position accurate)	6	View X	View point
(Da)	o	Fixed point (landmark, position approximate)	7		Datum point for grid of a plan
3	· 256	Summit of height (Peak) (when not a landmark)	8		Graphical triangulation point
(Db)	◉ 256	Peak, accentuated by contours	9	Astro	Astronomical
(Dc)	〰 256	Peak, accentuated by hachures	I0	Tri	Triangulation
(Dd)	〰	Peak, elevation not determined	(Df)	C of E	Corps of Engineers
(De)	⊙ 256	Peak, when a landmark	I2		Great trigonometrical survey station
			I3		Traverse station
			I4	Bdy Mon	Boundary monument
			(Dg)	◇	International boundary monument

E Units

I	hr, h	Hour	II	M, Mi / NMi, NM	Nautical mile(s)	21	′	Minute (of arc)
2	m, min	Minute (of time)				22	″	Second (of arc)
3	sec, s	Second (of time)	I2	kn	Knot(s)	23	No	Number
4	m	Meter	I2a	t	Tonne (metric ton equals 2,204.6 lbs)	(Ea)	St M, / St Mi	Statute mile
4a	dm	Decimeter				(Eb)	μsec, μs	Microsecond
4b	cm	Centimeter	I2b	cd	Candela (new candle)	(Ec)	Hz	Hertz (cps)
4c	mm	Millimeter	I3	lat	Latitude	(Ed)	kHz	Kilohertz (kc)
4d	m²	Square meter	I4	long	Longitude	(Ee)	MHz	Megahertz (Mc)
4e	m³	Cubic meter	I4a		Greenwich	(Ef)	cps, c/s	Cycles/second (Hz)
5	km	Kilometer(s)	I5	pub	Publication	(Eg)	kc	Kilocycle (kHz)
6	in, ins	Inch(es)	I6	Ed	Edition	(Eh)	Mc	Megacycle (MHz)
7	ft	Foot, feet	I7	corr	Correction	(Ei)	T	Ton (U.S. short ton equals 2,000 lbs)
8	yd, yds	Yard(s)	I8	alt	Altitude			
9	fm, fms	Fathom(s)	*I9	ht; elev	Height; Elevation			
I0	cbl	Cable length	20	°	Degree			

F Adjectives, Adverbs, Nouns, and Other Words

I	gt	Great	25	discontd	Discontinued	(Fe)	cor	Corner
2	lit	Little	26	prohib	Prohibited	(Ff)	concr	Concrete
3	Lrg	Large	27	explos	Explosive	(Fg)	fl	Flood
4	sml	Small	28	estab	Established	(Fh)	mod	Moderate
5		Outer	29	elec	Electric	(Fi)	bet	Between
6		Inner	30	priv	Private, Privately	(Fj)	Ist	First
7	mid	Middle	31	prom	Prominent	(Fk)	2nd, 2d	Second
8		Old	32	std	Standard	(Fl)	3rd, 3d	Third
9	anc	Ancient	33	subm	Submerged	(Fm)	4th	Fourth
I0		New	34	approx	Approximate	(Fn)	DW	Deep Water
II	St	Saint	35		Maritime	(Fo)	min	Minimum
I2	CONSPIC	Conspicuous	36	maintd	Maintained	(Fp)	max	Maximum
I3		Remarkable	37	aband	Abandoned	(Fq)	N'ly	Northerly
I4	D, Destr	Destroyed	38	temp	Temporary	(Fr)	S'ly	Southerly
I5		Projected	39	occas	Occasional	(Fs)	E'ly	Easterly
I6	dist	Distant	40	extr	Extreme	(Ft)	W'ly	Westerly
I7	abt	About	41		Navigable	(Fu)	Sk	Stroke
I8		See chart	42	N M	Notice to Mariners	(Fv)	Restr	Restricted
I8a		See plan	(Fa)	L N M	Local Notice to Mariners	(Fw)	Bl	Blast
I9		Lighted, Luminous	43		Sailing Directions	(Fx)	CFR	Code of Federal Regulations
20	sub	Submarine	44		List of Lights	(Fy)	COLREGS	Int'l Regulations for Preventing Collisions at Sea, 1972
21		Eventual	(Fb)	unverd	Unverified			
22	AERO	Aeronautical	(Fc)	AUTH	Authorized	(Fz)	IWW	Intracoastal Waterway
23		Higher	(Fd)	CL	Clearance			
23a		Lower						
24	exper	Experimental						

G Ports and Harbors

1	⚓	Anch	Anchorage (large vessels)
2	⚓	Anch	Anchorage (small vessels)
3		Hbr	Harbor
4		Hn	Haven
5		P	Port
6		Bkw	Breakwater
6a			Dike
7			Mole
8			Jetty (partly below MHW)
8a			Submerged Jetty
(Ga)			Jetty (small scale)
9		Pier	Pier
10			Spit
11			Groin (partly below MHW)
*12	ANCH PROHIBITED	ANCH PROHIB	Anchorage prohibited (Screen optional)
12a			Anchorage reserved
12b	QUARANTINE ANCHORAGE	QUAR ANCH	Quarantine anchorage
*12c			Quarantine Anchorage
*12d			Quarantine Anchorage
*12e		FISH PROHIB	Fishing prohibited
13	Spoil Area		Spoil ground (Dump Site)
(Gb)	Dumping Ground		Dumping ground (depths may be less than indicated) (Dump Site)
(Gc)	Disposal Area 92 depths from survey of JUNE 1972 85 90 87		Disposal area (Dump Site)
(Gd)	ⓟ		Pump-out facilities

14		Fsh stks	Fisheries; Fishing stakes
14a			Fish trap; Fish weirs (actual shape charted)
14b			Duck blind
15			Tunny nets
15a	Oys	Oys	Oyster bed
16		Ldg,Lndg	Landing place
17			Watering place
18		Whf	Wharf
19			Quay
20	Ⓐ ⑭		Berth
*20a	14 ⑭ Ⓑ		Anchoring berth
20b	3		Berth number
21	Dol		Dolphin
22			Bollard
22a	▪ S P M		Fixed single point mooring structure (lighted)
23			Mooring ring
24	⊙-		Crane
25			Landing stage
25a			Landing stairs
26	⊕	Quar	Quarantine
27			Lazaret
28	Harbor Master ⚓	Hbr Mr	Harbormaster's office
29	⊖	Cus Ho	Customhouse
30			Fishing harbor
31			Winter harbor
32			Refuge harbor
33		B Hbr	Boat harbor
34			Stranding harbor (uncovers at LW)
35			Dock
36			Drydock (actual shape on large scale charts)
37			Floating dock (actual shape on large scale charts)
38			Gridiron; Careening grid

G Ports and Harbors

39		Patent slip; Slipway; Marine railway
39a	Ramp	Ramp
40		Lock (point upstream)
41		Wetdock
42		Shipyard
43		Lumber yard
44	Health Office	Health officer's office
45	Hk	Hulk (actual shape on large scale charts)
45a		
* 46	PROHIBITED AREA / PROHIB AREA	Prohibited area (screen optional)

46a	10	Calling-in point for vessel traffic control
47		Anchorage for seaplanes
48		Seaplane landing area
* 49 / * 50	Under construction	Work in progress / Under construction
51		Work projected
(Ge)	Subm ruins	Submerged ruins
(Gf)	Dump site	Dump site

H Topography

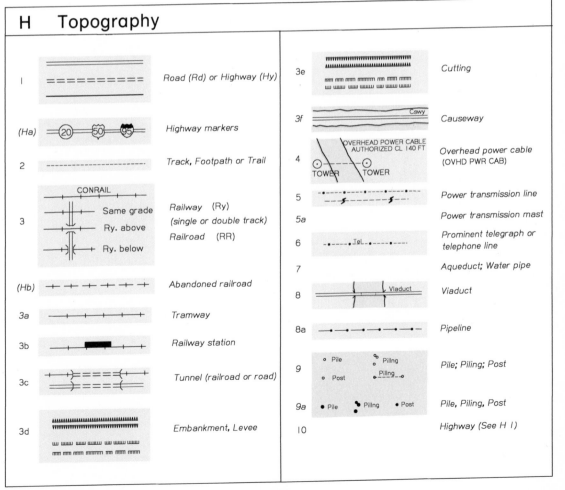

I		Road (Rd) or Highway (Hy)
(Ha)	20 50 95	Highway markers
2		Track, Footpath or Trail
3	CONRAIL / Same grade / Ry. above / Ry. below	Railway (Ry) (single or double track) / Railroad (RR)
(Hb)		Abandoned railroad
3a		Tramway
3b		Railway station
3c		Tunnel (railroad or road)
3d		Embankment, Levee

3e		Cutting
3f	Cswy	Causeway
4	OVERHEAD POWER CABLE AUTHORIZED CL 140 FT / TOWER TOWER	Overhead power cable (OVHD PWR CAB)
5		Power transmission line
5a		Power transmission mast
6	TgL	Prominent telegraph or telephone line
7		Aqueduct; Water pipe
8	Viaduct	Viaduct
8a		Pipeline
9	Pile Pilng / Post Pilng	Pile; Piling; Post
9a	Pile Pilng Post	Pile, Piling, Post
10		Highway (See H I)

295

H   Topography

11	_ _ _ _Sewer or Outfall_ _ _ _ _	Sewer
12		Culvert
13	Canal — Lock / Ditch — Sluice (Tidegate, Floodgate)	Canal, Ditch, Lock, Sluice
14		Bridge in general (BR)
(Hc)		Bridge under construction
14a		Stone, concrete bridge (same as H 14)
14b		Wooden brige (same as H 14)
14c		Iron bridge (same as H 14)
14d		Suspension bridge (same as H 14)
15		Drawbridge (in general)
16		Swing bridge (same as H 15)
16a		Lift bridge
16b		Weighbridge or Bascule bridge
17		Pontoon bridge
17a		Footbridge
18		Transporter bridge (same as H 14)
18a	VERT CL 6 FT	Bridge clearance, vertical
18b	HOR CL 28 FT	Bridge clearance, horizontal
19	Ferry Ferry On small-scale chart	Ferry (Fy)
(Hd)	Cable ferry	Cable ferry
20		Ford
21		Dam
22		Fence
23		Training wall
24	Log boom	Log boom

I Buildings and Structures

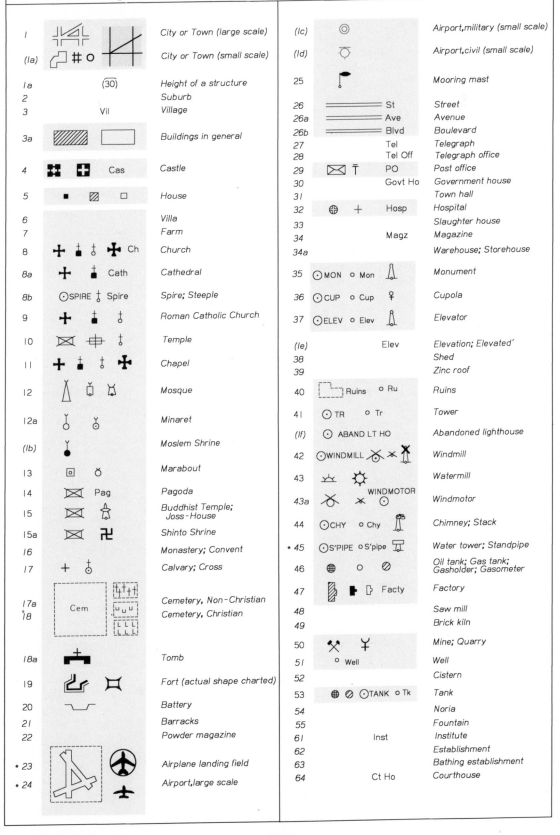

I		City or Town (large scale)
(Ia)		City or Town (small scale)
Ia	(30)	Height of a structure
2		Suburb
3	Vil	Village
3a		Buildings in general
4	Cas	Castle
5		House
6		Villa
7		Farm
8	Ch	Church
8a	Cath	Cathedral
8b	SPIRE Spire	Spire; Steeple
9		Roman Catholic Church
10		Temple
11		Chapel
12		Mosque
12a		Minaret
(Ib)		Moslem Shrine
13		Marabout
14	Pag	Pagoda
15		Buddhist Temple; Joss-House
15a		Shinto Shrine
16		Monastery; Convent
17		Calvary; Cross
17a	Cem	Cemetery, Non-Christian
18		Cemetery, Christian
18a		Tomb
19		Fort (actual shape charted)
20		Battery
21		Barracks
22		Powder magazine
*23		Airplane landing field
*24		Airport, large scale

(Ic)		Airport, military (small scale)
(Id)		Airport, civil (small scale)
25		Mooring mast
26	St	Street
26a	Ave	Avenue
26b	Blvd	Boulevard
27	Tel	Telegraph
28	Tel Off	Telegraph office
29	PO	Post office
30	Govt Ho	Government house
31		Town hall
32	Hosp	Hospital
33		Slaughter house
34	Magz	Magazine
34a		Warehouse; Storehouse
35	MON o Mon	Monument
36	CUP o Cup	Cupola
37	ELEV o Elev	Elevator
(Ie)	Elev	Elevation; Elevated′
38		Shed
39		Zinc roof
40	Ruins o Ru	Ruins
41	TR o Tr	Tower
(If)	ABAND LT HO	Abandoned lighthouse
42	WINDMILL	Windmill
43		Watermill
43a	WINDMOTOR	Windmotor
44	CHY o Chy	Chimney; Stack
*45	S'PIPE o S'pipe	Water tower; Standpipe
46		Oil tank; Gas tank; Gasholder; Gasometer
47	Facty	Factory
48		Saw mill
49		Brick kiln
50		Mine; Quarry
51	o Well	Well
52		Cistern
53	TANK o Tk	Tank
54		Noria
55		Fountain
61	Inst	Institute
62		Establishment
63		Bathing establishment
64	Ct Ho	Courthouse

I Buildings and Structures

65	🚩 Sch		School
(Ig)	🚩 HS		High school
(Ih)	🚩 Univ		University
66	■ ▨ □ Bldg		Building
67	Pav		Pavilion
68			Hut
69			Stadium
70	T		Telephone
71	⊕ ● ⊘		Gas tank; Gasometer
72	⊙ GAB o Gab		Gable
73			Wall

74		Pyramid
75		Pillar
76	⊙ ⧄ ⊡	Oil derrick
(Ii)	Ltd	Limited
(Ij)	Apt	Apartment
(Ik)	Cap	Capitol
(Il)	Co	Company
(Im)	Corp	Corporation
(In)	⊙	Landmark (position accurate)
(Io)	o	Landmark (position approximate)
(Ip)	⌐	Flare; Stack (on land)

J Miscellaneous Stations

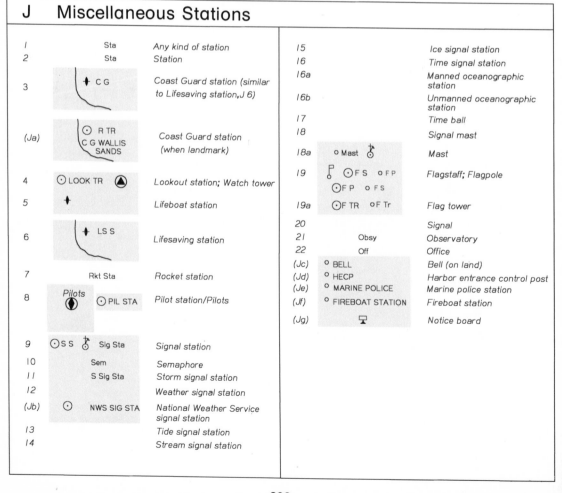

1	Sta	Any kind of station
2	Sta	Station
3	✦ C G	Coast Guard station (similar to Lifesaving station, J 6)
(Ja)	⊙ R TR C G WALLIS SANDS	Coast Guard station (when landmark)
4	⊙ LOOK TR ▲	Lookout station; Watch tower
5	✦	Lifeboat station
6	✦ LS S	Lifesaving station
7	Rkt Sta	Rocket station
8	Pilots ◖ ⊙ PIL STA	Pilot station/Pilots
9	⊙ S S ⚓ Sig Sta	Signal station
10	Sem	Semaphore
11	S Sig Sta	Storm signal station
12		Weather signal station
(Jb)	⊙ NWS SIG STA	National Weather Service signal station
13		Tide signal station
14		Stream signal station

15		Ice signal station
16		Time signal station
16a		Manned oceanographic station
16b		Unmanned oceanographic station
17		Time ball
18		Signal mast
18a	o Mast ⚲	Mast
19	⌐ ⊙ F S o F P ⊙ F P o F S	Flagstaff; Flagpole
19a	⊙ F TR o F Tr	Flag tower
20		Signal
21	Obsy	Observatory
22	Off	Office
(Jc)	o BELL	Bell (on land)
(Jd)	o HECP	Harbor entrance control post
(Je)	o MARINE POLICE	Marine police station
(Jf)	o FIREBOAT STATION	Fireboat station
(Jg)	⊡	Notice board

K Lights

I	✦ ● ☆ •	Position of light
*2	Lt	Light
(Ka)		Riprap surrounding light
3	Lt Ho	Lighthouse
4	AERO ☆ AERO	Aeronautical light
4a		Marine and air navigation light
*5	Bn	Light beacon
6		Light vessel; Lightship
8		Lantern
9		Street lamp
10	REF	Reflector
11		Leading light
11a	Lts in line 270°	Lighted range
12		Sector light
13	RED / GREEN	Directional light
14		Harbor light
15		Fishing light
16		Tidal light
17	Priv maintd	Private light (maintained by private interests; to be used with caution)
21	F	Fixed (steady light)
22	Oc; Occ	Occulting (total duration of light more than dark)
23	Fl	Single-Flashing (total duration of light less than dark)
(Kb)	L Fl	Long-Flashing (2 sec or longer)
	Fl (2+1)	Composite group-flashing
23a	Iso; E Int	Isophase (light and dark equal)
24	Q; Qk Fl	Continuous Quick Flashing (50 to 79 per minute; 60 in US)
	Q(3)	Group Quick
25	IQ; Int Qk Fl; I Qk Fl	Interrupted Quick Flashing
25a	S Fl	Short Flashing
(Kc)	VQ; V Qk Fl	Continuous Very Quick Flashing (80 to 159-usually either 100 or 120 per minute)
	VQ (3)	Group Very Quick
	IVQ	Interrupted Very Quick
	UQ	Continuous Ultra Quick (160 or more-usually 240 to 300 flashes per minute)
	IUQ	Interrupted Ultra Quick
26	Al; Alt	Alternating
27	Oc (2); Gp Occ	Group-Occulting
	Oc(2+3)	Composite group occulting
28	Fl (2); Gp Fl	Group Flashing
28a	S-L Fl	Short-Long Flashing
28b		Group-Short Flashing
29	F Fl	Fixed and Flashing
30	F Gp Fl	Fixed and Group Flashing
30a	Mo (A)	Morse Code light (with flashes grouped as in letter A)
31	Rot	Revolving or Rotating light
41		Period
42		Every
43		With
44		Visible (range)
(Kd)	M; Mi; N Mi	Nautical mile
(Ke)	m; min	Minutes
(Kf)	s; sec	Seconds
45	Fl	Flash
46	Oc; Occ	Occultation
46a		Eclipse
47	Gp	Group
48	Oc; Occ	Intermittent light
49	SEC	Sector
50		Color of sector
51	Aux	Auxiliary light
52		Varied
61	Vi	Violet
62		Purple
63	Bu; Bl	Blue
64	G	Green

K   Lights

65	Or; Y	Orange	72	Prov	Provisional light	80	Vert	Vertical lights
66	R	Red	73	Temp	Temporary light	81	Hor	Horizontal lights
67	W	White	(Kg)	D; Destr	Destroyed	(Kh)	VB	Vertical beam
67a	Y ; Am	Amber	74	Exting	Extinguished light	(Ki)	RGE	Range
(Ko)	Y	Yellow	75		Faint light	(Kj)	Exper	Experimental light
68	OBSC	Obscured light	76		Upper light			
68a	Fog Det Lt	Fog detector light	77		Lower light	(Kp)		Lighted offshore platform
70	Occas	Occasional light	78		Rear light			
71	Irreg	Irregular light	79		Front light	(Kq)		Flare (Flame)

L   Buoys and Beacons

•1	o	Approximate position of buoy	•21	Tel	Telegraph-cable buoy
•2		Light buoy	•22		Mooring buoy (colors of mooring buoys never carried)
•3	BELL BELL	Bell buoy	22a		Mooring
•3a	GONG GONG	Gong buoy	•22b	Tel Tel	Mooring buoy with telegraphic communications
•4	WHIS WHIS	Whistle buoy	•22c	T T	Mooring buoy with telephonic communications
•5	C	Can or Cylindrical buoy	•23		Warping buoy
•6	N	Nun or Conical buoy	•24	Y	Quarantine buoy
•7	SP	Spherical buoy	24a		Practice area buoy
•8	S	Spar buoy	•25	Explos Anch	Explosive anchorage buoy
•8a	P	Pillar or Spindle buoy	•25a	AERO	Aeronautical anchorage buoy
•9		Buoy with topmark (ball)	•26	Deviation	Compass adjustment buoy
•10		Barrel or Ton buoy	•27	BW	Fish trap (area) buoy (BWHB)
(La)		Color unknown	•27a		Spoil ground buoy
(Lb)	FLOAT	Float	•28	W	Anchorage buoy (marks limits)
•12	FLOAT FLOAT	Lightfloat	•29	Priv maintd	Private aid to navigation (buoy) (maintained by private interests, use with caution)
13		Outer or Landfall buoy	30		Temporary buoy
•14	RW BW	Fairway buoy (RWVS; BWVS)	30a		Winter buoy
•14a	RW BW	Midchannel buoy (RWVS; BWVS)	•31	HB	Horizontal bands
•15	R "2"	Starboard-hand buoy (entering from seaward – US waters)	•32	VS	Vertical stripes
•16	"1" "1"	Port-hand buoy (entering from seaward – US waters)	•33	Chec	Checkered
•17	RB BR RG GR RB	Bifurcation buoy	•33a	Diag	Diagonal bands
•18	RB BR RG GR BR	Junction buoy	41	W	White
•19	RB BR RG GR RG	Isolated danger buoy	42	B	Black
•20	RB BR RG GR G	Wreck buoy	43	R	Red
			44	Y	Yellow
•20a	RB BR RG GR G	Obstruction buoy	45	G	Green
			46	Br	Brown
			47	Gy	Gray

L Buoys and Beacons

48	*Bu*	*Blue*
48a	*Am*	*Amber*
48b	*Or*	*Orange*
* 51		*Floating beacon (and variations)*
* 52	Fixed beacons	(unlighted or daybeacons)

▲R Bn △RG Bn		*Triangular beacon*
▲Bn		*Black beacon*
■G Bn □GR Bn □W Bn		*Square and other shaped beacons*
□Bn ✦		*Color unknown*
		Variations

53	⊥ Bn □ ▲	*Beacon, in general*
54		*Tower beacon*
55		*Cardinal marking system*
56	△Deviation Bn	*Compass adjustment beacon*
57		*Topmarks*
58		*Telegraph-cable (landing) beacon*

* 59	Piles Piles Stumps	*Piles*
		Stumps
		Stakes, perches
(Lc)	MARKER Marker	*Private aid to navigation*
61	CAIRN Cairn △ ⚏	*Cairn*
62		*Painted patches*
63	⊙	*Landmark (position accurate)*
(Ld)	○	*Landmark (position approximate)*
64	REF	*Reflector*
65	MARKER	*Range targets, markers*
(Le)	W Or Y Y	*Special-purpose buoys*
	W Or Y Y	
66		*Oil installation buoy*
67	⊡	*Drilling platform*
70	NOTE: Refer to IALA Buoyage System description on page 48 for aids used in certain foreign waters.	
71		*LANBY (Large Auto. Nav. Buoy); Superbuoy*
72		*TANKER terminal buoy (mooring)*
73	ODAS	*ODAS (Oceanographic Data Aquisition System)*
(Lg)	Art	*Articulated light (floating light)*

M Radio and Radar Stations

1	○R Sta	*Radio telegraph station*
2	○RT	*Radio telephone station*
3	⊙ R Bn, Ro Bn	*Radiobeacon*
* 4	⊙ R Bn, RC	*Circular radiobeacon*
5	RD 072°30' ⊙ RD	*Directional radiobeacon; Radio range*
6	⊙ RW	*Rotating loop radiobeacon*
* 7	⊙ RDF, Ro DF, RG	*Radio direction finding station*
(Ma)	ANTENNA (TELEM) TELEM ANT	*Telemetry antenna*
(Mb)	R RELAY MAST	*Radio relay mast*
(Mc)	MICRO TR	*Microwave tower*
9	R MAST R TR	*Radio mast* / *Radio tower*

9a	TV TR Tr	*Television mast; Television tower*
10	R TR (WBAL) 1090 kHz	*Radio broadcasting station (commercial)*
* 10a	○R Sta	*QTG radio station*
11	⊙ Ra	*Radar station*
12	⊙ Racon ○Ra Sur	*Radar responder beacon*
13	Ra Ref	*Radar reflector*
14	Ra (conspic)	*Radar conspicuous object*
14a	⊙	*Ramark*
15	DFS ⊙R Telem	*Distance finding station (synchronized signals)*
16	AERO R Bn ⊙ 302 R C Aero	*Aeronautical radiobeacon*
17	○Decca Sta	*Decca station*
18	○ Loran Sta Venice	*Loran station (name)*

M Radio and Radar Stations

19	CONSOL Bn 190 kHz MMF ⋱⋱	Consol (Consolan) station
(Md)	AERO R Rge 342 ⋱⋱⋱	Aeronautical radio range
(Me)	Ra Ref Calibration Bn	Radar calibration beacon
(Mf)	LORAN TR SPRING ISLAND	Loran tower (name)
(Mg)	R TR F R Lt	Obstruction light
(Mh)	RA DOME DOME (RADAR) Ra Dome Dome (Radar)	Radar dome
(Mi)	uhf	Ultrahigh frequency
(MJ)	vhf	Very high frequency

N Fog Signals

1	Fog Sig	Fog-signal station	13	HORN	Air (foghorn)
2		Radio fog-signal station	13a	HORN	Electric (foghorn)
3	GUN	Explosive fog signal	14	BELL	Fog bell
4		Submarine fog signal	15	WHIS	Fog whistle
5	SUB-BELL	Submarine fog bell (action of waves)	16	HORN	Reed horn
6	SUB-BELL	Submarine fog bell (mechanical)	17	GONG	Fog gong
7	SUB-OSC	Submarine oscillator	18	⊙	Submarine sound signal not connected to the shore
8	NAUTO	Nautophone	18a	⊙	Submarine sound signal connected to the shore
9	DIA	Diaphone			
10	GUN	Fog gun	(Na)	HORN	Typhon
11	SIREN	Fog siren	(Nb)	Fog Det Lt	Fog detector light
12	HORN	Fog trumpet	(Nc)	Mo	Morse Code fog signal

O Dangers

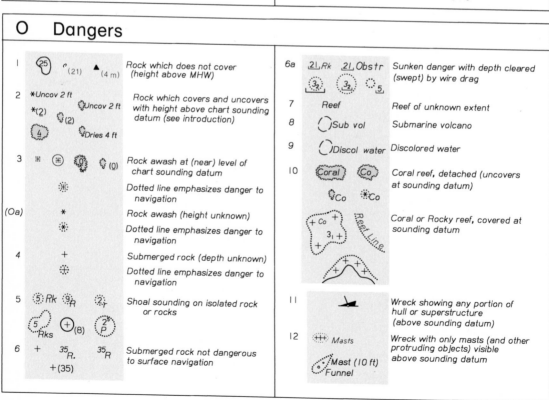

1	(25) (21) ▲ (4 m)	Rock which does not cover (height above MHW)
2	*Uncov 2 ft Uncov 2 ft *(2) (2) (4) Dries 4 ft	Rock which covers and uncovers with height above chart sounding datum (see introduction)
3	⋇ ⊛ (0) (0) ⊛	Rock awash at (near) level of chart sounding datum Dotted line emphasizes danger to navigation
(Oa)	* ⊛	Rock awash (height unknown) Dotted line emphasizes danger to navigation
4	+ ⊕	Submerged rock (depth unknown) Dotted line emphasizes danger to navigation
5	5 Rk 9 R 2 r 5 Rks (8) 2 P	Shoal sounding on isolated rock or rocks
6	+ 35 R. 35 R +(35)	Submerged rock not dangerous to surface navigation
6a	21 Rk 21 Obstr 3₂ 3₂ 5	Sunken danger with depth cleared (swept) by wire drag
7	Reef	Reef of unknown extent
8	Sub vol	Submarine volcano
9	Discol water	Discolored water
10	Coral Co Co Co	Coral reef, detached (uncovers at sounding datum)
	+ Co + 3₁ + Reef Line	Coral or Rocky reef, covered at sounding datum
11		Wreck showing any portion of hull or superstructure (above sounding datum)
12	Masts Mast (10 ft) Funnel	Wreck with only masts (and other protruding objects) visible above sounding datum

O   Dangers

13		Old symbols for wrecks
13a		Wreck always partially submerged
14		Sunken wreck dangerous to surface navigation (less than 11 fathoms over wreck)
14a		Sunken wreck covered 20 to 30 meters
15	Wk (9)	Wreck over which depth is known
15a	21 Wk 5 Wk	Wreck with depth cleared by wire drag
	5 Wk 21 Wk	
15b	8 Wk	Unsurveyed wreck over which the exact depth is unknown, but is considered to have a safe clearance to the depth shown
16	+++	Sunken wreck, not dangerous to surface navigation
17	Foul # fB	Foul ground, Foul bottom
17a		Mobil bottom (sand waves)
18	Tide rips	Overfalls or Tide rips
	Symbol used only in small areas	
19	Eddies	Eddies
	Symbol used only in small areas	
20	Kelp	Kelp, Seaweed
21	Bk	Bank
22	Shl	Shoal
23	Rf	Reef
23a		Ridge
24	Le	Ledge
25	Br or	Breakers
26		Submerged rock
27	5 Obstr	Obstruction
(Ob)	Obstr Well ✦	Submerged well
	Obstr Well	Submerged well (buoyed)
(Oc)	Obstruction (fish haven)	Fish haven (artificial fishing reef)
		(actual shape)

28		Wreck
29	Wreckage Wks	Wreckage
29a		Wreck remains (dangerous only for anchoring)
✴ 30	Subm piles Subm piling	Submerged piling
	Subm piles Stakes, Perches	
✴ 30a	Snags Stumps	Snags; Submerged stumps
31		Lesser depth possible
32	Uncov	Dries
33	Cov	Covers
34	Uncov	Uncovers
35	3 Rep (1983)	Reported (with date)
	Eagle Rk (rep 1983)	Reported (with name and date)
36	Discol	Discolored
37		Isolated danger
38		Limiting danger line
39	rky	Limit of rocky area
41	PA	Position approximate
42	PD	Position doubtful
43	ED	Existance doubtful
44	P Pos	Position
45	D	Doubtful
46	Unexam	Unexamined
(Od)	LD	Least Depth
(Oe)	Subm Crib	Crib
	Crib (above water)	
(Of)	⊡ ■ Platform (lighted) HORN	Offshore platform (unnamed)
(Og)	⊡ ■ Hazel (lighted) HORN	Offshore platform (named)

1		Leading line; Range line
2		Transit
3		In line with
4		Limit of sector
5		Channel, Course, Track recommended (marked by buoys or beacons)
5a	DW	Recommended track for deep draft vessels (defined by fixed marks)
5b	DW76 ft DW83 ft	Depth is shown where it has been obtained by the cognizant authority
(Pa)		Alternate course
6	Ra Ra	Radar-guided track
6a		Established traffic separation scheme. One-way traffic lanes (separated by line or zone)
6b		Established traffic separation scheme: Roundabout
		If no separation zone exists, the center of the roundabout is shown by a circle
6c	DW	Recommended direction of traffic flow
7		Submarine cable (power telegraph, telephone, etc.)
7a	Cable Area	Submarine cable area
7b		Abandoned submarine cable (includes disused cable)
8		Submarine pipeline
8a	Pipeline Area	Submarine pipeline area
8b		Abandoned submarine pipeline
9		Maritime limit in general
(Pb)	RESTRICTED AREA	Limit of restricted area
•10		Limits of national fishing zones
(Pc)		U.S. Harbor Line

11		Limit of dumping ground, spoil ground
12		Anchorage limit
•13		Limit of airport
•14		Limit of sovereignty (Territorial waters)
•15		Customs boundary
•16	++++++++	International boundary (also State boundary)
17		Stream limit
18		Ice limit
19		Limit of tide
20		Limit of Navigation
21		Recommended track (not marked by buoys or beacons)
21a	DW DW	Recommended track for deep draft vessels (track not defined by fixed marks)
21b	DW83 ft DW76 ft DW83 ft DW76 ft	Depth is shown where it has been obtained by the cognizant authority
22		District or province limit
23		Reservation line (Options)
24	COURSE 053° 00' TRUE MARKERS MARKERS	Measured distance
25	PROHIBITED AREA	Prohibited area (Screen optional)
(Pd)	SAFETY FAIRWAY	Shipping safety fairway (two-way traffic)
(Pe)		Limits of former mine danger area
(Pf)	17386	Reference larger scale chart
(Pg)		Limit of fishing areas (fish trap areas)
(Ph)		3-mile Territorial Sea Boundary 12-mile Contiguous Zone Boundary; headland to headland line
(Pi)		COLREGS demarcation line

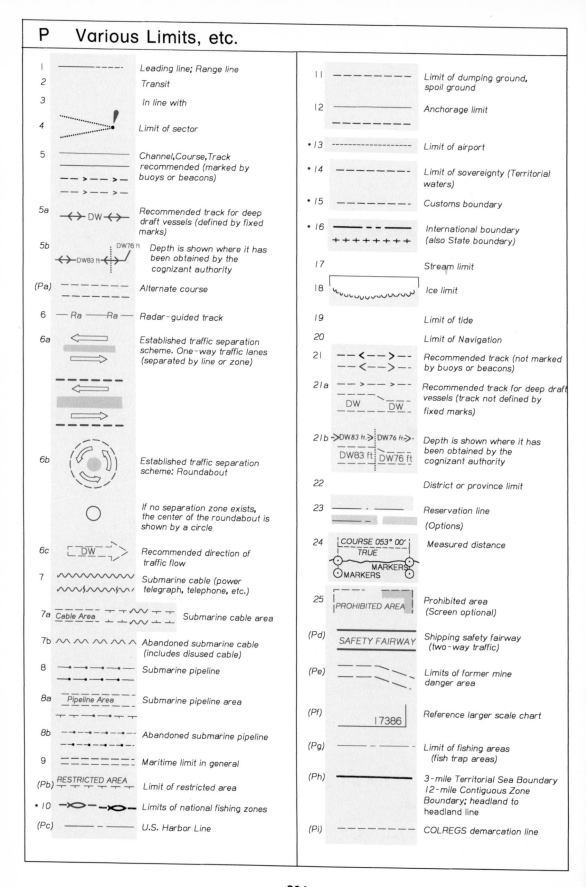

Q   Soundings

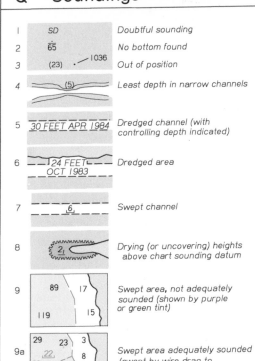

1	SD	Doubtful sounding
2	6̄5	No bottom found
3	(23) · 1036	Out of position
4	(5)	Least depth in narrow channels
5	30 FEET APR 1984	Dredged channel (with controlling depth indicated)
6	24 FEET OCT 1983	Dredged area
7	6	Swept channel
8	2₁	Drying (or uncovering) heights above chart sounding datum
9	89 17 119 15	Swept area, not adequately sounded (shown by purple or green tint)
9a	29 23 3 22 8 30 19 21 18 7	Swept area adequately sounded (swept by wire drag to depth indicated)

10		Hairline depth figures
10a	19 8₂ 7¾	Figures for ordinary soundings
11		Soundings taken from foreign charts
12	8₂ 19	Soundings taken from older surveys (or smaller scale charts)
13	8₂ 19	Echo soundings
14	8₂ 19	Sloping figures
15	8₂ 19	Upright figures
16	(25) (2)	Bracketed figures
17	6	Underlined sounding figures (drying)
18	3₂ 6₁	Soundings expressed in fathoms and feet
22		Unsounded area
(Qa)	6 5 2 ft	Stream

R   Depth Contours and Tints

Feet	Fm/Meters			
0	0			
6	1			
12	2			
18	3			
24	4			
30	5			
36	6			
60	10			
120	20			
180	30			
240	40			

Feet	Fm/Meters
300	50
600	100
1,200	200
1,800	300
2,400	400
3,000	500
6,000	1,000
Approximate depth contour	
Continuous lines, with values	

—— 5 —— (blue or black) —— 100 ——

S   Quality of the Bottom

1	Grd	Ground	11	Rk; rky	Rock; Rocky	(Sb)	Vol Ash	Volcanic ash
2	S	Sand	11a	Blds	Boulders	17	La	Lava
3	M	Mud; Muddy	12	Ck	Chalk	18	Pm	Pumice
4	Oz	Ooze	12a	Ca	Calcareous	19	T	Tufa
5	Ml	Marl	13	Qz	Quartz	20	Sc	Scoriae
6	Cy; Cl	Clay	13a	Sch	Schist	21	Cn	Cinders
7	G	Gravel	14	Co	Coral	21a		Ash
8	Sn	Shingle	(Sa)	Co Hd	Coral head	22	Mn	Manganese
9	P	Pebbles	15	Mds	Madrepores	23	Sh	Shells
10	St	Stones	16	Vol	Volcanic	24	Oys	Oysters

S   Quality of the Bottom

25	*Ms*	Mussels	42	*h; hrd*	Hard	60	*gn*	Green
26	*Spg*	Sponge	43	*stf*	Stiff	61	*yl*	Yellow
27	*K*	Kelp	44	*sml*	Small	62	*or*	Orange
28	*Wd*	Seaweed	45	*lrg*	Large	63	*rd*	Red
	Grs	Grass	46	*sy; stk*	Sticky	64	*br*	Brown
29	*Stg*	Sea-tangle	47	*bk; brk*	Broken	65	*ch*	Chocolate
31	*Spi*	Spicules	47a	*grd*	Ground (Shells)	66	*gy*	Gray
32	*Fr*	Foraminifera	48	*rt*	Rotten	67	*lt*	Light
33	*Gl*	Globigerina	49	*str*	Streaky	68	*dk*	Dark
34	*Di*	Diatoms	50	*spk*	Speckled	70	*vard*	Varied
35	*Rd*	Radiolaria	51	*gty*	Gritty	71	*unev*	Uneven
36	*Pt*	Pteropods	52	*dec*	Decayed	(Sc)	S/M	Surface layer and Under layer
37	*Po*	Polyzoa	53	*fly*	Flinty	76		Freshwater springs in seabed
38	*Cir*	Cirripedia	54	*glac*	Glacial			
38a	*Fu*	Fucus	55	*ten*	Tenacious	(Sd)		Mobile bottom (sand waves)
38b	*Ma*	Mattes	56	*wh*	White	(Se)	Si	Silt
39	*f; fne*	Fine	57	*bl; bk*	Black	(Sf)	Cb	Cobbles
40	*c; crs*	Coarse	58	*vi*	Violet	(Sg)	m	Medium (used only before S (sand))
41	*so; sft*	Soft	59	*bu*	Blue			

Foreign Bottoms   See glossary

T   Tides and Currents

1	HW	High water	*10*	ISLW	Indian spring low water
1a	HHW	Higher high water	*11*	HWF&C	High-water full and change (vulgar establishment of the port)
2	LW	Low water			
(Ta)	LWD	Low-water datum	*12*	LWF&C	Low-water full and change
2a	LLW	Lower low water	*13*		Mean establishment of the port
3	MTL	Mean tide level	*13a*		Establishment of the port
4	MSL	Mean sea level	*14*		Unit of height
4a		Elevation of mean sea level above chart (sounding) datum	*15*		Equinoctial
5		Chart datum (datum for sounding reduction)	*16*		Quarter; Quadrature
6	Sp	Spring tide	*17*	Str	Stream
7	Np	Neap tide	*18*	2 kn	Current, general, with rate
7a	MHW	Mean high water	*19*	2 kn	Flood stream (current) with rate
8	MHWS	Mean high-water springs	*20*	2 kn	Ebb stream (current) with rate
8a	MHWN	Mean high-water neaps	*21*	° Tide gauge	Tide gauge; Tidepole; Automatic tide gauge
8b	MHHW	Mean higher high water			
8c	MLW	Mean low water	*23*	vel	Velocity; Rate
9	MLWS	Mean low-water springs	*24*	kn	Knots
9a	MLWN	Mean low-water neaps	*25*	ht	Height
9b	MLLW	Mean lower low water	*26*		Tide

T   Tides and Currents

27	😊	New moon
28	●	Full moon
29		Ordinary
30		Syzygy
31	*fl*	Flood
32		Ebb
33		Tidal stream diagram

34	A B	Place for which tabulated tidal stream data are given
35		Range (of tide)
36		Phase lag
(Tb)		Current diagram, with *explanatory note*
(Tc)	CRD	Columbia River Datum
(Td)	GCLWD	Gulf Coast Low Water Datum

U   Compass

1	N	North
2	E	East
3	S	South
4	W	West
5	NE	Northeast
6	SE	Southeast
7	SW	Southwest
8	NW	Northwest
9	N	Northern
10	E	Eastern
11	S	Southern
12	W	Western
21	brg; My	Bearing
22	T	True
23	mag	Magnetic
24	var	Variation
25		Annual change
25a		Annual change nil

26	+15°	Abnormal variation; Magnetic attraction
27	deg	Degrees
28	dev	Deviation
29		Compass roses

The outer circle is in degrees with zero at true north. The inner circles are in points and degrees with the arrow indicating magnetic north.

ALBANIAN CHARTING TERMS

K, koder, kodra .... hill

ARABIC CHARTING TERMS

Dj, djebel .... mountain, hill
G, geb, gebel .... mountain, hill
J, jab, jabal, jabel .... mountain, hill
Jazirat, jazt .... island, peninsula
Jeb, jebel .... mountain, hill
Jez, jezirat .... island, peninsula
Jl .... mountain, hill
K, khawr .... inlet, channel
Si, sidi .... tomb
W, wad, wadi, wed .... valley, river, river bed

CHINESE CHARTING TERMS

Chg, chiang .... river, shoal, harbor, inlet, channel, sound

DANISH CHARTING TERMS

B .... bay, bight
Banke, bk .... bank
Bugt .... bay, bight
Fj, fjord .... inlet
Grd, grund .... shoal
Havn, havnen .... harbor
Hd .... headland
Hm .... islet
Hn .... harbor
Hne .... islets
Holm, holmene .... islet, islets
Hoved .... headland
K, kap .... cape
N, nord, nordre .... north, northern
One, oyane, oyene .... islands
Pt, pynt .... point
Skaer, skjaer, skr .... rock above water
Sd, sund, sundet .... sound

DUTCH (NETHERLANDS) CHARTING TERMS

As .... ash

B, baai .... bay
Bas .... woods in general
BaZ .... notice to mariners
Berg, bg .... mountain
Bk .... bank, broken
Br .... brown, latitude
Bu .... blue

CG .... coast guard station
Cy .... clay

D .... dark
Dia .... diaphone
Dm .... decimeter

Eb .... ebb

F .... fine
Fla .... chimney with flare

G .... gravel, shingle, green, gulf
Gb .... checkered, ground
GEB .... range
Gr .... great, large
Gs .... gray

H .... cape, hook, head, headland, hour
Hk .... reformed church
Horizontale lichten .... horizontal lights
HW .... high water

K .... creek, cape
Kenb, kenbaar .... conspicuous, remarkable
Kl, klein .... little, small
Kn .... knot
Krt .... see chart
Kt .... chalk

L .... light, longitude
LAT .... lowest astronomical tide
LW .... low water
LWS .... low water springs

M .... meter, mud, muddy, nautical mile
Min .... minute
Ml .... marl
Mond, monding .... mouth
Mosselen, Ms .... mussels

N .... north
No .... number
Nw, nwe .... new

O .... oysters
OMS .... submarine fog signal

Pas .... passage, pass
Pt .... point

R .... red, river, stream
Rf, rif .... reef
Ru .... ruins

S .... sand, south
Sh .... shells
Sk .... ooze
Sp .... speckled
SS .... signal station
St .... saint
Stn .... station

Tr .... tower

V .... volcanic
Vm .... fathom
Vs .... flagstaff, flagpole
Vss .... established traffic separation scheme
Vt .... foot, feet

W .... white

Y .... yellow, amber

ZW, zwart .... black

FINNISH CHARTING TERMS

K .... rock, reef
Kallio .... rock
Kari .... rock, reef
Kivi .... rock
Luodet, luoto, lu .... rock(s)
Ma, matala .... shoal
Sa, saaret, saari .... island(s)
Torni, tr .... tower

FRENCH CHARTING TERMS

A, Ae .... inlet, aeronautical
Ant .... front light
App, appr .... approach
Arch, archipel .... archipelago
Arg .... clay
+Astro .... astronomic station
Aux .... auxiliary light

V Abbreviations of principal foreign terms, (Glossary)

B	boulders, bay, shoal, white, blue
Baie	bay
Bal tel	telegraph-cable (landing beacon)
Bc	bank
B de sauv	lifeboat station
Bk, bks	bank, banks
Bl	blue
Blanc	white
Blds	boulders
Ble	battery
Blk	black
B M	low water
BM	bench mark
Bn	basin
Bo, boue	ooze, boulder
Br	broken, stream
Brk	broken
C	cape, cove, inlet
Ca	calcareous
Cal	calcareous, channel, narrows
Cap	cape
Carre, carriere	mine
Cath	cathedral
C de G	coast guard station
Ch, Chal, chan	channel, narrows, chimney, stack
Chap	chapel
Chat, chau	castle
Chau d'eau	water tower, standpipe
Chee	chimney stack
Chen	channel
Chk, chl	channel, narrows
Chlle	chapel
Chno	range
Cl	bell buoy, fog bell
Cler	steeple
Co, cor	coral
Coll, colline	hill
Coq	shells
Cq	shells
Cr	creek
Crl	coral
D	dark, hard
Dec	uncovers
Det, detr	strait, destroyed
Detroit	strait
Detruit	destroyed
Dist	district
Dk	dark
Dn	dolphin
Dne	customhouse
Duc d'Albe	dolphin
Dur	hard
E, eclat	flash
Egl	church
El	electric
Emb, embre	mouth
Ent	entrance, inlet
Entp	magazine
Entree	entrance, inlet
Env	about
Ep	wreck
Est	east, estuary
Ev	every, eventual
Ext	outer
F	ooze, light, unwatched light, fine
F Aero	aeronautical light
F b ae	fixed and group flashing light
F b e	fixed and flashing light
F det br	fog detector light
Fd	fjord, ground
F e	single flashing light

F el	isophase light
Feu	light
F f	fixed (steady light)
Fin	fine
Fjd	fjord
Fl	large river, single-flashing (total duration of light less than dark)
Fla	oil derrick
Fo	occulting light
Fque	factory
Fr	suburb
FS	flag tower
Ft	fort
FV	revolving, rotating light
G	coarse, gulf, large, great
Ga, gal	shingle, gravel
Gaz	gas tank
Gd	ground
Gde	large, great
Gg	fog gong
Gl	glacier
Glu	sticky
Gp	group
Gr	gravel, gray
Grd	ground
H	hard
Hal hop	hospital
Hd	head, headland
H de V	town hall
Hn	haven
Hr	height, harbor
Hrd	hard
Ht fd	shoal
Huit	oysters
I, is, Ile, ilot, it, its	island, islands, islet
Ineg	uneven
Inf	lower, lower light
Int	inlet, intensified, inner
Intd, interdit	prohibited
J	yellow
Jct	junction
L, lac	lake, loch, lough
Lag	lagoon
Lav	lava
Ldg	landing
Le, les	ledge, ledges
Lndg	landing
Lum	luminous
M	marl, mud, soft
Mad	madrepores
Mat	varied
ME	neap tide
Mge	anchorage
Mgne	mountain
Mid	middle
Mlg	anchorage
Mn	minute, mill
Mnt	monument
Mou	soft
Mouil, mouillage	anchorage
Moul	mussels
Moy	middle
Mt, mts	mount, mountain
Mus	mussels
MWL	low water datum
N	knots, black, north
NL	new moon

Nrs	narrows
Nv	new
0	annual change nil, occultation
Obs	obstruction
Obsc, obscd	obscured light
Occas	unwatched light, occasional light
Org	orange
P	port
Pas	passage
Pav	pavilion
Peb	pebbles
Pen	peninsula
Pi	stones
Pic	peak
Pierres	stones
Pit, pite	small, little
Pk	peak
Pl	full moon
Pla	tableland, sunken flat
Pm	high water
Pn	peak
Post	rear light
Pr I	peninsula
Prom	promontory
Pt, pte	landing, point
PTT	post office
Pyr	pyramid
Q	quay
R	river, rock, submerged rock, radio telegraph station, road, roadstead, red
Ra	radar responder beacon
RaRa	radar station
Rau	stream
Rav	ravine
RC	circular radiobeacon
RC Aero	aeronautical radiobeacon
Rd	road, roadstead, directional radiobeacon, radio range
Re	rock, submerged rock
Regl	standard
Relt	bearing
Rem	remarkable
Rer	rock, usually above water
Resr	oil tank
Rf, rfs	reef, reefs
RG	radio direction finding station
Rge	range of mountains
Rgl	standard
Riv	river
Rk, rks	rock, rocks
R Lta	riprap surrounding light
Ro Bn	radiobeacon
Ro Tel	radio telephone station
Ro Tr	radio tower
Rsv	oil tank
RT	radio telephone station
RW	rotating loop radiobeacon
S	sand, sector, south
Sal br	fog-signal station
Sante, ste	health officer's office
Sbrv	color of sector, white, red, green
Sect	sector
Sf	stiff
SG	sand and gravel
Sft	soft
Sh	shoal
Shin	shingle
Shl	shoal
Sif	fog whistle
Sif Gg	whistle buoy

Sir	siren
Sm	submarine, statute mile
Sm, sml	small
So, sft	soft
Som	summit
Sous-marin	submarine
Sp, spr	spring tide
St, ste	Saint
Stf	stiff
Stn	station
Str	strait
Stk	sticky
Sup	higher
Sy	sticky
Syz	syzygy
T	tufa
Tel	telegraph
Temp, tempre	temporary
The, tour, tr	tower
Tr	tufa
U	minutes
Us	factory
V	green, vertical lights
Var	varied
VC	see chart
VE	spring tide
Vig	lookout station, watchtower
Vio	violet
Vis	visible (range), conspicuous
Vol sm	submarine volcano
VPI	see plan
Vue	view point
Vx	old
W, wh	white
WT	watertower, standpipe
Y	nautical mile

GAELIC CHARTING TERMS

Bo, bogha	sunken rock
E, eilean, eileanen, en	island(s), islet(s)
Ru, rubha	point
Sg, sgeir, sgr	rock

GERMAN CHARTING TERMS

A, amt, anslalt	office, establishment
Ankpl	anchorage
Anl	jetty
Anst	approach
Auff	conspicuous
Aust	oysters
B	bay
Bntt	battery
Beabs	projected
Ber	correction
Bk	beacon in general, fixed beacon
Bl, blau	blue
Blink, blitz	long flash, short flash
Blk, blz	long flash, short flash
Blk (1&2)	composite group long flashing light
Blk (2)	group long flashing light
Blz	short flashing light
Blz-Blk	short-long flashing light
Blz (1&2)	composite group short flashing light
Blz (3)	group short flashing light
Bn	well
Bnt	varicolored
Br	brown, chocolate, latitude
Bucht	bay

Dev-Dlb	deviation dolphin
Dev-Tn	compass adjustment buoy
Di	diatoms
Dkl	dark
Drchf	passage, pass
Ehem	ancient
Einf	inlet
Eis-s	ice signal station
Enge	narrows
E-werk	electric works
Expl, explosiv	explosive
F	fine, fixed light
Fbr	factory
Fd	fjord
Fhrwss	channel
Fi, fi-hfn	fishing light, fishing harbor
Fj	fjord
Fkl	quick flashing light
Fkl unt	interrupted quick flashing light
Fl	river
Flgtm	flag tower
Fls, fls,	rock, rocky
FluB	river
Fr	foraminifera
G	yellow, gulf
Gas-T	gas tank, gasometer
Gb	coarse
GB	large, great
Gbg	range
GbK	shingle
Gd	ground
Gef-S	danger signal station
Gel	extinguished
Gelb	yellow
Gem	reported
Ger	lesser
Gez-F, gezeitenfeuer	tidal light
Ggf	eventual
Gl	globingerina
Glt	isophase light
Gl-Tn	bell buoy
Gn	green
Gr	gray
Grs	see-weed, grass, mattes
H	light
Hfn	haven, harbor, port
Hg	hill
H-I	peninsula
Hk	point
Hl-Tn	whistle buoy
Hs	house
Ht	hard, hut
HW	high water
I	institute
K	gravel
Ka	calcareous
Kai	quay
Kas	barracks
Kb-Bk	telegraph cable beacon
Kb-Tn	telegraph cable buoy
Kblg	cable length
Kl	little
Klippe, klp	rock, sunken rock
Kl St	pebbles
Kn	knot
Kor, koralle	coral
Kr	chalk
Krhs	hospital

Kst-W	coast guard station
Ku	cupola
Laz	lazaret
Lcht-TM	light
Ldg-Pl	landing, landing place
Lg	longitude
Lv	lava
M	oysterbed
Mdg	mouth
Meeresarm	loch, lough
Mgl	marl
Mk	marker
Ml-Bk	mile beacon
MNpHW	mean high water neaps
MNpNW	mean low water neaps
Mo(K)	morse code light, according to character
MSpHW	mean high water springs
MSpNW	mean low water springs
Mt	middle
Mt-F	middle light
Mun-Vers Gbt	explosives dumping ground
MW	mean tide level
N	north
NF	radio list
NfS	notices to mariners
N-L	fog light
N-lich, norlich	northern
NO	northeast
Not-S	distress signal station
Nr	number
N-S	fog signal station
N-Such-F	fog detector light
NW	low water
O	east
Ober feuer, ob-f	rear light, upper light
Obs	observatory
Od, oder	or
Ol-T, oltank	oiltank
Or	orange
P-A	post office
Pf, pfahle	pile
Priv	private
Pt	pteropods
Pyr	pyramid
Qm-F	cross light
Qrt-Tn	quarantine buoy
QTG kustenfunkstelle r	QTG-radio station
Qu	quartz
R	QTG-radio station, reef, red
Ra	radar, radar scanner, radar station
Ra-Ku	radar cupola, radar dome
Ra-Mst	radar mast
Ra-Tm	radar tower
RC	circular radiobeacon
RC (Aero)	aeronautical radiobeacon
Rcht-Bk	leading beacon
Rcht-F	leading light
RD	directional radiobeacon, radio range
Rf	reef
RG	radio direction finding station
Rgd	sand reef
R-S	lifeboat station, lifesaving station
Ru	ruins
RW	consol station, rotating loop radiobeacon
S	black, station, south
Sc	scoriae

Sch	shells
Sch-H	obstruction, depth unknown
Sch-H	sunken danger with depth, cleared by wire drag (in German waters by diver)
Schl	ooze, castle
Schp	shed
Schutt-S	spoil ground
Schw	sponge, faint
Sd	sand, sound
Sdkor	sabellaria
SFKI	very quick flashing light
Sgn	signal
Sgn-S	international signal station
Shb	sailing directions
Siel	sewer
Sk	mud, muddy
Skr	vertical
S-lich	southern
Sm	nautical mile
SMt	seamount
SO	southeast
Sp	summit, peak
St	flinty, stones
Stg	fucus, kelp
Str	strait
Strm-S	storm signal station
T	clay, and
Tafel	board
Tg-F	daytime light
Tlws	partly
Tm-Bk	tower beacon
Tst	tufa
Ton	clay
T-S	telegraph office
U	and
Ua	etc, other
Ubr	occulting light
Ubr(2)	group occulting light
Ubr(2&3)	composite group occulting light
U-F	front light, lower light
Und	and
Ungf	approximate
Unr gd	foul ground
Unr (mun)	foul (explosives)
Unr unrein	foul, wreck, remains dangerous only for anchoring
Unrein (munition)	foul (explosives)
Unterfeuer	front light, lower light
Untf	shoal
Usw	and so on
U-Wss-GI	submarine fog bell (action of waves)
U-Wss-GI	submarine fog bell (mechanical)
V	volcanic
Va	cinders
Vdklt	obscured
Verb	prohibited
Viol	violet
Vrd	concealed
Vrsdt	shoaled
Vrst	intensified
Vsw	experimented
W	lookout station, watch tower, white
Wache, wachtturm	lookout station, watch tower
Warn-F, warnfeuer	obstruction light
Wch	soft
Wchs Blk wr	long flashing light alternating, white-red
Wchs Blz wr	short flashing light alternating, white-red
Wchs Ubrwr	occulting light alternating white-red
Wchs(2)wgn	group alternating light, white-green

Wchs wr	alternating light, white-red
WeiB-rot	group alternating light, white-red
Wgr	horizontal
Wk	wreck
W-S	weather signal station
WSS-S	tide signal station
WSS-T	water tank
W-Tm	lookout station
Z, zah	tenacious, sticky
Zbr	broken
Zgl	brick kiln
Zolla-A, zoll-w	custom-house
Zrst	destroyed
Z-S	refuge for shipwrecked mariners
Zt-S	time signal station
Ztwl	temporary, temporary buoy
Ztws	occasional

GREEK CHARTING TERMS

Ak, akra, akrotirion	cape
Ali, angali	bight, open bay
Ang, angirovolion	anchorage
If, ifaloi, ifalos	reef(s)
Kolpos, ks	gulf
Limin, In	harbor
N	island
Nes	islet
Nis, nisidhes, nisis	islet(s)
Nisoi, nisos	island(s)
Noi	island
O, ormos	bay
Pot, potamos	river
Sk, skopeloi, skopelos	reef(s)
Ves	rocky islets
Vis	rocky islet
Voi, vos	rock(s)
Vrakhoi	rock(s)
Vrakhonisides	rocky islets
Vrakhonisis	rocky islet
Vrakhos	rock(s)

ICELANDIC CHARTING TERMS

Fjordhur, fjr	fjord
Gr, grunn	shoal

INDONESIAN CHARTING TERMS

Abu	cinders
Abu-abu	gray
Adjaib	remarkable
Arus	stream
Atap seng	zinc roof
Aum kabut	fog whistle
Badai	storm
Bagus	fire
Bahan peledak	explosive
Bajangan	shading
Bakobako	mangroves
Balai kota	town hall
Bandar	harbor, port
Bangunan	building
Barak	barracks
Baru	new
Barat	west
Barat daja	southwest
Barat laut	northwest
Baringan	bearing
Batre	battery
Batu	stones
Batu besar	boulder
Batu bulat-bulat	boulders
Batu karang	rock, rocky
Belukar	bushes
Berbatu-api	flinty

V Abbreviations of principal foreign terms, (Glossary)

Berduri	spicules	Kudus	saint
Berpasir	gritty	Kuning	yellow
Besar	great	Kwarsa	quartz
Besar (luas)	large		
Biara	monastery, convent	Laguna	lagoon
Biasa	ordinary	Lahar	lava, lava flow
Biru	blue	Lembah	valley
Budjur	longitude	Liat	tenacious, sticky
Bukit	hill	Limban	footbridge
Bukit pasir	sandhills, dunes	Lintang	latitude
Busuk	rotten	Listrik	electric
		Luar	outer
Dalam	inner	Luas	large
Dalam air	submarine	Lumpur	mud, muddy
Danau	lake	Lumut	mussels
Daratan	land		
Dengan	with	Madrepora	madrepores
Depa	fathom	Mata air	spring
Deradjat	degrees	Membusuk	decayed
Derwerga	wharf, quay	Menit	minute
Desimeter	decimeter	Merah	red
Detik	second	Mesdjid	mosque
Deviasi	deviation	Milimeter	millimeter
Diatoms	diatoma	Mil laut	nautical mile
Dibawah air	submerged	Millautdjam	knots
Dihentikan	discontinued	Muara	mouth
Dikenal	conspicuous	Muda	light
Dirusak	destroyed		
Diterangi	see plan	Nomor	number
Djalan masuk	entrance		
Djam	hour	P	white
Djarak	distant	Pagar	fence
Djingga	orange	Palem	tide gauge, tide pole
Dok	dock	Pasang	flood
		Pasir	sand
Galangan	shipyard	Patah	broken
Gamping	chalk	Paviljun	pavilion
Gelangkepil	mooring ring	Peg, pegunungan	range, mountain range
Gelap	eclipse	Pelabuhan	road, roadstead
Geredja	church, chapel	Pembetulan	correction
Gerundjal	uneven	Penataran	establishment
Glester	glacier	Pendaratan	landing place, leading stage, stairs
Gubug	shed, hut	Penerbitan	publication
Gudang	magazine	Pengeluaran	edition
Gunung	mountain	Penggaraman	saltpans
		Penggergadjian	saw mill
H	black	Penting	prominent
Halus	soft	Periode	period
Hidjau	green	Perseroan	company
Hitam	black	Pertambangan	mine, quarry
		Pertjobaan	experimental
Institut	institute	Peternakan	farm
Intji	inch	Pteropods	pteropoda
		Pulau	island
Kaki	foot	Puntjak	summit, peak
Kaku	stiff	Puri	castle
Kampung	village	Putih	white
Kasar	coarse		
Katedral	cathedral	Rabuk	marl
Kelompok	group	Rawa	marsh, swamp, slough
Kepulauan	archipelago	Rintangan	obstruction
Kapur	chalk	Ruangan	pillar
Keran	crane	Rumah	house
Kerang	shells, oysters	Rumput	grass
Kerangka	a number of sunken ships		
Keras	hard	Sawah	paddy fields
Kerikil	gravel, shingle, pebbles	Scoria	scoria
Ketjepatan	velocity, rate	Sedjati	true
Ketjil	little	Sekolah	school
Ketjil (sempit)	large	Sektor	sector
Kira-kira	approximate, about	Selat	strait
Koral	coral	Selatan	south
Kota	city or town	Semenandjung	peninsula
Kpn	archipelago	Sentimeter	centimeter
Kr, krueng	river	Seperempat	quarter, quadrature

Stadion	stadium
Suar atas	upper light
Suar bawah	lower light
Suar belakang	rear light
Suar darurat	auxiliary light
Suar depan	front light
Sumur	well
Sungai	river, stream
Surut	ebb
Taman	park, garden
Tandjung	cape
Tanggul	embankment, levee
Tangki	tank
Telegrap	telegraph
Telepon	telephone
Teluk	bay, gulf, creek, cove
Tembok	wall
Tempat	ground
Tengah	middle
Tenggara	southeast
Terbatas	pyramid
Tertutupes	glacial
Terusan	tunnel (railroad or road), cutting
Tiang	wreck masts visible
Tiangkepil	dolphin
Tiap-tiap	every
Timur	east
Tinggi	height, altitude
Tjandi	temple
Tjerlang	flash
Tjiatan	narrows
Tjoklat	brown, chocolate
Topan	typhoon
TPSB	great trigonometrical survey station
Triangulasi	triangulation
Tua	old
Tua, gelap	dark
Udjung	cape
Ug	cape
Unga	violet
Utara	north
Variasi	variation
W, wai	river

ITALIAN CHARTING TERMS

A	kelp
A band	flagstaff, flagpole
Aco	ancient
Aff	dries
Alb	hotel
All to, allineamento e rotta	leading line, range line
AM	high water
AN	notice to mariners
Anc, anco	anchorage
Ant	front
Appr	approximate
AR	antenna radio
Ar	orange
Arcgo	archipelago
A seg	signal mast
Astr	astronomical
Aum ann	increasing annually
Aus	auxiliary
A var nulla	annual change nil
Azz, azzurro	blue
B	bay
Ba, baia	bay
Banco	bank
Battigia	foreshore: strand (in general)
Bco	bank

Bianco	white
BM	low water
Bna	quay
Boc,	mouth
Bq	quarantine buoy
Br	dolphin
Briccole	dolphin
Bt	mooring buoy with telephonic communications
Btg	mooring buoy with telegraphic communications
BVB	compass adjustment buoy
C	cape
Cal	wharf
Cala	cove
Cam Neb	fog bell
Cam Stm	submarine fog bell
Can	channel
Cann Neb	fog gun
Capo	cape
Cas	castle
Cl	cove
Cle	hill
Cma	summit
Cna	range
Co	cape
Coclo, cocuzzolo	boulder
Colle	hill
Comlo	gable
Convto abb	monastery, convent
Cop	covers
Corne da nebbia	fog horn
Cosp, cospicuo	conspicuous
Cn	shells
Cr	coral
Czo	mountain, mount
Decl	variation
Depo	warehouse, storehouse
Dim ann	decreasing annually
Dir	directional
Dr	dredged
Eccl	occulation
EF	list of lights
Elettr	electric
Eso	estuary
Est	east
Eta	entrance
F	mud, muddy, river
Fca	factory
Fce	brick kiln
Fdo	fjord
F(Ff)	fixed light
Fi Neb	fog whistle
F Lam	fixed and flashing light
F Lam	fixed and group flashing light
Flo	chimney, stack
Fna	fountain
G	yellow
Gde	great
Grp	group
Gta	jetty
H	altitude
I	island, islands, islet, islets
IMAM	mean establishment of the port
Inf	lower, lower light
Int	intermittent light, occulting light
Irreg	irregular

ISAM	high water full and change (vulgar establishment of the port)
Iso	isophase
Ist, istitut	institute
Ito	islet
Lam	flash
Lam	group flashing light
Lam L	long flashing light
LM	mean sea level
LRS	chart datum (datum for sounding reduction)
Lum	lighted, luminous
M	madrepores, magnetic, mountain, mount
Mag	magazine
Mar	maritime
MAMQ	mean high water neaps
MAMS	mean high water springs
MBMQ	mean low water neaps
MBMS	mean low water springs
Mgna	mountain, mount
Mlo, molo	mole, pier
Mun	town-hall
Nauto	nautophone
NE	new edition
Nord	north
Nord Est	northeast
Nord Ouest	northwest
NP	new chart
Nvo	new
Occas	occasional light
Off	factory
Orriz	horizontal light
Osc	obscured light
Osp	hospital
Oss	observatory
Ost	obstruction
Ouest	west
P	square
Pali	piles, stakes, stumps, perches
Pass	pass, passage
Pco	peak
Per	period
PG	coast guard station
Pgio	mound, small hill
Pietre	stones
Pil	pillar
Pl	piles, stakes, stumps, perches
Pla	peninsula
Ple	groin
Plo	little, small
Po	hill
Port, portolano	sailing directions
Post	rear
PPTT	telegraph office
Pref	government house
Prio	promontory
Prog	projected
Provv	provisional light
P sb	landing
Pta	point, summit
Pte	bridge
Pto	haven
Pubbl	publication
Pzo	peak
R	radio telegraph station, rock, rocky, red
Ra	road, roadstead, radar station
Ra (cosp)	radar conspicuous
Ra Rifl	radar reflector
RC	circular radiobeacon
RC AERO	aeronautical radiobeacon
RD	directional radiobeacon, radio range
Rdf	radio broadcasting station
Rel	wreck showing any portion of hull or superstructure
RG	radio direction finding station
Rifl	reflector
Ril	bearing
Rist	reprint
RT	radio telephone station
R Telem	distance finding station (synchronized signals)
RW	rotating loop radiobeacon
S	sand, saint, south
Sc	quick flashing light, ridge of rocks
Sca	shoal
Scaf	wreck
Sci, sco	rock(s), reef(s)
Scogliera	ridge, ridge of rocks
Scogli, scoglio	rock(s), reef(s)
Scop Em	uncovers
Scra	ridge
Se, secca	shoal(s)
Seg	signal
Seg Neb	fog signal station
Segnale	signal
Sem, semaforo	semaphore
Serb, serbatolo bafta	oil tank
Serra	range of mountains
Set, settore	sector
SMt	seamount
Somm	submerged
Sopp, soppresso	discontinued
Sra	range of mountains
Stab	establishment
Ste	station
Ste Pt	pilot station
Ste Salo	lifeboat station
Ste Seg	signal station
Stm	submarine
Sto, stretto	strait
Sud	south
Sud Est	southeast
Sud Ouest	southwest
Sup	higher, upper
T	intermittent stream, torrent
Tam	temporary light
Tav	table-land
Tel	telegraph
Telno	telephone
Torre	tower
Tre	tower
Tre Ved	lookout station, watchtower
Tro Neb	fog trumpet
TV	television mast or tower
Uff	office
V	green, true, street
Va	villa
Vedi piano	see plan
Verde	green
Vert	vertical
Vietato	prohibited
Viol	violet
Vle	valley
Vno	volcano
Vo	old
VP	see plan
Vto	prohibited
Vto Anco	anchorage prohibited

V Abbreviations of principal foreign terms, (Glossary)

Zo .... elevation of mean sea level
above chart datum

JAPANESE CHARTING TERMS

B .... bay
Byoti .... anchorage
Cab .... cable length
Dake, de .... mountain, hill
Destd .... destroyed
Ga, gawa .... river
GTS .... great trigonometrical survey station
H .... height
Ha, hana .... cape, point
Hakuchi .... roadstead
Hi .... roadstead
Hn .... haven
Hr .... harbor
Irie .... loch, lough
Irikuti .... entrance
Ja, jima .... island
Ka .... river
Kaikyo, ko .... strait
Kaiwan .... gulf
Kako .... estuary
Kawa .... river
Ko .... loch, lough, strait
Koro .... passage
Kuti .... mouth
Kyoko .... fjord
LL .... list of lights
Ma .... village
Machi, mi .... town
Mi, misaki, mki .... cape
Mura .... village
Sa .... island
Saki .... cape, point
San .... mountain
Sankakusu .... delta
Sdo .... sound, pass, channel
Seto .... narrows, strait
Shima .... island
Si .... cape, point
Sn .... mountain
So .... narrows, strait
Suido .... sound, pass, channel
Take, te .... hill, mountain
Wan .... bay
Ya, yama .... mountain
Zaki .... cape, point
Zan .... mountain
Zi .... cape, point
Zn .... mountain

MALAY CHARTING TERMS

A, ayer .... stream
Bandar .... seaport
Batang .... river
Batu .... rock
Bdr, bendar .... seaport
Bt .... hill
Btg .... river
Bu .... rock
Bukit .... hill
Gg .... shoal, reef, islet
Gg .... mountain
Gosong, gosung .... shoal, reef, islet
Gunong .... mountain
Gusong .... shoal, reef, islet
Kali .... river
Kampong, kampung .... village
Karang .... coral reef, reef
Kg .... village
Kg .... coral reef, reef
Ki .... river
Kla, kuala .... river mouth

Lab, labn, labuan, labuhan .... anchorage, harbor
Ma, muara .... river mouth
Parit .... stream, canal, ditch
Pelabohan, pln .... roadstead, anchorage
PP .... group of islands
Prt .... stream, canal, ditch
Pu, pulau .... island
Pulau-pulau .... group of islands
Pulo, pulu .... island
Selat .... strait
SH .... strait
Si .... river
Sungai, sungei .... river
Tandjong, tandjung, tanjong .... cape
Telok, teluk .... bay
Tg .... cape
Tk .... bay

NORWEGIAN CHARTING TERMS

Aero .... aeronautical
Ankerplass .... anchoring berth, anchorage
Aust .... east
Austre .... eastern

Bae .... rock (submerged rock)
Bake .... tower beacon
Bakre fyr .... rear light
Banke .... bank
Bekk .... river, stream
Berg .... mountain, mount, hill
Bg .... mountain, hill
Bifyr .... auxiliary light
Bla .... blue
Bl, blink .... flash
Blokk .... boulder
Blot .... soft
Br .... brown
Br, bredde .... latitude
Bukt .... bay, bight
By .... city, town

Dia, diafon .... diaphone
Dm .... decimeter
Dnl .... sailing directions
Duc d'albe .... dolphin

Efs .... notice to mariners
Elv .... river, stream

F .... fine, rock, rocky, fixed
Famm .... fathom
Farbar .... navigable
Fast .... stiff, fixed
Favn .... fathom
Fd, fdn .... fjord
Fin .... fine
Flak .... rock (submerged rock), bank
Flo .... flood
Flu .... sunken rock
Fm .... occulting
Fm bl .... fixed and flashing
Fm n fm gpoc .... group occulting light
Fm oc .... occultation
Fremre .... front light
Fv .... fathom
Fyr .... light
Fyrskip .... light vessel, light ship

G .... green, coarse
Gl .... yellow
Gml .... old
Gn .... green
Godt synlig .... conspicuous
Gong .... fog gong
Gp .... group

316

V Abbreviations of principal foreign terms, (Glossary)

Gp hbl, gp hurtigbl	interrupted quick flashing light
Gp lynbl	group short flashing light
Gr	gravel
Gr, gra	gray
Grad	degree, degrees
Gron	green
Grov	coarse
Grunne	ground, rock (submerged rock), shoal
Grus	gravel
Gs	sea-weed, grass
Gul	yellow
H	white
Hamn	harbor
Havn, havneby	harbor, port
Hbl	quick flashing light
Hd	altitude, height
Hefte	ostruction
Hn	harbor
Holme	islet
Hv, hvert	every
Hvit	white
HW	high water
In	inner
K	cape
Kai	quay, wharf
Kbl	cable
Kil	cove
Klakk	rock (submerged rock)
Kn	knots
Kr	coral
L	little, small, clay
Lavere	lower, front light
Lav kyst	flat coast
Lbl	short flashing light
Ldg	longitude
Lei	channel
Lengde	longitude
Lille	little, small
Lop	channel
Los	pilot station always attended (on older charts)
Ls	pilot office
LW	low water
Lykt	light, light beacon
Lynbl, lynblink	short flashing light
Lys	light
Lysboye	light buoy
Lysende	lighted, luminous
Lysflate	light-float
M	middle, with, nautical
Magn	magnetic
Med	with
Mil	nautical mile
Myr	marsh, swamp
N	north, nautical mile, northern
Ned	lower
Nes	head, headland
No	northeast, number
Nord	north
Np	neap tide
Nr	number
Nut	peak
Ny	new
Os	estuary mouth
Ost	east
Ov	higher, upper, rear (light)
Ovre	upper
Oy	island

P	post office
Pilar	pile, pillar
Pir	pier
Pos	position
Pr	private
Puller	bollard
Pynt	point, head, headland
R	mooring ring
Rad mast	radio mast, radio tower
Rad st	radio broadcasting station
RC	circular radio beacon
RD	direction radiobeacon, radio range
Red	road, roadstead
Rev	reef, ridge
RG	radio directing finding station
Rod	red
RS	lifesaving station
R st	radio telegraph station
RT st	radio telephone station
RW	rotating loop radio beacon
S	second (of time), southern, south, black
Sec, sek	second (of time)
Sekt, sektor	sector
Sem, semafor	semaphore
Sg	shingle, pebbles
Sk	chimney, stack
Skall	rock (submerged rock)
Skalle	shoal
Skjaer, skjart	shoal
Skjer	rock (submerged rock)
Skjerane	rocks above water
Skolt	rock (submerged rock)
Sl	mud, muddy, ooze
So	southeast
Sor	south
Sp	spring tide
St	saint, large, stones
Stake	spar buoy
Stein	stones
Sto	ramp
Stor	large
St St	boulders
Svart	black
Syd	south
Sydlige	southern
Sydvest	southwest
Synl, synlig (rekkevidde)	visible (range)
T	ton
Taren	sunken rock
Tarn	tower
Tn	sunken rock
TS	telegraph station
Tydelig	conspicuous
Uren bunn	foul ground
Ute av bruk	abandoned
Ute av posisjon	out of position
Uv	submarine
Uv kl	submarine fog bell
V	cairn, tower beacon, western
Vest	west
Vg	bay, cove
Vks	alternating light
Vrak	wreck
Vorr	submerged jetty
Vs	weather signal station
Yt, ytre	outer
Zo	elevation of mean sea level above chart datum

V Abbreviations of principal foreign terms, (Glossary)

PERSIAN CHARTING TERMS

B, bandar . . . . harbor
Jab, jabal . . . . hill, mountain
Jazh, jazireh . . . . island, peninsula
K, khowr . . . . inlet, channel
R, rud . . . . river

POLISH CHARTING TERMS

Jez, jezioro . . . . lake
Kan, kanal . . . . channel
Miel, mielizna . . . . shoal
R, rzeka . . . . river
Wa, wyspa . . . . island
Zat, zatoka . . . . gulf, bay

PORTUGUESE CHARTING TERMS

A . . . . sand, yellow, range of tide
Aband . . . . abandoned
Aero . . . . aeronautical
Aerom. . . . . marine or air navigational light
Aeroporto . . . . airport
Aero RC . . . . aeronautical radiobeacon
AF, Af . . . . Shoal, fine sand
Afl . . . . muddy fine sand
Ag . . . . coarse sand, spire, steeple, compass
    adjustment buoy
Aguada . . . . watering place
Agua descorada . . . . discolored water
Al . . . . muddy sand
Alf . . . . custom-house
Alg . . . . kelp, sea-weed
Alt . . . . alternating light
Am . . . . neap tide, marine and air navigation
    light, yellow
Amarra . . . . cable length
Amarracao . . . . mooring
Amb, ambar . . . . amber
AN . . . . notices to mariners
Anc . . . . roadstead
Ang . . . . cove
Ant, anterior . . . . front, front light
Antigo . . . . ancient
Apito . . . . fog whistle
Apodrecido . . . . rotten
Aprox . . . . approximate
Ar . . . . clay
Arborizado . . . . woodland
Arco vis . . . . sector of visibility
Arg . . . . clay
Arm . . . . warehouse, storehouse
Arq . . . . archipelago
Arr . . . . brook
Ars . . . . Arsenal
At, AT . . . . lookout tower, power transmission line
Aux . . . . auxiliary light
Av . . . . spring tide, avenue
Az . . . . blue

B . . . . gravel, bay, white
Bal . . . . beacon, range targets (markers)
Barragem . . . . dam
Bat, bateria . . . . battery
Bb . . . . fathom
Bc, Bco . . . . bank
BM . . . . low water
BM AM . . . . mean low water neaps
BM AV . . . . mean low water springs
BM de AV da India . . . . Indian spring low water
Br . . . . white, fathom
Bt . . . . tenacious
Buz . . . . fog horn
Buz elt, buzina de nevoeiro electrica . . . . electric fog gun
Bxa, Bxas, Bxio, Bxo . . . . shoal

C . . . . strand, shingle, cape, gravel

Cabeco de poco . . . . wellhead
Cabo . . . . head, headland
Cabo submarino . . . . submarine cable
Cach . . . . waterfalls
Cada . . . . every
Cal . . . . inlet, calcareous
Calhau . . . . strand
Calvario . . . . calvary, cross
Camp . . . . spire, steeple
Can, canal . . . . channel
Canalizacoes subm . . . . submarine pipeline area
Canhao de nevoeiro (cerracao) . . . . fog gun
Cap . . . . chapel
Carta . . . . chart
Casa de campo . . . . villa
Cast . . . . castle, brown
Casuarina filao . . . . filao
Cat, catedral . . . . cathedral
Cem . . . . cemetery (non-christian)
Cerca de . . . . about
Ch, Chm, CH . . . . chimney, stack, elevation of top of
    building
Ci . . . . cirripeda
Cin, Cin F . . . . ash, cinders
Cin, cinzento . . . . gray
Cir . . . . cirripeda
Cist, cisterna . . . . cistern
Cl . . . . light (quality of bottom)
CM . . . . town-hall
Cme . . . . summit
Co . . . . shells, cape
Cob, cobre . . . . covers
Col . . . . hill
Com . . . . reported
Comp ag . . . . compass adjustment beacon
Compia . . . . company
Con . . . . shells, cape
Con, conv . . . . convent
Consp . . . . conspicuous
Cor . . . . coral, shoal
Cord . . . . range of mountains
Cor do sector . . . . color of sector
Corr . . . . correction
Corr, corrente . . . . stream
Cre . . . . chalk
Crista sub . . . . ridge
Cruz . . . . cross
CS . . . . wreck, telegraph cable buoy
Ct . . . . quick flashing light
Ctl(int) . . . . interrupted quick flashing light
CtR . . . . continuous very quick light
CtRI . . . . interrupted very quick light
CtR (n) . . . . group very quick light
CTT . . . . post and telegraph office
CtU . . . . continuous ultra quick light
CtUI . . . . interrupted ultra quick light
Cume . . . . summit
Cup . . . . cupola
Cz . . . . gray

D . . . . diatoms, distance, delta, doubtful, hard
DA . . . . water tower, stand-pipe
DC . . . . oil tank
Decl . . . . variation
Des . . . . landing
Des, descb, descobre
    (periodicamente) . . . . uncovers, dries
Descor, descolorido . . . . discolored
Desigual . . . . uneven
Dest . . . . destroyed
Det . . . . decayed
Dia . . . . diaphone
Dist . . . . distant
Distr . . . . district
D Mg . . . . variation

Dol	dolphin
Duque de alba	dolphin
Duro	hard
DW	deep water route
E	dark, red
Ecl	eclipse
Ed	edition
Electr, elt	electric
Elev, elevador	elevator
Empena	gable
En ch, enchente	flood
Enf	in line (range)
Ens	cove
Ent	entrance, magazine
Equin, equinocial	equinoctial
Erv mar	sea-weed, grass
ES	signal station
Esc	school, dark, scoriae
ES(corr)	stream signal station
ES(gelo)	ice signal station
ES(mare)	tide signal station
ES(meteor)	weather signal station
Esp	sponge
Espic, espiculas	spicules
Esponja	sponge
ES(temp)	storm signal station
Est	stadium
Estab	established
Estal, estaleiro	shipyard
Estb	established
Est salv	lifesaving station
Esto	creek
Estr	strait
Exp	experimental
Explos	explosive, explosive fog signal
Ext	outer
Extr, extremo	extreme
F	fixed light, branch, fine
Fab	factory
Fal	cliff, bluff
Faz	farm
Fc	fucus
Fd	fjord
F Gr Lp	fixed and group flashing light
F Lp	fixed and flashing light
For	foraminifera
Foz	mouth
Fr	faint light
FRI	fixed and flashing light
FRI Agr(n), FRI(n)	fixed and group flashing light
Ft, Fte	fort, small lighthouse
Fu	fucus
G	corse, gulf, phase lag
Gas	gasometer, gas tank
Gasoduto	gas pipeline
GC	Coast Guard station
Gde	great
Gelos	ice signal station
Ggo	fog gong, gong buoy
Gl	globigerina
Glac, glaciares	glacial, glaciers
Gol	narrows
Gongo	fog gong
Gov	government house
Gp	group
Gr	great, large
Gr Lp	group flashing light
Gr Lp C	group short flashing light
Gr Lp MR	group very quick flashing light
Gr Lp R	group quick flashing light
Gr Oc	group occulting light

GTS	great trigonometrical survey station
Gw	Greenwich
H	amplitude of a tidal component (half of the range), height, elevation, hour
Hor, Horiz	horizontal lights
Hora	time signal station, hour
I	intermittent light, island
Ig	church
Igp	creek
Igreja	church
IH	sale place of charts and nautical publications of the instituto hidrografico
Il	islet (islets)
Ilheu	islet
In	discontinued
Inf	lower, lower light
Inst	institute
Int	inner, intermittent light
Inter	intermediate, intermediate light
Irreg	irregular light
Is, Iso	isophase light
Ita	islet
J, jarda	yard
L	mud, muddy, lake, light
La, lama	sandy mud, lagoon, orange
LAN	list of navigational aids
LAR	list of radio navigational aids
L. aux	auxiliary light
Laz, lazareto	lazaret
LC	full moon
Ld	ooze
L det nev	fog detector light
L dir	directional light
LF	list of lights
L int	intermittent light
L irreg	irregular light
List, listrado	streaky
LN	new moon
L/ocas	occasional light
Lp	flash, flashing light
L part	private light
LpC	short flashing light
LpCl	short-long flashing light
Lp R	quick flashing light
L prov	provisional light
LpRln	interrupted quick flashing light
LpUR	ultra quick flashing light
LpURln	interrupted ultra quick flashing light
L rot	rotating light
L temp	temporary light
Lv	lava
Lz	light
M	meter, nautical mile, bearing
Mad	madrepore
Mag	magnetic
Mal, malhado	speckled
Mang	manganese
Mangal	mangrove
Manganesio	manganese
Mangrulho	dolphin
Mangue	mangrove
Mant, mantido	maintained
Marabuto	marabout
Mare	tide
Mareg, maregrafo	tide gauge
Mares	tide signal station
Mar, maritimo	maritime
Mas, MAST, Mast	mast, signal mast

MAST AT .... power transmission mast
MAST (Mas)TV .... television mast
MASTROS .... wreck of which only the masts are
                                    visible
Mat .... boulder
Md .... ground (shells), middle
Med .... middle
Merc .... magazine
Meteo .... weather signal station
Mex, mexilhao .... mussels
Mg .... magnetic, marl
Min .... minute of time
Ml .... soft
MN .... bench mark
Mo .... hill
Mo .... morse code light
MON, Mon .... monument
Most .... monastery, convent
Mt .... mountain, mount
Mt sub .... seamount
Mun .... municipality

Na sa .... our lady
Nauto .... nautophone
Navegavel, Navg .... navigable
NC .... new chart
NE .... new edition
NM, nm .... mean sea level, mean tide level
No .... number
No .... knot
Nora .... noria
Nos .... knots
Not, notavel .... remarkable
Novo .... new
NR .... chart datum (for sounding reduction)
NS .... our lord, our lady
NV .... unwatched light

Obs .... observatory
Obsc .... obscured light
Obst, Obstr .... obstruction
Oc .... occulting light
Ocas .... occasional, occasional light
Ocid, occidental .... western
Oc Agr (n), Oc (n) .... group occulting light
Oc (n t m) .... composite group-occulting light
Ocultacao .... occultation
Ol .... brick kiln
Oleoduto .... oil pipe
Ord ordinario .... ordinary
Os, Ost .... oysters
O sub .... submarine oscillator

P .... broken, black, harbor, port, haven,
                                    preliminary, stones
Pa .... branch
Pad, padrao .... standard
Pag, pagode .... pagoda
Par .... gray
Part .... private, privately, private aid to
                                    navigation
Pas, Pass .... passage, pass
Pc .... peak
Pcel .... bank, shelf
Pe .... foot
Ped, pedregoso .... flinty
Pen .... peninsula
Peq .... small, little
Per .... period
Pesq .... fish haven
PICO SUB .... seamount
Pil, pilar .... dolphin, pile, pillar
PILOTOS .... pilot station
Pinhal .... coniferous woodland
Pir, piramide .... pyramid

Plan .... tableland
PM .... high water
PM AM .... mean high water neaps
PM AV .... mean high water springs
Po .... peak, polyzoa
Poco .... well
Pod .... rotten
Pol .... polyparia, polyzoa, inch
Pontao .... hulk
Pos, posicao .... position
Post .... rear, rear light
Poste .... post
Pp .... pumice
Pr .... black, beach
Prados .... grass fields
Prel .... preliminary
Princ .... principal light
Proem, proeminente .... prominent
Prof menor, Profundidade minima em
    canais (ou rios) estreitos .... least depth in narrow channels
Proib .... prohibited
Proj .... projected
Prom .... promontory
Prov .... provisional, provisional light
Pt .... pteropods
Pta .... point
Ptal .... head, headland
Pub, Publ .... publication
Purp, purpura .... purple

Q .... quarantine, quarantine buoy, quarter,
                                    quadrature, broken, farm
QC .... 2nd quarter
QM .... 1st quarter
Qta .... farm
Qtel .... barracks
Qtz .... quartz

R .... true course, street, rock, rocky, river,
                                    stream, radiotelegraph station,
                                    measured distance, recommended
                                    track
Ra .... river, stream, radar guided track, radar
                                    station
Rad .... radiolaria
Rada .... road, roadstead
Rap .... ramp
Rc .... rock
Rch .... creek
Rd, RD .... radiolaria, directional radiobeacon
Red .... small
Res .... tank
Res met, Residuos metalicos .... mattes
Rest .... spit of land
RF .... radiobeacon
Rfe .... reef
RG .... radio direction finding station
Rl .... flashing light
Rl Agr(n) .... group flashing light
Rl L .... long-flashing light
Rl(n) .... group flashing light
Rl(n+m) .... composite group flashing light
Riba .... river, stream
Ric .... bar
Rio .... river
RN .... bench mark
Rocha .... rock, rocky
Rot .... revolving light, rotating light
RT .... radio telephone station
Ru, ruinas .... ruins
RV .... true course
RW .... rotating loop radiobeacon

S .... pebbles, saint, second of time, syzygy
Sa .... range of mountains, gritty

Sarg, sargaco	kelp, sea-tangle
Sc	small bay
Sem	semaphore
Ser	fog siren, saw mill
Set, setentrional	northern
Simb	symbol
Sin ac sub	submarine sound signal with platform
Sinais	signal station
Sinal	signal
Sino	fog bell
Sino sub	submarine fog bell
Sir	fog siren
Sizigia	syzygy
So	height of mean sea level above the zero of tidal staff
St, Sta, sto	saint, farm
Sub	submarine
Subm	submerged
Sup	higher, upper light
T	ton, temporary light, ground
Temp	temporary light
Temp bud, Templo budista	buddhist temple, joss-house
Tempo	storm signal station
Tf, TF	tufa, federal territory
T oleo	oil tank
Tr	tower, reed horn, power transmission mast
Trib	courthouse
Tufo	tufa
U	unit of height
Ur	bank, shelf
V	see, ooze, true, green
Va, vale	valley
Var	stranding harbor (dries at low-water)
Var, variado	varied
Vau	ford
Vaz, vazante	ebb
Vd	green
Velho	old
Vel, velocidade	velocity, rate
Ver	see
Vert	vertical, vertical lights
Vi	violet
Vig	lookout station, watch tower
Viol	violet
Vis, visivel (alcance)	visible (range)
Vm	red
Vsc	sticky
Vul	volcano
Vul sub	submarine volcano
X	schist
Z	azimuth (bearing)
Zero hidrografico (plano de reducao de sondas), ZH	chart datum (datum for sounding reduction)
Zo	elevation of MSL above chart datum
Zv	true bearing

ROMANIAN CHARTING TERMS

Br, brat, bratu, bratul	branch, arm
C, cap, capu, capul	cape
I, insula	island
L, lac, lacu, lacul	lake
O, ostrov, ostrovu, ostrovul	river island

RUSSIAN CHARTING TERMS

Banka, banki, bka, bki	bank(s)
B, bukhta	bay, inlet
G, gavan	harbor, basin
G, gora	mountain, hill
Ga, guba	bay, inlet, creek
Kam, kamen	rock
M, mys	cape
O, ostrov, ostrova	islands
Oz, ozero	lake
Pol, poluostrov	peninsula
Proliv, prv	channel, strait
R, reka	river
Z, zaliv	gulf, bay

SPANISH CHARTING TERMS

A	gritty, yellow, sand, orange, amber
À de Bda	flagstaff, flagpole
A de la M	range (of tide)
Abr, abra	haven
ACo	sand and gravel
Ad, aduana	customhouse
Aero	aeronautical
Aeromar	marine and air navigation light
AF	sand and mud
Agua	water tower, standpipe
Al	alternating
Alg	kelp, seaweed, seaweed grass
Alm	magazine
Alm	mussels
Alm Dep	warehouse, storehouse
Alr	about
Alt	alternating, alternating light
Am	yellow
An	narrows, orange
Anch	anchorage, bridge clearance (horizontal)
Ang, angostura	narrows
Ant	front light
Antg	ancient
Apag, apagdo, apagada	extinguished light
Aproxte	approximate
Ar	clay
Arc	clay
Arch, Archo, archipielago	archipelago
Arrf, arrecife	reef, rock (uncovers)
Arro	stream, creek
AS	signal mast
Aserr, aserrado	saw mill
Astrm, Astro	astronomical
At	faint light
Aux	auxiliary light
Av, avd	avenue
Az	blue
B	bay, white, reed horn
Ba, bahia	bay
Bajo	shoal
Bal	beacon (in general), bathing establishment
Banco	bank
Bat, bateria	battery
Bca	mouth
Bco	bank
Bd, bdo	soft
Bjmar	low water
BM	low water
Bo, boca	shoal
Br, braza	fathom
Bulevar	boulevard
Bzo	arm (of the sea), loch, lough, lake
C	cape, small, fog bell, shells, every, shingle, post office, quarter, quadrature
Ca	mountain range, shells, calcareous, house
Cable	cable length
Cabo	cape

Cada	every
Cal, caleta	inlet, tufa
Calle	street
Cam Send Hlla	track, footpath, trail
Can, canal	channel, sound, canal, fog gun
Canal	dredged channel
Carbon	cinders
Carta	chart
Cas, castillo	castle
Casa	house
Cat	cathedral
Cbl	cable length
Cc	shells
Cd	new candle
Ce	summit
Cerro	hill
Cha	chimney
Cima	summit
Cisterna	cistern
Cj	gravel
Cjo	farm
Cl	light (quality of bottom), coral
Claro	light (quality of bottom)
CN	post office
Co	gravel, hill
Col, colina	hill
Comp, agj	compass adjustment beacon
Con	with
Consp, conspicuo	conspicuous
Cord	range
Corr	correction
Corto	small
Cr	coral
Cra	mountain, range
Cre	summit
Crec	flood
Cruz	calvary, cross
Ct	quick flashing light
Cte	stream (current)
Ctl	int. quick flashing light
Cto	convent, quarter, quadrature
CtR	continuous very quick flashing light
Cu	quartz
Cubo, cubierto	covers
D	doubtful, delta, flash, flashing light, hard, diaphone
Darna	dock
De	drydock, floating dock
Demora, demarcacion	bearing
Des	flashing, landing place
Descol	discolored, discolored water
Descubo, descubierto	uncovers
Desemb	landing place
Dest, Desto	decayed, destroyed
Diaf	diaphone
DL	riprap surrounding light
Dm	city or town (small scale)
Do	reported
Dta	delta
E	east, sponge, explosive fog signal, station, eclipse
Ea	farm
E B Salv	lifeboat station
Ec	scoriae
E de I P	mean establishment of the port
Elec, Elect	electric
Enf	in line with
Ens, Ensa	bay, creek, cove
Ent, entr	entrance
Entrante	flood
Ep	sponge
Equin, equinicial	equinoctial
ES	signal station

E Salv	lifesaving station
Esc	scoriae, school
E S Horaria	time signal station
Esp	sponge
Esq	schist
Est	estuary
Estabdo, establecide	established
Este	east
Esto	stadium
Estr	strait
Event	eventual
Exp, Exper	experimental
Expl, Explos, Explosivo, Expvo	explosive
Ext, exterior	outer
F	fixed light, mud, muddy, fine
F D, Fd	fixed and flashing light, fjord
F Des	fixed and flashing
F Gp D	fixed and group flashing
F Gr D	fixed and group flashing
Fabca	factory
FC	railway
Fca	factory
Fdo	ground, fjord
Fdo So	foul ground, foul bottom
Fm	fathom
Fo	lighthouse, light
Fond	anchorage
Fondo Aero	anchorage for seaplanes
Fondo Prohdo	anchorage prohibited
Fondo sucio	foul ground, foul bottom
Freu	sound
Frontispicio, Fronton	gable
Fte	fort
G	phase lag, gray, coarse, revolving or rotating light, pebbles
Gde	great
Gj	flinty
Gja	farm
Gl	glacial, globigerina
Go	gulf, shingle
Gob, gobo	government house
Gong, Gongo, gong de niebla	fog gong
Gp	group
Gp Ct	group quick flashing light
Gp D	group-flashing light
Gp Oc	group occulting light
Gp Rp	group very quick flashing light
Gr	group, gray, large, gravel, shingle, chalk, Greenwich
Grd	great, large
Gr D	group flashing light
Gr Dc	composite group flashing light
Gr Des	group flashing light
Gris	gray
Gr Oc	group occulting light
Gru	coarse
Gs	coarse
H	hour, height, altitude, elevation, bridge clearance (vertical)
Hlla	track, footpath, trail
Hor, hort	horizontal light
I	island, islet
Ig, Igla	church
Inf, inferior	lower, lower light
Ins, instituto	institute
Int, Interior	inner
Ip	stiff
IQ	interrupted quick flashing
Irreg.	irregular light
Is	island
Iso, Isof	isophase light

Ite (Ites)	islet (islets)
IUQ	int. ultra quick flashing light
IVQ	int. very quick flashing light
Kn	knot, knots
L	light, large, ooze, lake, pond, loch, lough
La	lagoon
Laz, lazareto	lazaret
Lum	lighted, luminous
Lv	lava
M	meter, minute (of time), nautical mile, speckled
Ma	mountain
Mad, madrepora	madrepore
Mag	magnetic
Mal	jetty (partly below MHW)
Mco	magnetic
Md	middle, madrepores
M de S	mean low-water springs
M de Sic	spring tide
Mdo	speckled
Me, medio	middle, mole
Mejillon	mussels
Mj	mussels
Ml	ground (shells)
Mlle	mole
Mma	swamp
Mna	mountain, mount
Mo	morse code light, head, headland
Mol	windmill
Mon	monument
Mons, monasterio	monastery
Mr	marl
Ms de C	neap tide
Ms de S	spring tide
Mta	tableland (plateau)
Mte, Mtna	mountain, mount
Mte sub	seamount
Mto	monument
Municipio	town hall
Muralla, muro	wall
Mv	bearing
N	north, knot, nautophone, black, nautical mile
Naut, Nautof, nautofono	nautophone
Nav	navigable
N de R	bench mark
Nj, nja	orange
N M	mean tide level
NMM	mean sea level
No	number
Noray	dolphin
Not	notable
NRS	chart datum (datum for sounding reduction)
Ns	knots
Nv, nvo	new
O	dark, submarine oscillator
Obs, obsn	obstruction
Obso	observatory
Obston, obstr	obstruction
Obsv	observatory
Oc	occultation, occulting light
Ocas, Ocasl	occasional, occasional light
Occ	occulation, occulting light
Occidental	western
Of	office
Of de pto	harbormaster's office
Oficina	office
Of Tel	telegraph office
Ord, ordinario	ordinary
Os	oysters
Osc	submarine oscillator, dark, obscured light
Ost	oysters
O sub	submarine oscillator
P	peak, brown, stones, position, shoal sounding on isolated rock
Pal	wreck with only masts visible (above sounding datum)
Part	private, privately
Paso	passage, pass
Pblo	village
Pc	peak
Pd	pebbles
P de O	observation spot
P de R	point of reference
Penl, penla	peninsula
Peq	little
Pet	oil tank
Pg	sticky
Pie	foot, feet
Pil, pilares	pilar
Pl de R	chart datum (datum for sounding reduction)
Plmar	high water
Ply	beach
P M	high water
Pm M de S	mean high-water springs
PnFo	light vessel, lightship
Po	passage, pass
Pob	village
Pod, podrido	rotten
Pol, Polv, polvorin	powder magazine
Posn, posicion	position
Pq	small
Pr	brown
Prof men	lesser depth possible
Prof min	least depth in narrow channels
Proh, prohdo	prohibited
Prom, promo	promontory, prominent
Prov, Provo	provisional light
Proy, proyecto, proyo	projected
Pta	point
Pte	bridge
Pto	harbor, port
Publ, publicacion	publication
Pz	pumice
Q	stream, quick flashing light
Qb	broken
Queb, quebrada	stream
Quebrado	broken
R	red, radio fog-signal station, radio telegraph station, rock, river, qtg radio station
Ra	isolated rock, radar station, radar guided track
Rada	road, roadstead
Ram	ramark
RC	circular radiobeacon
Rcso	rocky
RD, Rd	directional radiobeacon, radiolaria
Rda	road, roadstead
Rdf	radio broadcasting station
Ref	hut
Regla	standard
RF, RFo	Radiobeacon
RG	radio direction finding station
Rga	ridge
Ria	creek
Rs	rocks, ruins
Rso	rocky

V   Abbreviations of principal foreign terms, (Glossary)

RT .... radio telephone station
Ru .... dumping ground
R Telem .... distance finding station
Rui, ruinas .... ruins
RW .... rotating loop radiobeacon

S .... station, south, fog signal, second (of
                                time), signal, syzygy, saint, sector
S A S .... submarine fog bell (mechanical)
S A So .... submarine fog bell (action of waves)
S C Bal .... cardinal marking system
S G .... unwatched light
S Mt .... seamount
S S .... signal station
Sa .... mountain range
SD .... doubtful sounding, lesser depth possible
Seg .... second (of time)
Sem .... semaphore
Senal .... signal
Sic .... syzygy
Sil, silb, silbato .... fog whistle
Sir, sirena .... fog siren
Sn .... saint
So .... low-water datum, sound
Son .... morse code fog signal
SSN .... submarine fog signal
Sta, sto .... saint
Subm .... submarine
sumdo, sumg, sumgd .... submerged
Sup .... higher, upper light
Supdo, Suprim .... discontinued

T .... temporary, temporary light, ton, tufa
Tba .... tomb
Te .... tower
Te bal .... tower beacon
Tejar .... brick kiln
Tel, telegrafo .... telegraph
Tem .... temporary, temporary light
Tfno .... telephone, mooring buoy with
                                telegraphic communications
Tn .... ton
Tri, Triang, triangulacion .... triangulation
Trompa, Trompeta de niebla .... fog trumpet
T V .... television mast, television tower

U .... unit of height, ultra quick flashing light
UQ .... ultra quick flashing light

V .... true, green, valley, volcano, ebb,
                                lookout tower
Valle .... valley
Var .... variation
Var an, Variacion anua .... annual change
VC .... see chart
Vc .... volcanic
V de T .... triangulation point (station)
Ve .... valley
Verd .... true
Vert .... vertical lights
Verto, Vertedero de draga ..... spoil area
Vi .... violet
Vigia .... lookout station, watch tower
Vis .... visible (range)
Vj .... old
Vo .... true
Vol, volcan .... volcano
VP .... see plan
VQ .... very quick flashing light
Vt .... streaky

Zo .... elevation of mean sea level above chart
                                sounding datum, chart datum (datum
                                for sounding reduction)

Zona C Subm .... submarine cable area
Zona Can Subm .... submarine pipeline area

SWEDISH CHARTING TERMS

B .... bay, bight
Berg, berget, bg, bgt .... mountain
Bk .... bank
Bukt .... bay, bight
Fj, fjard .... fjord
Grd, grund .... shoal
Hamm, hamnen .... harbor
Hd .... headland
Hm .... islet
Hn .... harbor
Holm, holmen .... islet
Huvud .... headland
K, kap .... cape
N, nord, norr, norra .... north, northern
Sk, skar, skaret .... rock above water

THAI CHARTING TERMS

Kh, khao .... hill, mountain
Laem, lm .... cape, point
Mae nam, MN .... river

TURKISH CHARTING TERMS

Ad, ada, adacik, adasi .... island, islet
Br, burnu, burun .... point, cape
Ca, cay, cayi .... stream, river
Da, dag, dagi .... mountain
De, dere, deresi .... valley, stream
Li, liman, limani .... harbor
Te, tepe, tepesi .... hill, peak

YUGOSLAV CHARTING TERMS

Br, brda, brdo .... mountain(s)
Gr, greben, grebeni .... rock, reef, cliff, ridge
Hr, hrid, hridi .... rock
L, luka .... harbor, port
O .... islet(s)
O, otoci .... island(s)
Otocic, otocici .... islet(s)
Otok .... island(s)
Pl, plicina .... shoal
Pr, prolaz .... passage
Sk, skolj, skoljic .... island, reef
U, uvala, uvalica .... inlet
Z, zaliv .... gulf, bay

Index of Abbreviations

A

aband	Abandoned	F 37
ABAND LT HO	Abandoned lighthouse	If
abt	About	F 17
AERO	Aeronautical	F 22; K 4
AERO R Bn	Aeronautical radiobeacon	M 16
AERO R Rge	Aeronautical radio range	Md
alt	Altitude	E 18
Al, Alt	Alternating (light)	K 26
Am	Amber	K 67a; L 48a
anc	Ancient	F 9
Anch	Anchorage	B 15; G 1, 2
Anch prohib	Anchorage prohibited	G 12
Ant	Antenna	Ma
approx	Approximate	F 34
Apprs	Approaches	Bg
Apt	Apartment	Ij
Arch	Archipelago	B 20
Art	Articulated light	Lg
Astro	Astronomical	D 9
AUTH	Authorized	Fc
Aux	Auxiliary (light)	K 51
Ave	Avenue	I 26a

B

B	Bay	B 2
B	Bayou	Ba
B	Beacon	L 54
B, b, bk	Black	L 42, S 57
Bdy Mon	Boundary monument	D 14
BELL	Fog Bell	N 14
bet	Between	Fi
B Hbr	Boat Harbor	G 33
Bk	Bank	O 21
Bkhd	Bulkhead	
Bkw	Breakwater	G 6
Bl	Blast	Fw
Bld, Blds	Boulder, Boulders	B 32; S 11a
Bldg	Building	I 66
Blvd	Boulevard	I 26b
BM	Bench Mark	D 5
Bn	Beacon (in general)	L 52, 53
BR	Bridge	H 14
Br, br	Brown	L 46; S 64
brg	Bearing	U 21
brk	Broken	S 47
Bu, bu	Blue	K 63; L 48; S 59
BWHB	Black and white horizontal bands	L 27
BWVS	Black and white vertical stripes	L 14, 14a

C

C	Can, Cylindrical (buoy)	L 5
C	Cape	B 22
C	Cove	B 5a
Ca	Calcareous	S 12a
Cap	Capitol	Ik
Cas	Castle	I 4
Cath	Cathedral	I 8a
Cb	Cobbles	Sf
cbl	Cable length	E 10
cd	Candela	E 12b
CFR	Code of Federal Regulations	Fx
C G	Coast Guard	J 3, Ja
ch	Chocolate	S 65
Ch	Church	I 8
Chan	Channel	B 10
Chec	Checkered (buoy)	L 33
CHY	Chimney	I 44
Cir	Cirripedia	S 38
Ck	Chalk	S 12
Cl	Clay	S 6
CL	Clearance	Fd
cm	Centimeter	E 4b
Cn	Cinders	S 21
Co	Company	Il
Co	Coral	S 14
Co Hd	Coral head	Sa
COLREGS	International Regulations for Preventing Collisions at Sea, 1972	Fy; Pi
concr	Concrete	Ff
conspic	Conspicuous	F 12
C of E	Corps of Engineers	Df
cor	Corner	Fe
Corp	Corporation	Im
Cov	Covers	O 33
corr	Correction	E 17
cps, c/s	Cycles per second	Ef
Cr	Creek	B 5
CRD	Columbia River Datum	Tc
crs	Coarse	S 40
Cswy	Causeway	H 3f
Ct Ho	Courthouse	I 64
CUP	Cupola	I 36
Cus Ho	Customhouse	G 29

D

D	Doubtful	O 45
D, Destr	Destroyed	F 14; Kg
dec	Decayed	S 52
deg	Degrees	U 27
dev	Deviation	U 28
Diag	Diagonal bands	L 33a
D F S	Distance finding station	M 15
Di	Diatoms	S 34
DIA	Diaphone	N 9
Discol	Discolored	O 36
discontd	Discontinued	F 25
dist	Distant	F 16
dk	Dark	S 68
dm	Decimeter	E 4a
Dol	Dolphin	G 21
DRDG RGE	Dredging Range	
DW	Deep Water	Fn

E

E	East, Eastern	U 2, 10
Ed	Edition	E 16
ED	Existence doubtful	O 43
elec	Electric	F 29
elev	Elevation	E 19
ELEV	Elevator	I 37

Index of Abbreviations

Index of Abbreviations

Lt	Light	K 2
lt	Light	S 67
Ltd	Limited	li
Lt Ho	Lighthouse	K 3
LW	Low water	T 2
LWD	Low water datum	Ta

M

M, Mi	Nautical mile	E 11; Kd
M	Mud, Muddy	S 3
m	Medium	Sg
m	Meter	E 4, d, e
m²	Square meter	E 4d
m³	Cubic meter	E 4e
m, min	Minute (of time)	E 2; Ke
Ma	Marsh	Bj.
Ma	Mattes	S 38b
mag	Magnetic	U 23
Magz	Magazine	I 34
maintd	Maintained	F 36
max	Maximum	Fp
Mc	Megacycle	Eh
Mds	Madrepores	S 15
Mg	Mangrove	Bk
MHHW	Mean higher high water	T 8b
MHW	Mean high water	T 7a
MHWN	Mean high-water neaps	T 8a
MHWS	Mean high-water springs	T 8
MHz	Megahertz	Ee
MICRO TR	Microwave tower	Mc
mid	Middle	F 7
min	minimum	Fo
Mkr	Marker	Lc
Ml	Marl	S 5
MLLW	Mean lower low water	T 9b
MLW	Mean low water	T 8c
MLWN	Mean low-water neaps	T 9a
MLWS	Mean low-water springs	T 9
mm	Millimeter	E 4c
Mn	Manganese	S 22
Mo	Morse code light, Fog signal	K 30a; Nc
mod	Moderate	Fh
MON	Monument	I 35
Ms	Mussels	S 25
μsec, μs	Microsecond (one millionth)	Eb
MSL	Mean sea level	T 4
Mt	Mountain, Mount	B 26
Mth	Mouth	B 13
MTL	Mean tide level	T 3
My	Bearing	U 21

N

N	North, Northern	U 1, 9
N	Nun, Conical (buoy)	L 6
N M, N Mi	Nautical mile	E 11
NAUTO	Nautophone	N 8
NE	Northeast	U 5
N'ly	Northerly	Fq
NM	Notice to Mariners	F 42
No	Number	E 23
Np	Neap tide	T 7

NW	Northwest	U 8
NWS	National Weather Service Signal Station	Jb

O

OBSC	Obscured (light)	K 68
Obs Spot	Observation spot	D 4
Obstr	Obstruction	O 27
Obsy	Observatory	J 21
Oc, Occ	Occulting (light), Occultation	K 22, 46
Oc, Occ	Intermittent (light)	K 48
Occas	Occasional (light)	F 39; K 70
ODAS	Oceanographic Data Aquisition System	L 73
Off	Office	J 22
Or, or	Orange	K 65; L 48b; S 62
OVHD PWR CAB	Overhead power cable	H 4
Oys	Oysters, Oyster bed	S 24; G 15a
Oz	Ooze	S 4

P

P	Pebbles	S 9
P	Pillar (buoy)	L 8a
P	Pond	Bb
P	Port	B 17; G 5
PA	Position approximate	O 41
Pag	Pagoda	I 14
Pass	Passage, Pass	B 9
Pav	Pavilion	I 67
PD	Position doubtful	O 42
Pen	Peninsula	B 21
PIL STA	Pilot station	J 8
Pk	Peak	B 29
Pm	Pumice	S 18
Po	Polyzoa	S 37
P O	Post Office	I 29
P, Pos	Position	O 44
priv	Private, Privately	F 30
Priv maintd	Privately maintained	K 17; L 29
Prohib	Prohibited	F 26
prom	Prominent	F 31
Prom	Promontory	B 23
Prov	Provisional (light)	K 72
Pt	Point	B 25
Pt	Pteropods	S 36
pub	Publication	E 15
P F	Pump-out facilities	Gd
PWI	Potable water intake	

Q

Quar	Quarantine	G 26
Q, Qk Fl	Quick flashing (light)	K 24
Qz	Quartz	S 13

R

R	Red	K 66; L 15, 43
R	River	Bd
R	Rocks	B 35
Ra	Radar station	M 11
Racon	Radar responder beacon	M 12

Index of Abbreviations

Ra (conspic)	Radar conspicuous object	M 14	Shl	Shoal	O 22
RA DOME	Radar dome	Mh	Si	Silt	Se
Ra Ref	Radar reflector	Lf; M 13	Sig Sta	Signal station	J 9
Ra Sur	Radio responder beacon	M 12	SIREN	Fog siren	N 11
RBHB	Red and black horizontal		Sk	Stroke, strike	Fu
	bands	L 17, 18, 19, 20, 20a	S-L Fl	Short-long flashing (light)	K 28a
R	Red beacon		Slu	Slough	Be; C 18
Bn		L 52	S'ly	Southerly	Fr
R Bn	Radiobeacon	M 3, 4, 16	sml	Small	F 4; S 44
RC	Circular radiobeacon	M 4	Sn	Shingle	S 8
Rd	Radiolaria	S 35	Sp	Spring tide	T 6
rd	Red	S 63	SP	Spherical (buoy)	L 7
Rd	Road, Roadstead	B 14; H 1	Spg	Sponge	S 26
RD	Directional Radiobeacon,		Spi	Spicules	S 31
	Radio range	M 5	S'PIPE	Standpipe	I 45
RDF, Ro DF	Radio direction finding		spk	Speckled	S 50
	station	M 7	S Sig Sta	Storm signal station	J 11
REF	Reflector	K 10; L 64	St	Saint	F 11
Rep	Reported	O 35	St	Street	I 26
Restr	Restricted	Fv	St	Stones	S 10
Rf	Reef	O 23	Sta	Station	J 1, 2
RG	Radio direction finding		std	Standard	F 32
	station	M 7	stf	Stiff	S 43
Rge	Range	B 27	Stg	Sea-tangle	S 29
RGE	Range	Ki	stk	Sticky	S 46
Rk	Rock	B 35	St M, St Mi	Statute mile	Ea
Rk, rky	Rock, Rocky	S 11	Str	Strait	B 7
Rky	Rocky	Bh	Str	Stream	Bc; T 17
R MAST	Radio mast	M 9	str	Streaky	S 49
Ro Bn	Radiobeacon	M 3	sub	Submarine	F 20
Rot	Rotating (light), Revolving	K 31	SUB-BELL	Submarine fog bell	N 5, 6
RR	Railroad	H 3	Subm, subm	Submerged	F 33; O 30
R RELAY			Subm Ruins	Submerged ruins	Gd
MAST	Radio relay mast	Mb	SUB-OSC	Submarine oscillator	N 7
R Sta	Radio telegraph station,		Sub Vol	Submarine volcano	O 8
	QTG Radio station	M 1, 10a	Subm W	Submerged Well	Ob
RT	Radio telephone station	M 2	SW	Southwest	U 7
rt	Rotten	S 48	sw	Swamp	Bl
R TR	Radio tower	M 9			
Ru	Ruins	I 40	**T**		
RW	Rotating loop radiobeacon	M 6	t	Tonne	E 12a
RW			T	Ton	Ei
Bn	Red and white beacon	L 52	T	Telephone	I 70; L 22c
RWVS	Red and white vertical		T	True	U 22
	stripe	L 14, 14a	T	Tufa	S 19
Ry	Railway	H 3	TB	Temporary buoy	L 30
			Tel	Telegraph	I 27; L 22b
S			Telem Ant	Telemetry antenna	Ma
S	Sand	S 2	Tel Off	Telegraph office	I 28
S	South, Southern	U 3, 11	Temp	Temporary (light)	F 38; K 73
S	Spar (buoy)	L 8	ten	Tenacious	S 55
Sc	Scoriae	S 20	Thoro	Thoroughfare	B 9
Sch	Schist	S 13a	Tk	Tank	I 53
Sch	School	I 65	TR, Tr	Tower	I 41; L 63; Ld; M 9a
Sd	Sound	B 8	Tri	Triangulation	D 10
SD	Sounding doubtful	Q 1	T.T.	Treetop	C 11
SE	Southeast	U 6	TV TR	Television tower (mast)	M 9a
sec, s	Second (time; geo, pos.)	E 3; Kf			
SEC	Sector	K 49	**U**		
Sem	Semaphore	J 10	uhf	Ultra high frequency	Mi
S Fl	Short flashing (light)	K 25a	Uncov	Uncovers; Dries	O 2, 32, 34
sft	Soft	S 41	Univ	University	Ih
Sh	Shells	S 23			

Index of Abbreviations

unverd	Unverified	Fb
unev	Uneven	S 71
μsec. μs	Microsecond (one millinoth)	Eb
UQ	Continuous Ultra Quick	Kc

V

var	Variation	U 24
vard	Varied	S 70
VB	Vertical beam	Kh
vel	Velocity	T 23
Vert	Vertical (lights)	K 80
VERT CL	Vertical clearance	H 18a
vhf	Very high frequency	Mi
Vi, vi	Violet	K 61; S 58
View X	View point	D 6
Vil	Village	I 3
Vol	Volcanic	S 16
Vol Ash	Volcanic ash	Sb
VQ, V Qk Fl	Very quick flashing (light)	Kc
VS	Vertical stripes	L 32

W

W	West, Western	U 4, 12
W, wh	White	K 67; L 41; S 56
W Bn	White beacon	L 52
Wd	Seaweed	S 28
Whf	Wharf	G 18
WHIS	Fog whistle	N 15
Wk	Wreck	O 15, 15a, 28
Wks	Wrecks, Wreckage	O 29
W Or	White and orange	Le
W'ly	Westerly	Ft

Y

Y, yl	Yellow	L 24, 44; S 61
yd, yds	Yard(s)	E 8

1st	First	Fj
2nd, 2d	Second	Fk
3rd, 3d	Third	Fl
4th	Fourth	Fm

°	Degree	E 20
'	Minute (of arc)	E 21
"	Second (of arc)	E 22

AIDS TO NAVIGATION

IN

UNITED STATES WATERS

MODIFIED U.S. AID SYSTEM
LATERAL AIDS AS SEEN ENTERING FROM SEAWARD

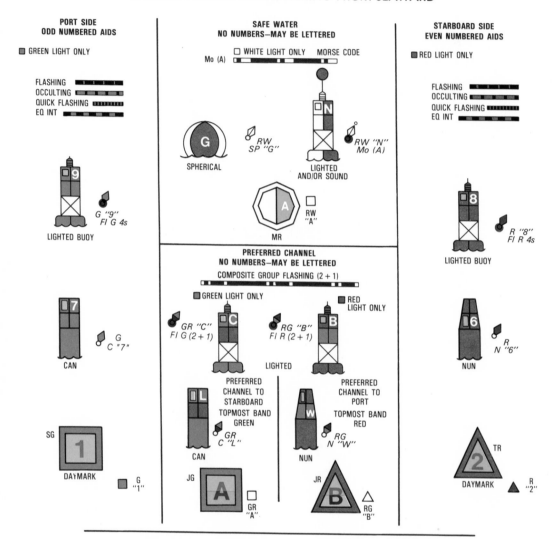

PORT SIDE
ODD NUMBERED AIDS

■ GREEN LIGHT ONLY

FLASHING
OCCULTING
QUICK FLASHING
EQ INT

G "9"
Fl G 4s

LIGHTED BUOY

7

G
C "7"

CAN

SG

1

DAYMARK

G "1"

SAFE WATER
NO NUMBERS—MAY BE LETTERED

□ WHITE LIGHT ONLY   MORSE CODE
Mo (A)

G
SPHERICAL

RW
SP "G"

N

RW "N"
Mo (A)

LIGHTED
AND/OR SOUND

A
MR

RW "A"

PREFERRED CHANNEL
NO NUMBERS—MAY BE LETTERED
COMPOSITE GROUP FLASHING (2 + 1)

■ GREEN LIGHT ONLY   ■ RED LIGHT ONLY

C
GR "C"
Fl G (2 + 1)

B
RG "B"
Fl R (2 + 1)

LIGHTED

L
PREFERRED
CHANNEL TO
STARBOARD
TOPMOST BAND
GREEN

GR
C "L"

CAN

JG

A

GR "A"

W
PREFERRED
CHANNEL TO
PORT
TOPMOST BAND
RED

RG
N "W"

NUN

JR

B

RG "B"

STARBOARD SIDE
EVEN NUMBERED AIDS

■ RED LIGHT ONLY

FLASHING
OCCULTING
QUICK FLASHING
EQ INT

8

R "8"
Fl R 4s

LIGHTED BUOY

6

R
N "6"

NUN

TR

2

DAYMARK

R "2"

NOTE: WHEN USED ON THE INTRACOASTAL WATERWAY, THESE AIDS ARE
ALSO EQUIPPED WITH SPECIAL YELLOW STRIPS, TRIANGLES, OR SQUARES.
WHEN USED ON THE WESTERN RIVERS (MISSISSIPPI RIVER SYSTEM), THESE
AIDS ARE NOT NUMBERED. (MISSISSIPPI RIVER SYSTEM ABOVE BATON
ROUGE AND ALABAMA RIVERS)

AIDS TO NAVIGATION ON NAVIGABLE WATERS
except Western Rivers and Intracoastal Waterway

LATERAL SYSTEM AS SEEN ENTERING FROM SEAWARD

PORT SIDE
ODD NUMBERED AIDS
■ GREEN OR □ WHITE LIGHTS

FIXED
FLASHING
OCCULTING
QUICK FLASHING
EQ INT

"9"
Fl G 4sec

LIGHTED BUOY

"C"7"

CAN 7

SG

1

DAYMARKS

G
"1"

MID CHANNEL
NO NUMBERS—MAY BE LETTERED
□ WHITE LIGHT ONLY

MORSE CODE

Mo(A)

BW"N"
Mo(A)

BW
C"T"

CAN   **LIGHTED**

BW
N"B"

NUN

MB   A   BW "A"   **DAYMARK**

JUNCTION
MARK JUNCTIONS AND OBSTRUCTIONS
NO NUMBERS—MAY BE LETTERED
INTERRUPTED QUICK FLASHING

□ WHITE OR ■ GREEN   □ WHITE OR ■ RED

BR "M"
I Qk Fl G

RB "D"
I Qk Fl R

LIGHTED

PREFERRED
CHANNEL TO
STARBOARD
TOPMOST BAND
BLACK

BR
C"N"

CAN

PREFERRED
CHANNEL TO
PORT
TOPMOST BAND
RED

RB
N"L"

NUN

JG   A   GR "A"

JR   B   RG "B"

STARBOARD SIDE
EVEN NUMBERED AIDS
■ RED OR □ WHITE LIGHTS

FIXED
FLASHING
OCCULTING
QUICK FLASHING
EQ INT
GROUP FLASHING (2)

R"8"
Fl R 4sec

LIGHTED BUOY 8

R
N"6"

NUN 6

TR

DAYMARK 2   R "2"

BUOYS HAVING NO LATERAL SIGNIFICANCE—ALL WATERS

SHAPE HAS NO SIGNIFICANCE
NO NUMBERS—MAY BE LETTERED
MAY BE LIGHTED
ANY COLOR LIGHT EXCEPT
RED OR GREEN

W Or
C

SPECIAL PURPOSE

FIXED
FLASHING
OCCULTING

W
C"N"

ANCHORAGE

BW
C

FISH NET AREA

GW
C

DREDGING

UNLIGHTED

DANGER

EXCLUSION AREA

DAYMARKS HAVING NO LATERAL SIGNIFICANCE
MAY BE LETTERED

NW   SUBMERGED DANGER JETTY   W □ Bn

NR   A   RW □ Bn

NG   A   GW □ Bn

NB   M   BW □ Bn

AS SEEN ENTERING FROM NORTH AND EAST—PROCEEDING TO SOUTH AND WEST

PORT SIDE
ODD NUMBERED AIDS
■ GREEN OR □ WHITE LIGHTS

FIXED ▬▬▬▬ OCCULTING ▬ ▬ ▬
FLASHING ▬ ▬ ▬ QUICK FLASHING ▪▪▪▪▪▪▪
EQ INT ▬ ▪ ▬ ▪

3
"3"
Fl G 4sec

LIGHTED BUOY

9
C"9"

CAN

SG-I
1
■G "1"

DAYMARKS

SG-SY
5
G "5"■

DUAL PURPOSE DAYMARKS

5
C"5"

DUAL PURPOSE BUOYS

TR-SY
6
R "6"

6
R N"6"

JG-SY
A
GR "A"

JR-SY
D
RG "D"

SN 7530-01-GF2-5560

JUNCTION
MARK JUNCTIONS AND OBSTRUCTIONS
NO NUMBERS—MAY BE LETTERED
INTERRUPTED QUICK FLASHING

□ WHITE OR ■ GREEN LIGHTS □ WHITE OR ■ RED LIGHTS

J
BR"J"
I Qk Fl G

N
RB"N"
I Qk Fl R

PREFERRED CHANNEL
A TO STARBOARD TOPMOST BAND BLACK
TO PORT TOPMOST BAND RED **S**

CAN
BR C"A"

RB N"S"
NUN

JG-I
A
□ GR "A"

JR-I
B
RG "B"

MID CHANNEL MORSE CODE NO NUMBERS—MAY BE LETTERED
□WHITE LIGHT ONLY

MB-I
B
DAYMARK

BW
"B"

T
BW C"T"

N
BW Mo(A) "N"

N
BW N"8"

CAN LIGHTED NUN

STARBOARD SIDE
EVEN NUMBERED AIDS
■ RED OR □ WHITE LIGHTS

FIXED ▬▬▬▬ OCCULTING ▬ ▬ ▬
FLASHING ▬ ▬ ▬ QUICK FLASHING ▪▪▪▪▪▪▪
EQ INT ▬ ▪ ▬ ▪
GROUP FLASHING (2) ▬ ▪▪ ▬ ▪▪

8
R"8"
Fl R 4sec

LIGHTED BUOY

6
R N"6"

NUN

TR-I
2
R "2"

DAYMARK

TR-TY
6
R "6"

6
R N"6"

DUAL PURPOSE DAYMARKS

DUAL PURPOSE BUOYS

SG-TY
5
G "5"■

5
C"5"

JG-TY
C
□ GR "C"

JR-TY
B
RG "B"

DUAL PURPOSE MARKING USED WHERE THE ICW AND OTHER WATERWAYS COINCIDE

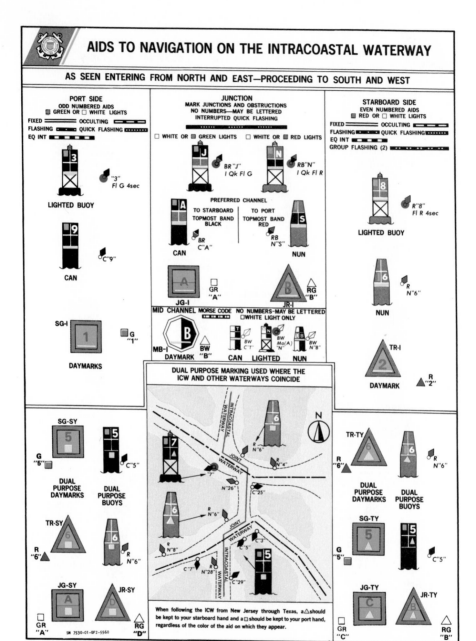

INTRACOASTAL WATERWAY

6
R N"6"

N

7
"7"
R N"4"

N"26"

C"25"

6
R N"6"

JOINT WATERWAY

6
R N"8"

C"3"

C"5"

5
C"7"
N"28"
C"29"

JOINT WATERWAY

INTRACOASTAL WATERWAY

When following the ICW from New Jersey through Texas, a △ should be kept to your starboard hand and a ◻ should be kept to your port hand, regardless of the color of the aid on which they appear.

334

AIDS TO NAVIGATION ON WESTERN RIVERS
(MISSISSIPPI RIVER SYSTEM)

AS SEEN ENTERING FROM SEAWARD

PORT SIDE	JUNCTION	STARBOARD SIDE

PORT SIDE

☐ GREEN OR ☐ WHITE LIGHTS
FLASHING

LIGHTED BUOY

CAN

SG

PASSING DAYMARK

CG

CROSSING DAYMARK

176.9

MILE BOARD

JUNCTION

MARK JUNCTIONS AND OBSTRUCTIONS
INTERRUPTED QUICK FLASHING

PREFERRED CHANNEL TO STARBOARD	PREFERRED CHANNEL TO PORT
TOPMOST BAND BLACK	TOPMOST BAND RED
☐ WHITE OR ☐ GREEN LIGHTS	☐ WHITE OR ☐ RED LIGHTS

LIGHTED

CAN	NUN

JG	JR

STARBOARD SIDE

☐ RED OR ☐ WHITE LIGHTS
GROUP FLASHING (2)

LIGHTED BUOY

NUN

TR

PASSING DAYMARK

CR

CROSSING DAYMARK

123.5

MILE BOARD

RANGE DAYMARKS AS FOUND ON

NAVIGABLE WATERS — EXCEPT — ICW — MAY BE LETTERED

KGW	KWG	KWB	KBW	KWR	KRW	KRB	KBR	KGB	KBG	KGR	KRG

INTRACOASTAL WATERWAY — MAY BE LETTERED

KGW-I	KWG-I	KWB-I	KBW-I	KWR-I	KRW-I	KRB-I	KBR-I	KGB-I	KBG-I	KGR-I	KRG-I

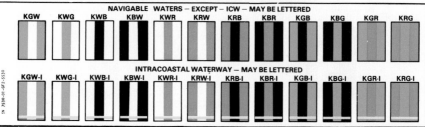

SN 7530-01-GF2-5530

UNIFORM STATE WATERWAY MARKING SYSTEM

STATE WATERS AND DESIGNATED STATE WATERS FOR PRIVATE AIDS TO NAVIGATION

REGULATORY MARKERS

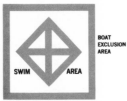

BOAT EXCLUSION AREA

SWIM AREA

EXPLANATION MAY BE PLACED OUTSIDE THE CROSSED DIAMOND SHAPE, SUCH AS DAM, RAPIDS, SWIM AREA, ETC.

DANGER

ROCK

THE NATURE OF DANGER MAY BE INDICATED INSIDE THE DIAMOND SHAPE, SUCH AS ROCK, WRECK, SHOAL, DAM, ETC.

CONTROLLED AREA

SLOW

NO WAKE

TYPE OF CONTROL IS INDICATED IN THE CIRCLE, SUCH AS SLOW, NO WAKE, ANCHORING, ETC.

MULLET LAKE

BLACK RIVER

INFORMATION

FOR DISPLAYING INFORMATION SUCH AS DIRECTIONS, DISTANCES, LOCATIONS, ETC.

BUOY USED TO DISPLAY REGULATORY MARKERS

MAY SHOW WHITE LIGHT
MAY BE LETTERED

5 MPH

AIDS TO NAVIGATION

MAY SHOW WHITE REFLECTOR OR LIGHT

MOORING BUOY

WHITE WITH BLUE BAND
MAY SHOW WHITE REFLECTOR OR LIGHT

RED-STRIPED WHITE BUOY

MAY BE LETTERED
DO NOT PASS BETWEEN BUOY AND NEAREST SHORE

BLACK-TOPPED WHITE BUOY

MAY BE NUMBERED
PASS TO NORTH OR EAST OF BUOY

RED-TOPPED WHITE BUOY

PASS TO SOUTH OR WEST OF BUOY

CARDINAL SYSTEM

MAY SHOW GREEN REFLECTOR OR LIGHT

MAY SHOW RED REFLECTOR OR LIGHT

SOLID RED AND SOLID BLACK BUOYS

USUALLY FOUND IN PAIRS
PASS BETWEEN THESE BUOYS

3

PORT SIDE ———— LOOKING UPSTREAM ———— STARBOARD SIDE

4

LATERAL SYSTEM

SN 7530-01-GF2-5540

336

INTERNATIONAL ASSOCIATION OF LIGHTHOUSE AUTHORITIES (I.A.L.A.) MARITIME BUOYAGE SYSTEM

L 70   Buoys and Beacons IALA* Buoyage System

Where in force, the IALA System applies to all fixed and floating marks and occasionally lighthouses, sector lights, range marks, lightships and LANBY's (large automatic navigational buoys).

The standard **buoy shapes** are cylindrical (can) ⊏⊐, conical △, spherical ○, pillar (including high focal plane) ⬧, and spar ∤, but variations may occur, for example: lightfloats ⬭. In the illustrations below, only the standard buoy shapes are used. In the case of fixed **beacons** (lighted or unlighted), only the shape of the topmark is of navigational significance.

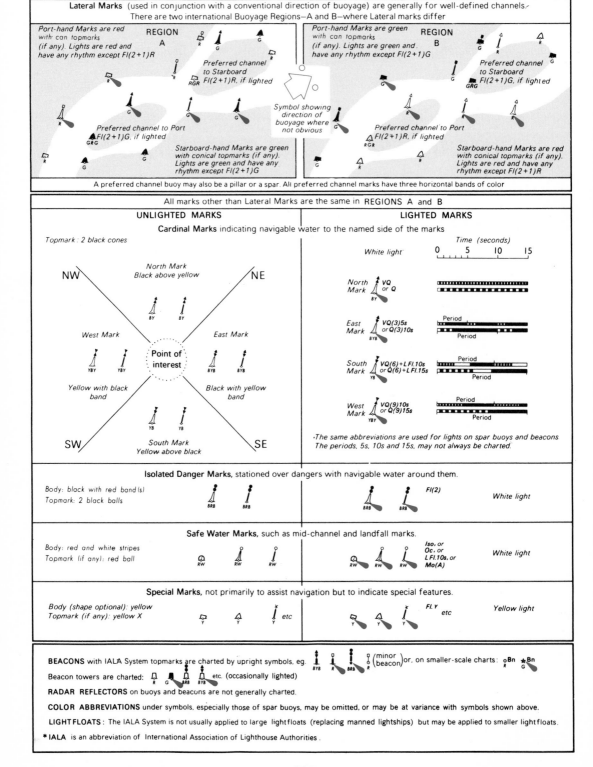

Lateral Marks (used in conjunction with a conventional direction of buoyage) are generally for well-defined channels.
There are two international Buoyage Regions—A and B—where Lateral marks differ

REGION A

Port-hand Marks are red with can topmarks (if any). Lights are red and have any rhythm except Fl(2+1)R

Preferred channel to Starboard Fl(2+1)R, if lighted

Preferred channel to Port Fl(2+1)G, if lighted

Starboard-hand Marks are green with conical topmarks (if any). Lights are green and have any rhythm except Fl(2+1)G

REGION B

Port-hand Marks are green with can topmarks (if any). Lights are green and have any rhythm except Fl(2+1)G

Preferred channel to Starboard Fl(2+1)G, if lighted

Preferred channel to Port Fl(2+1)R, if lighted

Starboard-hand Marks are red with conical topmarks (if any). Lights are red and have any rhythm except Fl(2+1)R

Symbol showing direction of buoyage where not obvious

A preferred channel buoy may also be a pillar or a spar. All preferred channel marks have three horizontal bands of color

All marks other than Lateral Marks are the same in REGIONS A and B

UNLIGHTED MARKS

LIGHTED MARKS

Cardinal Marks indicating navigable water to the named side of the marks

Topmark : 2 black cones

White light

Time (seconds)
0   5   10   15

North Mark
Black above yellow

NW   NE

West Mark   East Mark

Point of interest

Yellow with black band   Black with yellow band

SW   SE

South Mark
Yellow above black

North Mark   VQ or Q   BY

East Mark   VQ(3)5s or Q(3)10s   BYB   Period

South Mark   VQ(6)+L Fl.10s or Q(6)+L Fl.15s   YB   Period

West Mark   VQ(9)10s or Q(9)15s   YBY   Period

The same abbreviations are used for lights on spar buoys and beacons. The periods, 5s, 10s and 15s, may not always be charted.

Isolated Danger Marks, stationed over dangers with navigable water around them.

Body: black with red band(s)
Topmark: 2 black balls

BRB   BRB   Fl(2)   White light

Safe Water Marks, such as mid-channel and landfall marks.

Body: red and white stripes
Topmark (if any): red ball

RW   Iso. or Oc. or L Fl.10s, or Mo(A)   White light

Special Marks, not primarily to assist navigation but to indicate special features.

Body (shape optional): yellow
Topmark (if any): yellow X

Y   etc   Fl.Y etc   Yellow light

BEACONS with IALA System topmarks are charted by upright symbols, eg. ∤ (minor beacon) or, on smaller-scale charts: ○Bn ★Bn

Beacon towers are charted: ⬚ ■ ⬚ etc. (occasionally lighted)

RADAR REFLECTORS on buoys and beacons are not generally charted.

COLOR ABBREVIATIONS under symbols, especially those of spar buoys, may be omitted, or may be at variance with symbols shown above.

LIGHT FLOATS : The IALA System is not usually applied to large lightfloats (replacing manned lightships) but may be applied to smaller lightfloats.

✱ IALA is an abbreviation of International Association of Lighthouse Authorities.

IALA MARITIME BUOYAGE SYSTEM
LATERAL MARKS REGION A

PORT HAND

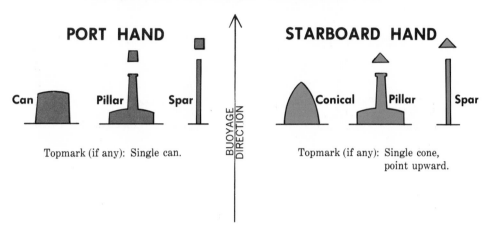

Can   Pillar   Spar

Topmark (if any): Single can.

STARBOARD HAND

Conical   Pillar   Spar

Topmark (if any): Single cone,
point upward.

BUOYAGE DIRECTION

Lights, when fitted, may have any phase
characteristic other than that used
for preferred channels.

Examples

Quick Flashing
Flashing
Long Flashing
Group Flashing

PREFERRED CHANNEL
TO STARBOARD

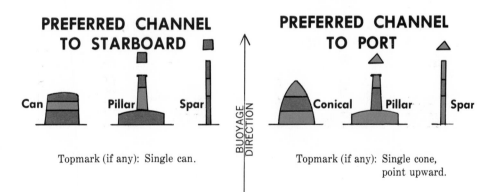

Can   Pillar   Spar

Topmark (if any): Single can.

PREFERRED CHANNEL
TO PORT

Conical   Pillar   Spar

Topmark (if any): Single cone,
point upward.

BUOYAGE DIRECTION

Lights, when fitted, are composite
group flashing Fl (2 + 1).

IALA MARITIME BUOYAGE SYSTEM
LATERAL MARKS REGION B

PORT HAND

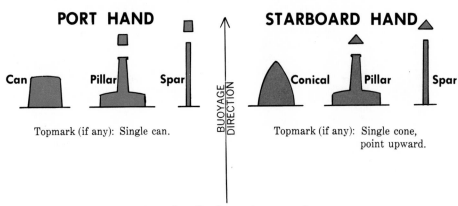

Can   Pillar   Spar

Topmark (if any): Single can.

BUOYAGE DIRECTION

STARBOARD HAND

Conical   Pillar   Spar

Topmark (if any): Single cone, point upward.

Lights, when fitted, may have any phase
characteristic other than that used
for preferred channels.

Examples

Quick Flashing
Flashing
Long Flashing
Group Flashing

PREFERRED CHANNEL
TO STARBOARD

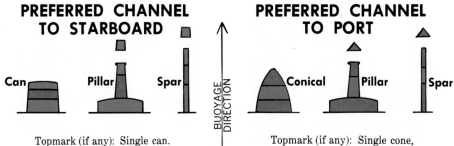

Can   Pillar   Spar

Topmark (if any): Single can.

BUOYAGE DIRECTION

PREFERRED CHANNEL
TO PORT

Conical   Pillar   Spar

Topmark (if any): Single cone, point upward.

Lights, when fitted, are composite
group flashing Fl (2+1).

IALA MARITIME BUOYAGE SYSTEM
CARDINAL MARKS REGIONS A AND B

Topmarks are always fitted (when practicable).
Buoy shapes are pillar or spar.

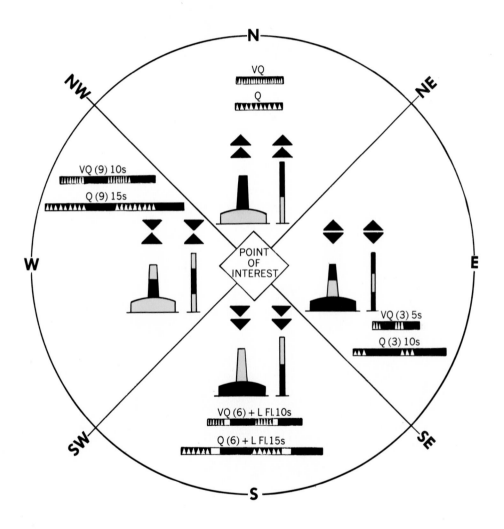

Lights, when fitted, are **white** . Very Quick Flashing
or Quick Flashing; a South mark also has a
Long Flash immediately following the quick flashes.

IALA MARITIME BUOYAGE SYSTEM
REGIONS A AND B
ISOLATED DANGER MARKS

Topmarks are
always fitted
(when practicable).

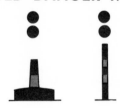

Light, when fitted, is
white
Group Flashing (2)

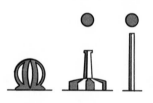 Fl (2)

Shape: Optional, but not
conflicting with lateral
marks; pillar or spar
preferred.

SAFE WATER MARKS

Topmark (if any):
Single sphere.

Light, when fitted,
is **white**
Isophase or Occulting,
or one Long Flash
every 10 seconds or
Morse "A".

Iso

Occ

L Fl.10s

Morse "A"

Shape: Spherical
or
pillar or spar.

SPECIAL MARKS

Topmark (if any):
Single X shape.

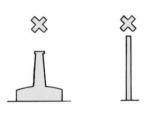

Light (when fitted) is
yellow and may have
any phase characteristic
not used for white lights.

Examples
 Fl Y
Fl(4) Y

Shape: Optional, but not
conflicting with
navigational marks.

Appendix B.   Abbreviations and Symbols Commonly Used in Piloting

ATP	Allied Tactical Publication	GMT	Greenwich Mean Time
Bn	beacon	IALA	International Ass'n of Lighthouse Authorities
C	course		
CIC	combat information center	JA	captain's battle sound-powered telephone circuit
CO	commanding officer		
COG	course over the ground	JW	navigation sound-powered telephone circuit
D	drift		
DB	danger bearing	kn	knot (of speed)
DG	degaussing	L	latitude
DMA	Defense Mapping Agency	λ	longitude
		LMT	local mean time
DMAHTC	Defense Mapping Agency Hydrographic/ Topographic Center	LOGREQ	logistics requirement message
		LOP	line of position
DMAODS	Defense Mapping Agency Office of Distribution Services	LPO	leading petty officer
		M	magnetic
DR	dead reckoning	m	meter
E	east	Mi	mile
ec.	eclipse (of a light)	MHW	mean high water
ECM	electronic countermeasures	MLW	mean low water
		N	north
EP	estimated position	1-N	DMAHTC *Catalog of Nautical Charts*
ETA	estimated time of arrival		
		NEOC	Naval Eastern Oceanography Center
ETD	estimated time of departure	NOS	National Ocean Survey
fm	fathom		
ft	foot	NWOC	Naval Western Oceanography Center

OOD	officer of the deck	SOA	speed of advance
OTSR	optimum track ship routing	SOG	speed over ground
		SOP	standard operating procedure
P_n	north pole		
pgc	per gyro compass	SOPA	senior officer present afloat
PIM	position of intended movement	T	true
		TR	track
PMP	parallel motion protractor	W	west
		XO	executive officer
PPI	plan position indicator	yd	yard (unit of measure)
psc	per standard compass		
pstgc	per steering compass	ZD	zone description
P_S	south pole	ZT	zone time
QMOW	quartermaster of the watch		

R   relative

RF   radiofrequency

RPM   revolution per minute

S   speed; south; set

SI   International System of Units

Miscellaneous Symbols

°	degrees
′	minutes of arc
″	seconds of arc
□	estimated position
⊙	fix, running fix
⌂	DR position
λ	longitude

Quartermasters, 6–7
  of watch (QMOW), 6–7

Radar, 171–82
  characteristics of, 171–73
  scope presentation, 173–79
    interpretation of, 176–79
  use of, during piloting, 179–81
Radar Navigation Manual, 75
Radar navigation team, 16
Radar operator, of piloting team.
  See Piloting team
Radio broadcast warnings, 47–50
Radio Navigation Aids, 75
Range lights, 85–86
Rhumb line, defined, 35
Running fix, 140–44
  described, 140
  determination of, in piloting, 140–44
  labeling of, 141–42, 149

Sailing Directions 60–64, 270, 272–73
  Enroute volumes, 60–62, 272
  Planning Guides, 60–62
  World Port Index, 60, 63
Secondary lights, described, 78
Sector lights, 84–85
Set, of current, defined, 208
Shaft RPM,
  use of, to estimate speed, 117–18
Ship's Deck Log, 6, 14, 138
Ship's position report, 280–81
  illustrated, 281
Side-lobe effect, radar, 178–79
Six-minute rule, 125
Small circle, defined, 19
Speed of advance (SOA), defined, 146
Stadimeter, 113–14
  Brandon, 114
  Fisk, 113
Stand, of tide, 185
Standard compass, 152
Steering compass, 152
Symbols, piloting, standard, 149
  See also Appendix B
Summary of Corrections, 47
  See also Notice to Mariners

Tactical data folder, 244–45
Telescopic alidade, 112
Terrestrial coordinate system, 18–23
Three-arm protractor, 123–24
Three-minute rule, 125
Tidal Current Diagrams, 227

Tidal Current Tables, 69–71, 210–27,
    228, 230–31
  layout of, 210–17
  use of, 215–24
  See also Current, tidal
Tide, 183–205
  bridge problem, 200–203
  causes of, 183–85
  cycle, defined, 184
  effect of unusual meteorological con-
    ditions, 204
  neap, 185
  predicting height of, 189–200
  reference planes for, 187–89
  shoal problem, 203–4
  spring, 184–85
  table, construction of, 194–200
  types of, 185–86
Tide Tables, 69, 189–205, 227, 272
  layout of, 190–94
  use of, during voyage planning, 272
  use of, to predict tide, 194–205
Time, 260–67
  conversions, 264–67
  Greenwich mean (GMT), 262–66
  zone, 261–67
  See also Voyage planning
Transfer. *See* Precise piloting

U.S. Lateral System:
  beacons of, 100–102
  buoys of, 97–100, 103, 104–7
  described, 96–97
  See also Chart No. 1

Variation, 153–56
  defined, 153–54
  isogonic lines of, 155
Visual navigation aids, 77–109
Voyage planning, 260–83
  Captain's Night Orders, 281–82
  departure/arrival dates, determina-
    tion of, 270–73
  miscellaneous considerations, 280–83
  process of, 267–70
  SOAs, determination of, 277–78
  time, 260–67
    See also Time
  track, plotting, 273–78
  using OTSR services, 278–80

Weems Navigation Plotter, 121–22
Weems Parallel Plotter, 121
World Port Index, 60, 63
Worldwide Navigational Warning Sys-
  tem, 48